Fluctuations in top predator populations in lakes can cascade through food webs to alter nutrient cycling, algal biomass and primary production. Trophic cascades may interact with nutrients and physical factors to explain most of the variance in lake ecosystem process rates.

In this book, a multidisciplinary research team tests this idea by manipulating whole lakes experimentally, and coordinating this with paleolimnological studies, simulation modeling, and small-scale enclosure experiments.

Consequences of predator–prey interactions, behavioral responses of fishes, diel vertical migration of zooplankton, plankton community change, primary production, nutrient cycling and microbial processes are described. Paleolimnological techniques enable the reconstruction of trophic interactions from past decades. Hypotheses from the authors' simulation models are evaluated, and prospects for further modeling of trophic cascades are assessed. Prospects for analysing the interaction of food web structure and nutrient input in lakes are explored.

This book will be of interest to graduate students and researchers in ecology, aquatic ecology, resource management and limnology.

Cascading trophic interactions

Cambridge Studies in Ecology presents balanced, comprehensive, up-to-date, and critical reviews of selected topics within ecology, both botanical and zoological. The Series is aimed at advanced final-year undergraduates, graduate students, researchers, and university teachers, as well as ecologists in industry and government research.

It encompasses a wide range of approaches and spatial, temporal, and taxonomic scales in ecology, including quantitative, theoretical, population, community, ecosystem, historical, experimental, behavioral and evolutionary studies. The emphasis throughout is on ecology related to the real world of plants and animals in the field rather than on purely theoretical abstractions and mathematical models. Some books in the Series attempt to challenge existing ecological paradigms and present new concepts, empirical or theoretical models, and testable hypotheses. Others attempt to explore new approaches and present syntheses on topics of considerable importance ecologically which cut across the conventional but artificial boundaries within the science of ecology.

CAMBRIDGE STUDIES IN ECOLOGY

ALSO IN THE SERIES

The trophic cascade in lakes

edited by

STEPHEN R. CARPENTER and
JAMES F. KITCHELL

CAMBRIDGE
UNIVERSITY PRESS

CAMBRIDGE UNIVERSITY PRESS
Cambridge, New York, Melbourne, Madrid, Cape Town,
Singapore, São Paulo, Delhi, Tokyo, Mexico City

Cambridge University Press
The Edinburgh Building, Cambridge CB2 8RU, UK

Published in the United States of America by Cambridge University Press, New York

www.cambridge.org
Information on this title: www.cambridge.org/9780521566841

First published 1993
First paperback edition 1996

A catalogue record for this publication is available from the British Library

Library of Congress Cataloguing in Publication data

The trophic cascade in lakes/edited by Stephen R. Carpenter and
James F. Kitchell.
   p.   cm. – (Cambridge studies in ecology)
Includes bibliographical references and index.
ISBN 0-521-43145-X
1. Lake ecology. 2. Biological productivity. 3. Food chains
(Ecology) I. Carpenter, Stephen R. II. Kitchell, James F.
III. Series.
QH541.5.L3T76 1993
574.5'26322–dc20 92-36737 CIP

ISBN 978-0-521-43145-3 Hardback
ISBN 978-0-521-56684-1 Paperback

To our mentor Arthur Hasler,
pioneer ecosystem experimenter;
and
in memory of David Benkowski,
whose good humor and skill live on among
the legends of the northern lakes

Contents

Contributors

David Benkowski (Deceased)

Nicolaas Bouwes, *Department of Fisheries and Wildlife, Utah State University, Logan, UT 84322, U.S.A.*

Stephen R. Carpenter, *Center for Limnology, University of Wisconsin, Madison, WI 53706, U.S.A.*

Philip A. Cochran, *Division of Natural Sciences, St. Norbert College, DePere, WI 54115, U.S.A.*

Michael L. Dini, *Department of Biological Sciences, Texas Tech University, Box 43131, Lubbock, TX 79409, U.S.A.*

James J. Elser, *Department of Zoology, Arizona State University, Tempe, AZ 85287, U.S.A.*

Monica M. Elser, *1843 East Laguna Drive, Tempe, AZ 85282, U.S.A.*

Xi He, *Center for Limnology, University of Wisconsin, Madison, WI 53706, U.S.A.*

James R. Hodgson, *Department of Biology, St. Norbert College, De Pere, WI 54115, U.S.A.*

James F. Kitchell, *Center for Limnology, University of Wisconsin, Madison, WI 53706, U.S.A.*

Peter R. Leavitt, *Zoology Department, Biological Sciences Building, University of Alberta, Edmonton, Alberta, Canada K6G 2E9*

David M. Lodge, *Department of Biological Sciences, University of Notre Dame, Notre Dame, IN 46556, U.S.A.*

Neil A. MacKay, *Zoology Department, Arizona State University, Tempe, AZ 85287, U.S.A.*

Susan M. Moegenburg, *Department of Zoology, 233 Bartram, University of Florida, Gainesville, FL 32611, U.S.A.*

John A. Morrice, *Department of Biology, University of New Mexico, Albuquerque, NM 87131, U.S.A.*

Michael L. Pace, *Institute for Ecosystems Studies, Cary Arboretum, Box AB, Millbrook, NY 12545, U.S.A.*

Patricia R. Sanford, *Center for Limnology, University of Wisconsin, Madison, WI 53706, U.S.A.*

Mark D. Scheuerell, *Cornell Biological Field Station, R.D. #1, Bridgeport, NY 13030, U.S.A.*

Patricia A. Soranno, *Center for Limnology, University of Wisconsin, Madison, WI 53706, U.S.A.*

Ann L. St. Amand, *Phycotech, Inc., P.O. Box 218, Baroda, MI 49101, U.S.A.*

Russell Wright, *Center for Limnology, University of Wisconsin, Madison, WI 53706, U.S.A.*

Preface

This book attempts a bridge between ecosystem and population ecology. These fields have gradually separated in recent decades as ecosystem science has focused on production and biogeochemical processes at large scales, while population biology has emphasized biotic interactions, often at small scales and typically excluding feedbacks with the physical–chemical environment. The trophic cascade concept has elements of both. It emphasizes the consequences of population interactions for production processes. Population dynamics influence nutrient cycles and must therefore be considered in an ecosystem context. These questions entail concerted work at several different scales, using theoretical principles and practical tools derived from several of the subdisciplines of ecology.

Our approach focused intensive effort on three experimental lakes for seven years. A coordinated, multidisciplinary team is mandatory for this kind of research. We are fortunate to have worked with a remarkable group of collaborators, postdocs, graduate students, technicians, and undergraduates in the course of this project. Their contributions include participation as coauthors of this volume. We also wanted to produce an integrated synthesis of the ecosystem experiments, while minimizing gaps and redundancies that sometimes arise in multiauthored collections. In an attempt to avoid these difficulties, one or both of us is among the authors of most chapters. Our job was made easier by the efforts of Xi He on chapters about fishes and Pat Soranno on chapters about zooplankton.

Individual chapters acknowledge the helpful contributions of reviewers. We thank them all. Mike Pace (New York Botanical Garden), Bob Paine (University of Washington) and John Birks (University of Bergen) read the manuscript carefully and made numerous helpful suggestions.

We appreciate the wise and helpful leadership of those affiliated with the University of Notre Dame Environmental Research Center (UNDERC) near Land o' Lakes, Wisconsin. Early constraints to the use of the lakes for this project were overcome through the foresight of Father Theodore Hesburgh and Professor Robert Gordon, at that time President and Vice President, respectively, of the University of Notre Dame. The existence and growth of UNDERC owe to their vision and energy. More recently, we have enjoyed an excellent working environment due to the efforts of Dr Ronald Hellenthal, Gillen Director of UNDERC, and Dr Martin Berg, Assistant Director. Like all users of UNDERC, we have benefitted from the generosity of the Hank family, for which we are deeply grateful. We thank all members of the Notre Dame community for the opportunity to work at this remarkable field facility.

We have been helped on countless occasions by Tom Frost and the staff of the University of Wisconsin's Trout Lake Station, Boulder Junction, Wisconsin. Our approaches and interpretations have been enhanced significantly by close association with the Northern Lakes Long-Term Ecological Research and the Little Rock Lake Acidification programs sited there. We thank Tim Kratz for many helpful discussions of year-to-year variability in lakes of the region, and Walt Haag, Tim Meinke, and Carl Watras for sound advice about equipment and methods.

Linda Holthaus of the Center for Limnology kept track of text, figures, and tables; cajoled authors; compiled references; and maintained a tenor of organization that was essential for the production of this book. We marvel at her skills as a coordinator, and are grateful for her efforts. Thanks are due to Liz Krug and Bonnie Throgmorton for preparing the final reference list. Paul Carpenter, Bill Feeny, John Morrice and Mark Scheuerell provided valuable help on computer graphics.

Financial support for this project came from the Ecosystems Studies Program of the U.S. National Science Foundation.

Stephen R. Carpenter
James F. Kitchell

Madison, Wisconsin
June 1992

1 · *Cascading trophic interactions*

James F. Kitchell and Stephen R. Carpenter

Introduction

The extent to which physical–chemical or biotic factors influence community structure and ecosystem function continues as one of the fundamental issues of ecology. The action and interaction of abiotic and biotic factors was recognized in early concepts of plant succession (McIntosh, 1985) and continues in the most contemporary reviews of plant–animal interaction (Strong, 1992). In animal community ecology, there have been several recent syntheses of the effects of multiple controlling factors (Menge & Sutherland, 1976, 1987; Fretwell, 1977; Oksanen *et al.*, 1981; Power, 1992; Strong, 1983, 1992). Vigorous debate has surrounded the relative roles of predation and competition (Hairston, Smith & Slobodkin, 1960; Murdoch, 1966). Predation has been viewed from the standpoints of predator control of prey communities (Oksanen, 1983, 1990) and of prey constraints on predator communities (Price *et al.*, 1980; Kareiva & Sahakian, 1990; Hunter & Price, 1992).

Like the other branches of ecology, limnology has evolved through debates about the roles of abiotic and biotic factors (Edmondson, 1991). In some respects, lakes are ideal systems for the study of multifactor interactions at the ecosystem scale (Carpenter, 1988*a*, pp. 4–5). Boundaries are clear and the difficulties of system definition that plague some areas of ecology (McIntosh, 1985) are lessened. Lakes are amenable to experimentation on a variety of scales, including whole-lake manipulations (Frost *et al.*, 1988). At a global scale, insolation and climate have dominant effects on lake ecosystems (Brylinsky & Mann, 1973). At scales ranging from lake districts to individual lakes, nutrient input rate, water renewal rate and lake morphometry are prominent abiotic factors (Schindler, 1978; Fee, 1979; Carpenter, 1983). At these scales, biotic effects are also evident and may contribute to the substantial variability observed in basic ecosystem processes such as primary production.

Over the past several decades, two general lines of inquiry have emerged in the efforts of aquatic scientists. The seminal observations of Hrbacek *et al.* (1961) and the pioneering work of Brooks & Dodson (1965) set the stage for an immense diversity of studies designed to clarify the unknowns of grazing, competition and predation as interactions of primary importance in regulating aquatic populations and community structure (Kerfoot & Sih, 1987). A second pursuit emerged from general interest in water quality, flourished during the limiting nutrient debate, and remains fully established as the first order of consideration in limnological textbooks (e.g. Wetzel & Likens, 1991).

Each of the topics identified in the preceding paragraphs is complex; a contemporary synthesis would be a book-length monograph in itself. Our goals are more modest and specific: to present the results of a large-scale experimental test of the trophic cascade hypothesis which coupled abiotic and biotic effects and was proposed to explain the unaccounted for variability of primary production rates in lakes (Carpenter, Kitchell & Hodgson, 1985).

The trophic cascade hypothesis states that nutrient input sets the potential productivity of lakes and that deviations from the potential are due to food web effects (Carpenter *et al.*, 1985). Nutrient and food web effects are complementary, not contradictory, but they act at different time scales (Carpenter, *et al.*, 1985; Carpenter, 1988*a*). Food web effects stem from variability in predator–prey interactions and their effects on community structure (Carpenter & Kitchell, 1987). Acting through selective predation, variability at the top of the food web cascades through zooplankton and phytoplankton to influence ecosystem processes (Carpenter *et al.*, 1985). This is the definition of trophic cascades used throughout this book. Our usage derives from Paine (1980). It is distinct from the cascade model used by theorists to describe the statistical distribution of links and other static properties of food webs (Cohen, 1989).

In this chapter, we first summarize the roots of the trophic cascade idea for lakes. We consider the major kinds of studies that have been used to learn about lake ecosystems: comparisons, long-term studies, simulation models, mesocosms and ecosystem experiments. We then explain our approach, in which ecosystem experiments are central but all five methods of study are represented. Finally, we explain the goals and structure of the remainder of the book.

The trophic cascade concept in lakes

The ideas about a trophic cascade in lakes derive from two primary sources. First is the extension of thermodynamic principles to ecology, which yields the expectation that organic production in lakes should be a function of nutrient status. Strong correlations exist between nutrient loading or nutrient concentration and primary production (Schindler, 1978). Flow of energy and matter upward through food chains is a central paradigm of the ecosystem approach (Lindeman, 1942; LeCren & Lowe-McConnell, 1981; Odum, 1969). Although the correlations confirm a logical expectation, the log–log regressions used in these empirical analyses account for only part of the variance observed when primary production rates or their surrogates are considered as a function of nutrient loading rates (Carpenter et al., 1991). At any given level of nutrient loading or concentration, algal concentration or production may differ among lakes by an order of magnitude or more. Measurement error may account for some of that variance, but a mechanistic alternative is also plausible.

The second major element of the cascade idea derives from the evolutionary principles widely employed in contemporary population biology and community ecology. Consumers are typically selective in the types and sizes of resources they consume (Hall et al., 1976; Kerfoot & Sih, 1987). This tenable extension of natural selection theory is embodied in aquatic ecology as the principles of size-selective predation (Hrbacek et al., 1961; Brooks & Dodson, 1965), the keystone predator concept (Paine, 1966) and theories of optimal foraging and habitat usage (Werner, 1986). In lake ecosystems, the result of selective predation plays a major role in community composition at each trophic level (de Bernardi, 1981). Piscivores determine the size and species composition of the planktivorous fish assemblage beneath them in a food web (Tonn & Magnuson, 1982). Selective planktivory by fishes and invertebrate predators profoundly influences the community of herbivorous zooplankton which, in turn, regulates the amount and kinds of phytoplankton that compete for nutrients (Brooks & Dodson, 1965; Sommer, 1989). Much of the available nutrient pool derives from recycling through excretion processes that are strongly size-dependent (Kitchell et al., 1979; Peters, 1983). Thus, the rates of primary production can be substantially influenced by a trophic cascade of size-selective predation processes that start at the top of the food web.

The components of the cascade argument have a crucial nexus at the zooplankton (Fig. 1.1). Larger herbivores are consumed selectively by planktivorous fishes (Brooks & Dodson, 1965). Carnivorous zooplankton feed most heavily on smaller zooplankton (Hall *et al.*, 1976). Abundant planktivorous fishes shift the zooplankton composition toward dominance by smaller individuals. When planktivorous fishes are absent, predation by planktivorous invertebrates and competition among herbivores shift the zooplankton toward larger individuals (Brooks & Dodson, 1965; Hall *et al.*, 1976). Large herbivores such as *Daphnia* have a greater impact on phytoplankton because they consume a broad range of sizes and morphologies of algae (Burns, 1968; Gliwicz, 1980; Bergquist, Carpenter & Latino, 1985). Owing to their size, large zooplankton have lower mass-specific rates of nutrient excretion (Peters, 1983). In comparison with a small-bodied zooplankton assemblage of equal biomass, an assemblage dominated by large *Daphnia* should graze a broader spectrum of algae but recycle nutrients at lower rates. Thus, algal biomass and primary production should be less in *Daphnia*-dominated lakes than in lakes dominated by small zooplankton such as *Bosmina*, small calanoid copepods, or rotifers. Zooplankton biomass should be directly related to nutrient concentration but, for equivalent zooplankton biomass, algal biomass and production should be inversely related to mean zooplankton size (Carpenter & Kitchell, 1984).

Evidence for trophic cascades stems from a variety of sources as represented in recent reviews (Northcote, 1988; Power, 1992). Although each case has its idiosyncracies, the central idea about the effects of a top predator are documented through experimental studies in ponds (Hurlbert & Mulla, 1981; Spencer & King, 1984), lakes (Henrikson *et al.*, 1980; Shapiro & Wright, 1984; Carpenter *et al.*, 1987), rivers (Power, 1990) and intertidal (Paine, 1980) and subtidal marine communities (Estes & Palmisano, 1974; Mann & Breen, 1972). Important evidence has also come from studies of variability in lakes spanning gradients of nutrient richness and food web structure (Carpenter *et al.*, 1991; Persson *et al.*, 1992).

A comparative analysis of these state variables in 25 lakes sampled from 2 to 6 years each revealed both nutrient and predation effects (Fig. 1.2) (Carpenter *et al.*, 1991). Total phosphorus concentration during spring (a surrogate for nutrient input) was positively correlated with summer mean chlorophyll ($r^2 = 0.40$, $p < 0.001$). Zooplankton mean length, an indicator of size-selective predation and the intensity of grazing, was negatively correlated with summer mean chlorophyll ($r^2 = 0.45$, $p < 0.001$). A multiple regression combining food web and

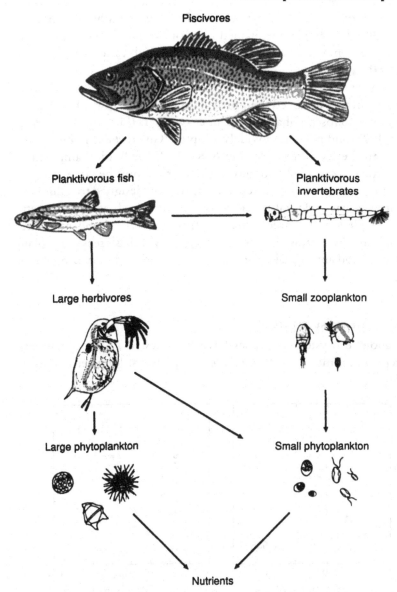

Fig. 1.1. Major interactions of the trophic cascade illustrated with selected organisms from our experimental lakes. Nutrients (mainly inorganic phosphorus and nitrogen) are provided by inputs from the watershed and recycling by animals. The relationship of this food web to the microbial food web is detailed in Fig. 14.1. Organisms are not to scale; for scale drawings see Figs. 4.1, 8.1, 11.1, and 14.2.

nutrient variables to predict chlorophyll was highly significant ($R^2 = 0.68$, $p < 0.001$) and explained far more variance than regressions based on nutrient variables alone or food web variables alone (Carpenter et al., 1991).

Analogous reasoning forms the basis for 'biomanipulation' as a management tool designed to improve water quality. Many of the ideas for that owe to Joseph Shapiro and his co-workers (Shapiro, Lamarra & Lynch, 1975) and have been widely adopted (Gulati et al., 1990). As we have detailed elsewhere (Carpenter & Kitchell, 1992), biomanipulation and the trophic cascade hypothesis are similar but not the same. Biomanipulation draws from a diverse set of approaches designed to reduce the expression of an undesirable ecological attribute such as hypolimnetic oxygen depletion, bluegreen algae blooms or overly abundant littoral macrophytes. In contrast, the trophic cascade hypothesis seeks to explain the within- and among-lake variability in the basic primary production process.

Ecosystem epistemology

Ideas about ecosystem processes have been tested in five essentially different ways: mesocosms, interlake comparisons, long-term studies,

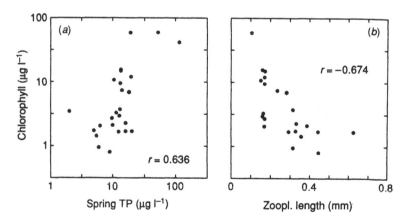

Fig. 1.2. Scatterplots relating photic zone chlorophyll concentration to (a) total phosphorus concentration during spring and (b) mean zooplankton length. Each data point is the mean of 2–6 summer stratified seasons for a lake in the Great Lakes region of North America (Carpenter et al., 1991). Correlations are highly significant ($p < 0.001$). From Carpenter et al. 1991. Copyright © Springer-Verlag New York, Inc. Used by permission.

simulation models, and ecosystem experiments. While these categories are somewhat arbitrary and have some obvious overlaps, all ecosystem studies could be assigned to at least one of them. Each approach has different strengths. All five approaches have been used in our research on the trophic cascade, although we view ecosystem experiments as the core of our work. We will review these before explaining why we think that ecosystem experimentation is central to understanding the trophic cascade.

Comparisons usually involve analysis of how some ecosystem property of interest (such as primary production or plankton community structure) changes along gradients of potential causal factors' (such as nutrient supply or fish density). Comparative studies can easily be scaled to whole ecosystems, often at relatively low cost (Cole, Lovett & Findlay, 1991). They quickly reveal the range of possible system states and relationships among key variables. On the other hand, it can be very difficult to detect dynamics and infer responses to perturbation from comparisons of static properties of ecosystems (Carpenter & Kitchell, 1988; Carpenter et al., 1991).

Long-term studies involve analysis of ecosystem dynamics through direct observation or paleoecological surrogates (Strayer et al., 1986; Likens, 1989). Despite their relative rarity, long-term records have made remarkable contributions by reducing the constraints of time scale that limit much ecological research (Strayer et al., 1986; Likens, 1989; Magnuson & Bowser, 1990; Jassby, Powell & Goldman, 1990; Edmondson, 1991). Because the population dynamics of fishes play out over decades, long-term studies have been useful in studies of trophic cascades (Jassby et al., 1990; Kitchell, 1992) (Chapter 15). The main disadvantage of long-term studies is the slow rate at which new insights develop (Walters, 1986; Carpenter, 1988a). One must wait for rare, unpredictable informative changes in order to learn about the system (Kitchell, 1992; Edmondson, 1991). Where good paleolimnological indicators exist, this difficulty may be circumvented. As detailed in Chapter 15, the documented history of manipulations in Peter, Paul and Tuesday Lakes, plus the calibration of neolimnological responses due to our experimental work, allow us to create a unique combination of paleolimnological observations that cover the entire range of trophic conditions in these lakes.

Simulation analyses are based on models designed to serve as a simplified surrogate for the complexities of natural systems. Models of ecosystem dynamics depend on the rationale that the major components and interactions have been included, and that forcing variables and parameters can be known rather precisely (e.g. Chapra & Reckhow, 1983). Simulation models have proven valuable for understanding the possible outcomes of postulated interactions in ecosystems (e.g. DeAngelis & Waterhouse, 1987). However, developing model structures that can cope with the variability of ecosystems remains a challenging problem (Beck, 1983; Walters, 1986). Simulation models are usually calibrated and tested using field data, and are therefore used with one or more of the other approaches. Models and their evaluation have played a central role in the design of our experimental work. As candidly detailed in Chapter 16, the mixture of successes and failures has been very informative.

Mesocosms are laboratory or field enclosures used for experimental studies of selected ecosystem components or interactions. Experimental ponds (Hall, Cooper & Werner, 1970; Hurlbert & Mulla, 1981) are mesocosms that are relatively large and complex but, due to their depth, offer environments that simulate only the littoral zone of lakes. Because mesocosm experiments can be controlled and replicated, they are a powerful experimental tool for studying ecological mechanisms (Hairston, 1989; Mazumder et al., 1990; Soto & Hurlbert, 1991). Because of their typically small size, mesocosms necessarily exclude some processes that may be important for predicting ecosystem dynamics (Frost et al., 1988; Carpenter & Kitchell, 1988). For example, inshore–offshore migrations of fishes and hydrodynamic fluxes of nutrients cannot easily be included in enclosure experiments. Therefore, mesocosm results may not be directly transferable to the ecosystem scale and can, in fact, yield erroneous extrapolations (Frost et al., 1988). Recognizing these limitations, we have used mesocosm experiments to study selected interactions between zooplankton and their predators (Chapter 9) and between herbivorous zooplankton and the algae they graze (Chapter 12), and the complexities of interrelationships among microbes, protozoa and zooplankton (Chapter 14).

Experimentation at the ecosystem scale has made important contributions to ecology in general, and limnology in particular, over the past several decades (Likens, 1985; Schindler, 1988). Such experiments lack the scale problems of mesocosms and, unlike most comparative studies, can be designed to directly address ecosystem dynamics and responses to perturbation. Ecosystem manipulations can significantly increase learning rate, as measured by rate of reduction of error in model predictions (Walters, 1986; Carpenter, 1988a). Because ecosystem experiments simulate management actions at the appropriate scale, they have been a convincing means of resolving uncertainty in management controversies (Walters, 1986; Schindler, 1988; Kitchell, 1992). The main disadvantage of ecosystem experiments is that replication is difficult or impossible (Matson & Carpenter, 1990). Access to experimental sites often limits ecosystem experimentation. In fact, the University of Notre Dame Environmental Research Center, where we conducted our work, is among the few places on earth where controlled access to lakes allows for whole-lake experimental studies that are unaffected by perturbations of the watershed, anglers and/or a curious public. Despite these difficulties, deliberate or inadvertent perturbations of lake food webs have yielded valuable information about trophic cascades (Henrikson et al., 1980; Shapiro & Wright, 1984; Carpenter et al., 1987; McQueen et al., 1989; Benndorf, 1990; Jeppeson et al., 1990; Reinertsen et al., 1990; Kitchell, 1992).

Our approach

The central idea of this book – that effects derived from the fish community cascade through the food web to influence primary production processes – is obviously testable by whole-lake experimentation. More importantly, evidence from other approaches leaves gaps that can only be resolved by whole-lake experiments. Several comparative studies are consistent with the cascade hypothesis, but offer less direct insight about response capacity of perturbed lakes (Mills & Schiavone, 1982; Pace, 1984; Quiros, 1990; Carpenter et al., 1991; Persson et al., 1992). Mesocosm studies reveal a complex array of interactions that could dampen trophic cascades (Kerfoot & Sih, 1987; Frost et al., 1988; Gulati et al., 1990; DeMelo, France & McQueen, 1992). For example, compensatory shifts in species composition (Strong, 1992) (Chapter 8) or migrations of key predators or their prey (Chapters 5 and 9) could override cascade effects. Fishes and plankton, their functional and numerical responses to food resources, their habitat selection behaviors, their life history strategies and the integrated result in population and trophic dynamics all operate at the scale of entire lakes. Those variables function interactively with other large-scale processes such as mixing, diffusion and light extinction. Thus, the trophic cascade embodies a set of processes demanding the whole ecosystem as an experimental unit and the growing season as a minimum duration. As noted above, our ecosystem experiments were coordinated with comparative, long-term, simulation and mesocosm studies where these alternatives were appropriate and advantageous. Frost *et al.* (1988) and Tilman (1989) have discussed the utility of combining approaches at several scales for ecosystem studies.

When we initiated this project in the early 1980s, only two whole-lake food web experiments addressing fish, zooplankton, phytoplankton and nutrients were known to us (Henrikson *et al.*, 1980; Shapiro & Wright, 1984). Now, results are published from several more whole-lake experiments (Benndorf, 1990; Jeppeson *et al.*, 1990; Sanni & Waervagen, 1990; van Donk *et al.*, 1990; Giussani, de Bernardi & Ruffoni, 1990; Reinertsen *et al.*, 1990) and a large number of inadvertent perturbations of food webs in whole lakes. Like our work, most of these studies examined a comprehensive array of response variables ranging from fish to nutrients, and from population to ecosystem processes. We believe that our experiments are a unique and valuable addition to this literature for several reasons.

(1) Unlike several other whole-lake experiments, we employed a reference ecosystem (Likens, 1985) as a check for trends unrelated to our manipulations.
(2) Our experimental lakes lie in protected watersheds closed to the public. As a result, we were spared the confounding effects of unplanned or unknown human disturbance (Chapter 2).
(3) We employed statistical approaches that compensate for the lack of replicability inherent in large-scale experimentation (Chapter 3).
(4) We coordinated our experiments with paleolimnological studies to gain a long-term view (Chapter 15) and with simulation models to conduct specific tests of theory (Chapter 16).

It is clear from the literature that a wide range of responses can develop from perturbations of lake food webs (Carpenter & Kitchell, 1988). We do not yet know the conditions that cause one particular sequence of events to occur rather than the alternatives. Each ecosystem experiment is a valuable datum toward developing that understanding.

The seven intensive years of this program involved two major manipulations. The first (1984–6) was a reciprocal transplant of the tops of two contrasting food webs. Largemouth bass were removed from Peter Lake and planted in Tuesday Lake after its minnow population had been removed. Those minnows were then planted in Peter Lake. This experiment was designed to test maximum response capacity; i.e., to determine the greatest possible contrast of food web effects on community structure and ecosystem processes. The second set of manipulations (1987–90) sought to test for rate-response effects and evaluate time lags due to recruitment fluctuations commonly observed in natural populations of planktivorous or piscivorous fishes. Again, Peter and Tuesday Lakes were manipulated by altering their fish communities. In both cases, we evaluated the prospect of a cascade of responses by monitoring each trophic level, by conducting independent experiments on key interactions and by calibrating the response signals as they entered the sedimentary record. Throughout the period of both experiments, a similar and nearby system, Paul Lake, was monitored as an unmanipulated reference and the source of evidence on background variation. During 1986–9, a fourth system, Bolger Bog, was added as a site for experimental evaluation of behavioral responses that became a major feature of ecosystem responses (Chapter 5).

Our specific hypotheses and experimental tests pertain to the pelagic zones of lakes. This focus was largely dictated by the pragmatisms of limited resources and the fact that littoral zones are small and sparsely vegetated in our study lakes. We fully recognize the importance of the littoral zone in many of the world's lakes (Wetzel, 1990), the strong interactions that could drive cascades in the littoral zone (Carpenter & Lodge, 1986); Lodge *et al.*, 1988), and the feedbacks between littoral and pelagic zones that affect pelagic fish populations and trophic cascades (Gulati *et al.*, 1990; Boers, van Ballegooijen & Uunk, 1991; Persson *et al.*, 1992; Carpenter *et al.*, 1992*b*).

Although food web structure is a central element of our work, we have not pursued the kinds of analyses expressed in the work of Pimm (1982) and Cohen, Briand & Newman (1990), or the debates stemming from interest in a general set of principles that might emerge from studies of food web structure (Paine, 1988; Sugihara, Schoenly & Trombia, 1989; Schoenly & Cohen, 1991). Our interest in food webs emphasizes function and dynamic feedback. We recognize and document the trophic ontogeny and omnivory (Chapters 4, 6, and 7) and behavioral shifts (Chapters 5 and 9) that are beyond the pale of static, structural approaches. Temporal variability is a central theme of our work.

Response to the trophic cascade argument has been and continues to be surprising. Ecologists of all sorts, and limnologists in particular, responded with a barrage of ideas, results and criticisms. Apparently, the discipline of aquatic ecology was predisposed to a novel notion. One of our limnological colleagues offered some insight when he pointed out that water quality issues had become the province of chemists and engineers. He reasoned that the trophic cascade idea was of substantial interest because it '. . . puts biology back in limnology'. People tended to choose sides in the unfortunately simplified dichotomy represented by 'Is it top-down or bottom-up?' (McQueen, Post & Mills, 1986; Northcote, 1988). That debate continues (Power, 1992; Strong, 1992).

We did not invent the main ideas of the trophic cascade but did combine them in a way that we hoped would help resolve the contradictions or shortcomings of singular views and that might offer a means for evaluating unexplained variability. We can take some credit for helping advance the conceptual framework and we can offer the detailed results of our theoretical and experimental work.

Many of the primary results from this program have been and will continue to be published in the outlets most appropriate to a specific subdiscipline of ecology. In our experience, many readers are not familiar with (i.e. don't have enough time to read) the breadth of specialized evidence developed in the growing diversity of journals and books such as those employed as primary outlets by theorists, water chemists, experimental ecologists, fisheries scientists, plankton ecologists, paleolimnologists, resource managers, and so forth. That observation led us to believe that we needed a single, comprehensive airing of our findings. We offer this volume as synthesis and documentation of the detail and diversity of results that emerged from seven years of experimental and theoretical work. In addition to new and comprehensive analyses, it contains our candid assessment of successes, failures and what we believe to be the most insightful ways to proceed with the next set of unknowns.

Overview of the book

The next two chapters provide details of our experimental design and our statistical methods for analyzing and interpreting the ecosystem experiments. Chapter 2 also provides information on the surroundings, history, and general limnology of our experimental site. Chapter 3 introduces analytical methods that are not well known to ecologists and may be applicable to a wide range of large-scale ecological investigations (Matson & Carpenter, 1990).

Fishes, the independent variables of our ecosystem experiments, are treated in the next trio of chapters. The fish populations are described in Chapter 4. We found that behavioral responses of fishes to their predators were as important as consumption in transmitting effects to lower trophic levels. Behavioral interactions of fishes are analyzed in more detail in Chapter 5. Chapter 6 presents diet data and consumption rates that form the basis for analyses of fish effects on zooplankton.

Zooplankton, the nexus of the cascade in pelagic systems, are treated in the next four chapters. The carnivorous insect larva *Chaoborus*, which proved pivotal in some of our manipulations, is the subject of Chapter 7. Chapter 8 details the population densities and community structure of the omnivorous and herbivorous zooplankton. Diel vertical migration proved to be an important adaptive response of *Daphnia*, the keystone grazer in our experiments. An analysis of migration by *Daphnia* appears in Chapter 9. In Chapter 10, we present the biomass and size structure of the zooplankton, which form the basis for expected responses by the phytoplankton.

Phytoplankton responses are analyzed in the next three chapters. Composition and biomass of epilimnetic phytoplankton are presented in Chapter 11. We found that the deep-dwelling phytoplankton of the metalimnion were distinctive in their community structure and responses to nutrients and grazers. The metalimnetic phytoplankton are analyzed in Chapter 12. Chapter 13 presents the evidence of ecosystem responses that are central to our hypotheses: the dynamics of chlorophyll, primary production, and their relationships to light and nutrients.

Recently, aquatic scientists have discovered a remarkable complex of interactions among bacteria and protozoa in both oceans and lakes (Stockner & Porter, 1988; Porter *et al.*, 1988). The opportunity provided by our experiments attracted a fruitful collaboration and the consequent additional evidence of microbial responses to food web manipulation as summarized in Chapter 14.

Our experimental lakes have long histories of limnological study and annually varved sediments that provide unusual opportunities in paleolimnology. Whole-lake experiments allowed the calibration of paleolimnological indicators through analyses of changes in deposition during the course of the experiment. Our efforts to perform such calibrations and reconstruct the histories of these lakes are described in Chapter 15.

The concept of the trophic cascade in lakes, the hypotheses tested in this book, and the design of these ecosystem experiments, were largely based on simulation models. Our models and the expectations we derived from them are summarized and critically evaluated in Chapter 16.

Chapter 17 closes the book with a synthesis of progress to date, speculations about the generality of trophic cascades in aquatic and terrestrial systems, and evaluation of new directions.

We suggest the following guidelines for readers who wish to read the book selectively. The main features of the experimental design are presented in Table 2.3 and accompanying text. Chapter 3 is most important for readers who wish to follow statistical details of several chapters. Chapters 4–15, which contain the detailed results of our experiments, each end with summaries of the main points. Our essential findings can be gleaned from a reading of those summaries. Chapter 16 takes stock of models of the trophic cascade, and candidly evaluates our hypotheses. Chapter 17 offers a synthesis of our findings in relation to the current status of the trophic cascade concept.

Acknowledgements

We thank Tom Frost, Tom Martin, Mike Pace and Lars Rudstam for helpful advice on this chapter.

2 · *Experimental lakes, manipulations and measurements*

Stephen R. Carpenter and James F. Kitchell

The experimental lakes

Our studies were conducted in Paul, Peter, and Tuesday Lakes (Fig. 2.1). These lakes lie on the grounds of the University of Notre Dame Environmental Research Center (UNDERC) near Land o' Lakes, Wisconsin, U.S.A. (89°32′ W, 46°13′ N). UNDERC occupies more than 2800 ha of land donated to the University of Notre Dame in the 1940s 'for the scientific purposes of Forestry, Botany, Biology and allied sciences' (Gillen, 1939). The limnological potential of the property had been recognized in the 1920s, when E. A. Birge, C. Juday and associates sampled most of its lakes (Beckel, 1987).

Because the UNDERC facility is privately owned and protected, it offers remarkable opportunities for field experimentation. Fish populations in the experimental lakes are unexploited, and can be manipulated without the complications of sport or commercial fishing. The drainage basins are undeveloped and lie entirely within the UNDERC property, so disturbances or chemical inputs to the lakes that might confound ecosystem experiments are minimized. UNDERC is about 40 km from the University of Wisconsin's Trout Lake Station, a national center of limnological activity for more than 50 years (Magnuson & Bowser, 1990). The array of intensively studied lakes in the region provides additional reference (or 'control') systems for manipulative experiments (Carpenter et al., 1989) as well as opportunities for comparative studies (Carpenter et al., 1991).

The lakes lie in the Northern Highland Lake District. The geological province is characterized by primarily Precambrian bedrock capped by a thin layer of sedimentary rocks left by the Paleozoic seas. On top of this are glacial deposits left by the Woodfordian and Valderan substages of the Wisconsinan glaciers (Paull & Paull, 1977). The soils are infertile glacial outwash with reduced cation exchange capacity, leaving them susceptible to acidification and bog formation. As the Valderan glaciation (approximately 12000 years ago) receded, the kettle basins of Paul, Peter, and Tuesday Lakes arose from the melting of blocks of ice in the glacial drift.

The regional climate has cool summers and cold winters, with no dry season. Annual precipitation is as much as 100 cm of snow and rain. Mean January temperatures range from −10 °C to −20 °C, and ice thickness on the lakes can exceed 60 cm. The lakes are generally free of ice by late April or early May. In most springs, the lakes do not mix completely. Mean date of the last killing frost is 3 June. Average July temperatures range from 16 °C to 21 °C. Frosts occur regularly by late September. The lakes become isothermal by the end of October, and usually freeze in November. Paul and Peter Lakes can mix in the autumn.

(a)

Fig. 2.1. Aerial photographs of the experimental lakes taken in July 1988. (a) Tuesday Lake (foreground) and paired Paul and Peter Lakes (background). (b) Tuesday Lake. (c) Paul (foreground) and Peter Lakes. Boat landings and transect lines used for sampling can be seen in all three lakes. The large bags visible in Peter Lake were used for experiments on diel vertical migration of *Daphnia* (Chapter 9). Photographs taken by Tony Aloi.

(b)

(c)

Table 2.1. *Common names and Latin names of fishes in this book*

Common name	Latin name
creek chub	*Semotilus atromaculatus*
fathead minnow	*Pimephales promelas*
finescale dace	*Phoxinus neogaeus*
golden shiner	*Notemigonus crysoleucas*
largemouth bass	*Micropterus salmoides*
mudminnow	*Umbra limi*
rainbow trout	*Oncorhynchus mykiss*
redbelly dace	*Phoxinus eos*
yellow perch	*Perca flavescens*

Drainage basins of the lakes were clear cut early in the twentieth century (UNDA, 1988). Forests of the uplands around the lakes correspond to the northern hemlock-hardwoods described by Curtis (1959). These woods are now dominated by sugar maples (*Acer saccharum*), which appear even-aged and probably date from the time of logging. Red maple (*Acer rubrum*), yellow birch (*Betula lutea*), eastern hemlock (*Tsuga canadensis*), balsam fir (*Abies balsamea*) and white spruce (*Picea glauca*) also occur in the uplands. The *Sphagnum* bogs around the lakes harbor black spruce (*Picea mariana*), white cedar (*Thuja occidentalis*) and tamarack (*Larix laricina*) trees as well as bog rosemary (*Andromeda polifolia*), labrador tea (*Ledum groenlandicum*) and other ericad shrubs typical of northern Wisconsin lowlands (Curtis, 1959). *Potentilla palustris* and *Utricularia* spp. often grow at the bog edges. Littoral zones of the lakes are sparsely vegetated with *Isoetes braunii*, *Potamogeton epihydrus* and water lilies (*Nuphar luteum* and *Nymphaea odorata*).

Paul and Peter Lakes were connected by a narrow channel prior to 1951. In that year, Professor Arthur Hasler and his associates divided them for experimental use by constructing an earthen dike across the channel. A culvert allowed water to flow from Paul Lake to Peter Lake. Their study was the first ecosystem experiment to use a reference (or 'control') ecosystem (Likens, 1985). The goal of the experiment was to test the idea that brown-water lakes could be made more productive and suitable for fishes by liming. The largemouth bass and yellow perch assemblages of both lakes were removed with rotenone, and both lakes were restocked with rainbow trout (Table 2.1). Peter Lake received a series of limings which cleared the water, while Paul remained as the reference system. Although the last liming was in 1976, effects of the

Table 2.2. *Morphometric and chemical characteristics of the study lakes*

Chemical characteristics are means measured from 1984 through 1990.

Characteristic	Paul Lake	Peter Lake	Tuesday Lake
maximum depth (m)	15.0	19.6	18.5
mean depth (m)	3.9	6.0	6.9
volume (m³)	57 823	137 352	83 337
area (m²)	14 804[a]	22 886	12 102
pH	6.1	6.6	5.6
alkalinity (μeq l^{-1})	55	173	21
dissolved inorganic carbon (mg l^{-1})	1.3	2.7	0.7
conductivity (μmhos cm^{-1}):			
epilimnion	18	29	17
hypolimnion	48	42	26

Note:
[a] Area excludes small bog islands found in shallow (< 1 m) water, which are not shown on the map.

limings were apparent during our studies. Peter Lake consistently had higher pH, higher alkalinity, higher dissolved inorganic carbon concentrations, and greater transparency than Paul Lake (Elser, Elser & Carpenter, 1986b) (Table 2.2). Details of the limings and fish manipulations that antedated our study are provided by Kitchell & Kitchell (1980), Elser *et al.* (1986b), Leavitt, Carpenter & Kitchell (1989), and Chapter 15. By 1980, the fish communities of Paul and Peter Lakes consisted of dense populations of largemouth bass.

Tuesday Lake is located about 0.5 km from Paul and Peter Lakes (Fig. 2.1). Prior to initiation of our research, Tuesday Lake had been studied intermittently since 1956 (Schmitz, 1958; UNDA, 1988). Because the lake is small, deep and sheltered from the wind, its thermocline is shallow and water below about 8–10 m depth is always anoxic. To our knowledge, the lake has never mixed naturally to the bottom. For most of the lake's history, the fish fauna has been typical of winterkill lakes of the region and includes northern redbelly dace, finescale dace, and mudminnow. Tuesday Lake was also part of the program of whole-lake experimentation established by Hasler. It was artificially destratified by aeration in 1956, but the effects were short-lived (Schmitz, 1958). In 1961, Tuesday Lake was stocked with rainbow trout (UNDA, 1988).

However, the trout probably lived no more than one summer. Dace and mudminnows had regained dominance by 1970, and possibly earlier (UNDA, 1988).

Each of the three lakes is relatively deep for its surface area (Figs 2.2 and 2.3; Table 2.2). All three lakes exhibit the hyperboloid geometry typical of kettle lakes (Carpenter, 1983). The lake water has a low ion content. Paul and Tuesday Lakes are mildly acidic, whereas Peter Lake is circumneutral because of its long history of liming. All three lakes have higher conductivities in the hypolimnion, suggestive of spring meromixis.

Experimental manipulations

Three major sets of whole-lake manipulations were conducted (Table 2.3): a reciprocal exchange of fish assemblages between Peter and Tuesday Lakes in 1985; reversal of the manipulation in Tuesday Lake beginning in 1987; and a series of planktivore additions to Peter Lake beginning in 1988. Paul Lake remained an undisturbed reference ecosystem throughout our experiments.

The reciprocal exchange of fish assemblages in 1985 involved nearly equal biomasses of piscivores and planktivores (Carpenter *et al.*, 1987) (Table 2.3). Piscivorous largemouth bass were removed from Peter Lake by angling and electrofishing and held behind blocking nets before

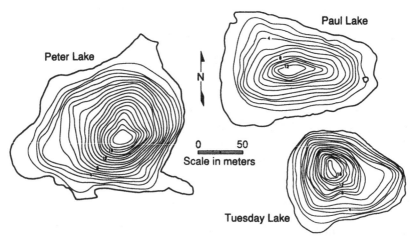

Fig. 2.2. Bathymetric charts of Paul, Peter, and Tuesday Lakes drawn to the same scale.

Table 2.3. *Manipulations performed in Peter and Tuesday Lakes, 1984–90*

Paul served as an undisturbed reference system throughout this period.

Time	Peter Lake	Tuesday Lake
23–31 May 1985	375 bass (45.7 kg) removed	44 901 minnows (56.4 kg) removed
	44 901 minnows (56.4 kg) added	375 bass (47.5 kg) added
27 July 1985	91 bass (10.1 kg) removed	91 bass (10.1 kg) added
20 September 1986	—	761 bass (177.9 kg) removed
22 May 1987	—	1390 minnows (3.3 kg) added
17 May 1988	3000 trout (149.3 kg) added	—
23 May 1989	3000 trout (62.9 kg) added	—
28 August 1989	490 bass (43.7 kg) removed	—
	52 trout (11.4 kg) removed	—
23 May 1990	20 000 shiners (94.4 kg) added	—
May–August 1990	1367 bass (7.1 kg) removed	—
	37 trout (18.5 kg) removed	—
5 July 1990	—	4000 bass (2 kg) added

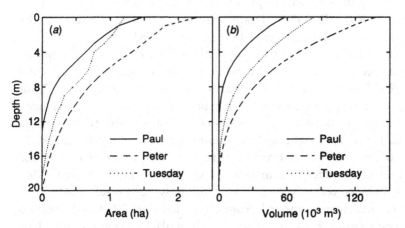

Fig. 2.3. Hypsometric curves of Paul, Peter, and Tuesday Lakes. Plots versus depth show (*a*) area (m^{-2}) at depth and (*b*) volume (thousands of m^3) below depth.

release into Tuesday Lake. An additional group of bass was moved to Tuesday Lake later in 1985. Overall, about 90% of the bass biomass of Peter Lake was removed in 1985. Planktivores were removed from Tuesday Lake by minnow trapping and held behind blocking nets before release into Peter Lake. These represented about 90% of the planktivore biomass of Tuesday Lake. By species, planktivores moved to Peter Lake were: 39654 redbelly dace, 2692 finescale dace, and 2655 mudminnows. Mortality of all fishes due to handling was less than 3% (Carpenter *et al.*, 1987).

The piscivore introduction to Tuesday Lake was reversed after the 1986 field season. Angling, electroshocking, and finally rotenone were used to remove the bass. Small populations of adult planktivores were reintroduced in the spring of 1987 in approximately the same species ratio present in 1984: 1250 redbelly dace, 70 finescale dace and 70 mudminnows. Planktivore populations were allowed to grow without further manipulation until 1990 (Chapter 4). On 5 July 1990, 4000 young of the year largemouth bass were introduced to study trophic ontogeny of this species in competition with planktivores.

Manipulations in Peter Lake from 1988 through 1990 were intended to simulate exploitation of a piscivore population in a system that contained both piscivores and planktivores. We hypothesized that variability in zooplankton and phytoplankton biomass would increase as exploitation drove piscivore populations to intermediate densities (Carpenter & Kitchell, 1987). Testing the hypothesis required both the introduction of planktivores and the removal of piscivores.

To establish the planktivore populations needed for this experiment, planktivore introductions to Peter Lake occurred each spring from 1988 to 1990. Rainbow trout were introduced in 1988 and 1989. Trout became piscivorous very rapidly (Chapter 6), so in 1990 we introduced an obligate planktivore, golden shiner. Trout and shiners were purchased from fish hatcheries and transported to the site by truck. In addition, a few planktivores were trapped from other waters on the UNDERC property and introduced to Peter Lake in August 1990. These included redbelly dace, fathead minnows, and creek chubs.

Piscivores were removed from Peter Lake in 1989 and 1990 to facilitate expansion of the planktivore populations. Largemouth bass were removed by electroshocking and angling. Rainbow trout were removed by gill netting and angling.

Routine measurements

The methods of fish ecology and limnology are complex technical topics in their own right (Bagenal, 1978; Wetzel & Likens, 1991). In this book, we have chosen to present a limited amount of methodological detail. We attempt to present what was measured and its ecological significance, while omitting particulars that would be intelligible only to aquatic ecologists. We hope that policy makes the book acceptably brief and accessible to a broad ecological audience. Consequently, some details that will be interesting and important for specialists have been omitted. In all cases, however, specialized information about the methods is available in our journal articles and/or a separately printed methods manual (Soranno, 1990). This section presents the overall design of the sampling program. Further methodological information, and citations of papers that present methods in greater detail, are found in subsequent chapters.

Through the summer stratified seasons of 1984–90, we monitored fishes, zooplankton, phytoplankton, and selected physicochemical variates in all three lakes (Table 2.4). Microbial measurements were made from 1988 through 1990 (Chapter 14). Sampling focused on summer stratification because that is the period of most intense metabolic activity in these ecosystems. In most years of the study, one or two samples were taken during fall mixis, and one or two samples were taken through the ice.

Piscivore populations were censused by intensive angling and electrofishing at the beginning and end of each field season (late May and late August). A smaller number of piscivores was sampled by angling every two weeks from May through September, to measure changes in diets. Measurements made routinely on piscivores were length, mass, age, and diet. These were determined by nonlethal methods, and handling mortality was very low throughout the study. Regular and continued tagging of adult fish resulted in multiple recaptures, which allowed studies of individual variation in diets and growth. After 1985, more than two thirds of the bass in Paul and Peter Lakes and all the bass in Tuesday Lake carried tags.

Planktivore populations were assessed by minnow trapping every two weeks from May through September. Measurements made routinely on planktivores were length, mass and diet by species. The invertebrate predator *Chaoborus* was sampled by night tows every two weeks from May through September.

Table 2.4. *Measurements performed routinely in Paul, Peter and Tuesday Lakes, 1984–90*

Frequency of measurements pertains to the period of summer stratification, May–September.

Variable	Method	Frequency
piscivore populations	mark–recapture by electroshocking	twice per summer
planktivore populations	minnow traps calibrated by depletion estimates	every two weeks
piscivore diets	stomach flushing of fish captured by angling	every two weeks
Chaoborus populations	vertical net hauls at night	every two weeks
zooplankton populations	vertical net hauls by day	weekly
zooplankton vertical migration	diel profiles with plankton trap	monthly
phytoplankton populations	pooled samples from epilimnion	weekly
profiles of chlorophyll *a*, pheopigments, dissolved inorganic C, pH, alkalinity, alkaline phosphatase activity, light, temperature, dissolved oxygen	Van Dorn samples at 6–10 depths	weekly
profiles of primary production	^{14}C fixation	weekly or every two weeks
ammonium enhancement response	pooled samples from epilimnion	weekly
N and P enrichment of chlorophyll *a* growth	pooled samples from epilimnion	every two weeks
total nitrogen, phosphorus	pooled samples from epilimnion, hypolimnion	weekly
surface irradiance	pyrheliometer	continuous

Zooplankton, phytoplankton, chemistry and physical variables were sampled weekly from May through September. Samples were taken routinely for zooplankton enumeration, phytoplankton enumeration, chlorophyll *a*, pheopigments, carbon fixation, total nitrogen, total phosphorus, physiological indicators of nutrient deficiency (alkaline phosphatase activity and ammonium enhancement response), pH and dissolved inorganic carbon. Growth bioassays to determine effects of nutrient enrichment were conducted every two weeks in the laboratory. Profiles were measured *in situ* of temperature, dissolved oxygen and photosynthetically active radiation. Microbial measurements (Chapter 14) were coordinated with the routine limnology after 1988.

This core sampling program enabled us to assess the responses of all trophic levels, from fish to primary producers, to ecosystem manipulations. Additional measurements and experiments were also conducted to address specific questions addressed in the chapters to follow.

Acknowledgements

We thank Tom Frost and Paul Rasmussen for helpful reviews.

3 · Statistical analysis of the ecosystem experiments

Stephen R. Carpenter

Introduction

Experimentation at the ecosystem scale has made important contributions to ecology in general, and limnology in particular, over the past several decades (Likens, 1985; Schindler, 1987). Unlike some alternative approaches, large experiments are appropriately scaled for direct, strong inference about ecosystem dynamics and responses to perturbation (Chapter 1). The main disadvantage of ecosystem experiments is that replication is difficult or impossible (Matson & Carpenter, 1990).

By emphasizing whole lake experiments, we attain the appropriate scale but sacrifice replication. We have compensated for this shortcoming in several ways.

First, some of our manipulations have been strong and sustained ones, in the sense that changes in the independent variates (the fishes) were near the extremes of the natural range, and maintained for many generations of the zooplankton and phytoplankton populations that were the dependent variates (Carpenter, 1989). Such manipulations attempt to cause changes that are large enough to be evident without resorting to statistics, and would be viewed as ecologically significant by most practitioners. Strong sustained manipulations have been used in most ecosystem experiments, with the consequence that subtle responses and interactions are usually not detected (Likens, 1985; Schindler, 1987). For a variety of views on the utility of such 'sledgehammer' experiments, see Hurlbert (1984), Schindler (1987), Crowder et al. (1988), Kitchell et al. (1988) and Carpenter (1989).

Second, in some cases we have used data from many reference lakes to test for responses using conventional statistics (Carpenter et al., 1989). This approach was possible because our study lakes are near the Northern Lakes Long-Term Ecological Research Site (Magnuson & Bowser, 1990), which offered comparable data from multiple reference lakes for certain comparisons.

Third, we have combined other approaches with ecosystem experimentation where possible. Since the approaches can be complementary, a polythetic program that draws on all of them can offer strengths that are not achieved by any single method (Kitchell *et al.*, 1988). We have used mesocosm experiments to study mechanisms of selected interactions (Chapters 9, 12 and 14), simulation models to explore the implications of certain ideas (Chapter 16), comparative studies to test our hypothesis across a wider range of lakes (Carpenter *et al.*, 1991) and paleolimnological studies to examine the long-term dynamics of the trophic cascade (Chapter 15).

Finally, to test our hypotheses we have applied statistical techniques that are new to ecosystem ecology. Even unreplicated experiments are susceptible to insightful statistical analysis (Matson & Carpenter, 1990).

Interpretation of an ecosystem manipulation reduces to two essential questions. Did the system change? If so, did the manipulation cause the change? In a replicated, randomized experimental design, statistical analysis resolves both questions simultaneously. In an unreplicated experiment, only the first question can be addressed statistically (Frost *et al.*, 1988; Carpenter *et al.*, 1989). To answer the second question affirmatively, one must show that the manipulation is the most plausible reason for the change. Many ecosystem experiments employ reference ecosystems (like Paul Lake) to check the possibility that regional weather or environmental factors caused changes that might mistakenly be attributed to manipulation (Likens, 1985). Answers to the second question also depend on the ecological interpretation and significance of the changes (Frost *et al.*, 1988; Carpenter *et al.*, 1989).

The following sections describe the statistical methods employed to search for nonrandom changes in our experimental systems. These include filtering, transfer functions, and intervention analyses. The review is intended for limnologists, ecologists, and fisheries biologists who are not familiar with time series techniques. For these scientists, I hope to provide the minimal statistical background necessary to understand analyses in later chapters. For a deeper treatment of time series techniques, readers are directed to Chatfield (1984) and Wei (1990).

All time series analyses reported in this book were performed with the ARIMA procedure of the Statistical Analysis System (SAS Institute, 1988). I wrote the programs for the stochastic simulations reported below, calculations of impulse weights (Wei, 1990), and randomized intervention analysis (Carpenter *et al.*, 1989). All stochastic simulations and the randomized intervention analyses employ the pseudorandom number generators of Press *et al.* (1989).

Overview

Time series data from ecosystems are often highly variable. The fluctuations are not entirely random, but consist of a mixture of deterministic dynamics, serial dependency, seasonality, and patternless variation. We want to quantify the deterministic dynamics that underlie ecosystem function. The other three components (serial dependency, seasonality and patternless variation) can produce spurious apparent relationships that seriously mislead our search for dynamic mechanisms (Carpenter *et al.*, 1991). By serial dependency, we mean the tendency for each observation in the time series to depend on past observations. By seasonality, we mean the tendency for observations to follow a pattern that changes regularly with time of the year. Patternless variation, often called 'white noise' or 'error', is the background against which mechanistic dynamics must be discerned.

This book uses two approaches to separate deterministic change from the other kinds of variation in time series. First, transfer functions quantify the relationship between an input time series (e.g. a predator's population) and a response time series (e.g. a prey's population). Transfer functions can be used when the fluctuations of input and response are lagged in time. The delay between input and response fluctuations was typically a few weeks or less in our applications of transfer functions. Second, intervention analyses quantify the change in a response variable (e.g. a prey's population) following a manipulation (e.g. of a predator's population). Intervention analysis proved most useful when manipulations were step changes that were sustained for relatively long time periods, typically two or more years in our study. We used two kinds of intervention analysis, one derived from time series techniques (Box & Tiao, 1975) and the other a randomization test (Carpenter *et al.*, 1989). Filtering is the first step in fitting a transfer function or performing an intervention analysis by the time series method.

Filtering

Deterministic patterns in time series are far easier to detect when serial dependency and seasonality have been removed by filtering. A family of models called Autoregressive Integrated Moving Average (ARIMA) models can be fitted to time series to remove, or filter, serial dependency and seasonality. The identification and fitting of the appropriate ARIMA model for a given series is an iterative process that is well described in textbooks and will not be reviewed here (Chatfield, 1984; Wei, 1990).

All series analyzed for this book followed autoregressive models, a subset of the larger ARIMA family. In most cases presented in this book, the simplest autoregressive model, known as the AR(1) model, was sufficient. The equation for the AR(1) model is

$$(1 - \phi B)X(t) = a(t), \qquad [3.1]$$

where $X(t)$ is the original time series indexed by time t, and $a(t)$ is a series of independent residuals from a distribution that does not change through time. B is the backshift operator, with the property that $B_C X(t) = X(t - C)$; the backshift at lag C refers to the observation of X at C time intervals in the past. The parameter ϕ is the autoregressive parameter that accounts for the effects of serial correlation. Using the definition of the backshift operator, equation 3.1 expands to

$$X(t) = \phi X(t - 1) + a(t). \qquad [3.2]$$

Thus, in the AR(1) model $X(t)$ is simply a constant (ϕ) times the past value of X, plus a noise term $a(t)$.

In some cases, our data required more elaborate autoregressive models. Sometimes the value of $X(t)$ depended on the immediate past value $X(t-1)$ and the value before that, $X(t-2)$, leading to the AR(2) model:

$$(1 - \phi B - \phi_2 B_2)X(t) = a(t). \qquad [3.3]$$

Here, ϕ_2 is the coefficient for serial dependency of X on values at two time steps in the past. In other cases, the value of $X(t)$ depended on the immediate past value $X(t-1)$ and the value the previous year, $X(t-C)$, where C is the number of time intervals in an annual cycle. In this case, we fit the seasonal AR(1) model

$$(1 - \phi B)(1 - \phi_C B_C)X(t) = a(t), \qquad [3.4]$$

where ϕ_C is the coefficient for serial dependency of X on the previous year's value. Seasonal AR(2) models combine the features of equations 3.3 and 3.4:

$$(1 - \phi B - \phi_2 B_2)(1 - \phi_C B_C)X(t) = a(t). \qquad [3.5]$$

Even in this relatively complicated case, X is represented as a linear function of past values and noise. This point may be more obvious if equation 3.5 is expanded using the definition of the backshift operator:

$$X(t) = \phi X(t-1) + \phi_2 X(t-2) + \phi_C X(t-C)$$
$$- \phi\phi_C X(t-C-1) - \phi_2\phi_C X(t-C-2) + a(t). \qquad [3.6]$$

That is, $X(t)$ is simply a weighted combination of past values of X (where the ϕs are the weights), plus a random component $a(t)$. All autoregressive models express time series as weighted combinations of past values, plus noise.

Filtering a time series produces a series of uncorrelated residuals (Fig. 3.1). The original series contains an obvious seasonal trend, and a less obvious serial dependency. A series like this might arise for a population that increased during the course of each summer, with the population biomass at any time t dependent on the biomass at the previous time $t-1$. Sampling during seven summer seasons produces the sawtooth pattern with seven upwardly sloping segments. Each segment contains 15 data points, representing 15 samples in the course of each summer. Fitting the seasonal AR(1) model

$$(1 - \phi B)(1 - \phi_C B_C) X(t) = a(t) \qquad [3.7]$$

with C set equal to 15 leads to the parameter estimates $\phi = -0.209$ and $\phi_{15} = 0.840$, with residuals $a(t)$ shown as the filtered series in Fig. 3.1. Note that the seasonal pattern has disappeared. The filtered series also has no significant serial dependencies, as determined by plots of autocorrelation and partial autocorrelation functions (Wei, 1990).

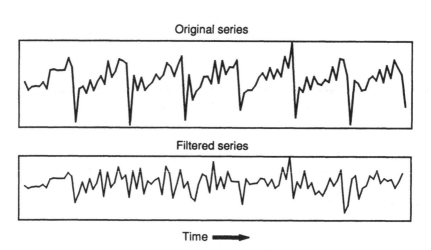

Original series

Filtered series

Time ➡

Fig. 3.1. Simulated time series. The original series contains serial dependency and a seasonal dependency. These are removed in the filtered series, which resembles patternless noise.

Time series models like equations 3.1–3.7 are usually fitted to unbroken sequences of data (Wei, 1990). The time series analyzed in this book consist of sequences of summer observations broken by winter periods with no observations. We coped with missing winter observations as follows. By truncating the longer series, all summer series were adjusted to the same number of observations (C), corresponding as closely as possible to the same days of the year. The sequence of these series was then analyzed as if it was an unbroken series of seasonal observations with cycle length C. In our data, C was 12–18 observations, depending on the variate analyzed. In some simulated examples of this chapter (see Figs 3.3 and 3.4), I have deliberately introduced strong trends within each summer season to show that the time series models are capable of filtering them. In our study, the purpose of filtering was to remove autocorrelations that might confound predator–prey relationships (see below). The predator–prey responses are expected to be evident at relatively short lags of 1–6 weeks (samples). Any autocorrelations introduced by the procedure for handling missing values should be removed by seasonal terms at lags of C (from 12 to 18) or a multiple of C, and have no effect on the analysis for predator–prey effects. In most of the actual applications found in later chapters, autocorrelation and partial autocorrelation functions did not indicate the need for any seasonal term, and simple AR(1) models like equation 3.1 were used.

Filtering and dynamic relationships

Filtering removes some of the difficulties of interpreting correlations among dynamic variables in time series. Serial correlation and seasonality can lead to spurious correlations that mislead interpretations of unfiltered series. Suppose the time series of Fig. 3.1 represents biomass of a predator which has a strong negative effect on the biomass of prey measured in the next sample. Time series measured for predator and prey show increasing biomass of the predator and decreasing biomass of the prey during each summer sampling period (Fig. 3.2). A naive analysis might simply correlate biomasses of predator and prey. The value is -0.107, which is not statistically significant. More detailed ecological examples of potentially misleading correlations are discussed by Carpenter & Kitchell (1987, 1988) and Carpenter et al. (1991).

The cross correlation function shows how the correlation changes as a function of time lag (Fig. 3.2). Each cross correlation is computed by shifting the series by the indicated lag, and calculating the product-moment correlation coefficient (r) between the lagged series. The cross correlations of the original series show results obtained without filtering to remove serial dependencies or seasonal trends. At positive lags S, we see the correlation of the predator at time 0 with the prey at time $t + S$ in the future. At negative lags $-S$, we see the correlation of the predator at time 0 with the prey at time $t - S$ in the past. Many correlations are significant, including the direct effect of predator on prey at lag $+1$, an artifactual seasonal signal at lag -15, and numerous artifacts of serial correlation at both positive and negative lags. The cross-correlations of the filtered series were obtained from the residuals after fitting each series

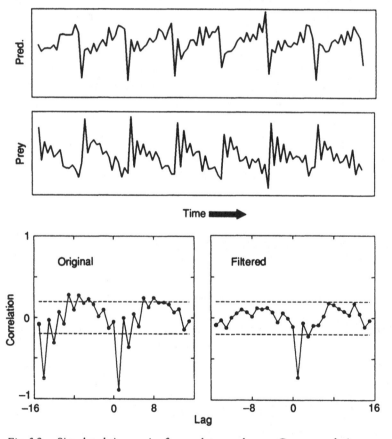

Fig. 3.2. Simulated time series for predator and prey. Cross correlations versus lag (in units of the interval between samples) are shown for the original time series, and for the time series after filtering to remove serial dependency and seasonality. Cross correlations outside the dashed lines are greater than their standard error.

to equation 7. The only significant correlation is that at lag +1, which represents the direct negative effect of predator biomass on the prey one sample later. By filtering the original series, spurious correlations were removed, leaving only the signal derived from the mechanistic relationship between predator and prey.

Transfer functions

We have seen that cross correlations of filtered series can identify relationships between two time series. Transfer functions extend the analysis by fitting models to such relationships. Identification and fitting of transfer functions is an iterative process, well described in textbooks, which will not be recounted here (Wei, 1990).

Transfer functions can be viewed as relationships among anomalies. After the ARIMA filter has removed serial dependencies and seasonal trends, the remaining fluctuations may be viewed as anomalies, in the sense that they are not explainable by serial dependency or seasonality. If anomalies in the input variable are followed by anomalies in the response variable, then a transfer function model can be fit.

Most of our transfer functions used the model

$$(1 - \phi B)Y(t) = \omega_0 X(t - S) + a(t), \qquad [3.8]$$

where $Y(t)$ is the time series for the response (dependent) variable, $X(t)$ is the time series for the input (independent) variable, and $a(t)$ is the time series of the residual. B is the backshift operator introduced previously. The left hand side of the equation is simply the AR(1) model introduced above. The right hand side of the equation contains the transfer model $\omega_0 X(t - S)$. The parameter ω_0 represents the effect of X on the current value of Y at S time units in the past. When expanded using the definition of the backshift operator, we have

$$Y(t) = \phi Y(t - 1) + \omega_0 X(t - S) + a(t). \qquad [3.9]$$

That is, Y depends on its immediate past value, the value of X at S time units in the past, and a noise term.

In a few cases, more elaborate transfer function models were needed. A relatively complex example is

$$(1 - \phi B - \phi_2 B_2)(1 - \phi_C B_C)Y(t) = \\ [(\omega_0 - \omega_1 B)X(t - S)/(1 - \delta B)] + a(t). \qquad [3.10]$$

Here, a second autoregressive parameter ϕ_2 and a seasonal parameter ϕ_C

are included in the autoregressive model on the left hand side. The transfer function is the bracketed terms on the right hand side. This transfer function contains two numerator parameters ω and a denominator parameter δ. Additional numerator parameters account for effects of X on Y at additional lags. Denominator parameters modify the time course of the change in Y following a change in X. Effects of the denominator parameter are illustrated by the example below.

Transfer functions may be easier to interpret when converted to the form

$$
\begin{aligned}
(1 - \phi B - \phi_2 B_2)(1 - \phi_C B_C)Y(t) = \\
(\nu_0 + \nu_1 B_1 + \nu_2 B_2 + \ldots)X(t) + a(t).
\end{aligned}
\tag{3.11}
$$

Here, the transfer function (right-hand side of equation 10) has been converted to a weighted combination of past values of the input series X. The weights ν_j are called impulse response weights. Each ν_j measures the effect of the value of X at j time units in the past on the current value of Y. Thus a plot of impulse weights ν_j versus time lag j shows the shape of the historical dependency of Y on X. The impulse weights can be calculated directly from the ω and δ coefficients (Wei, 1990, p. 291).

Some features of transfer functions can be illustrated by example. I simulated four sampling seasons of 15 samples each, where the input X followed the seasonal autoregressive model

$$
(1 - \phi B)(1 - \phi_{15} B_{15})X(t) = a(t) + \alpha M(t)
\tag{3.12}
$$

with $\phi = 0.3$ and $\phi_{15} = 0.3$. The residuals $a(t)$ were normally distributed with mean 0 and standard deviation 0.1. A very small standard deviation was used so the deterministic changes in the series would be evident graphically. The term $\alpha M(t)$ represents a manipulation that causes X to shift abruptly (see section below on intervention analysis). M is a series set to 0 prior to the manipulation, and 1 afterwards. Therefore M represents a step change shift at the time of manipulation. The parameter α sets the magnitude of the change in X following manipulation. In our example, $\alpha = 1$ and M changed value at the beginning of the third cycle. Thus, for cycles 1 and 2 (of 15 samples per cycle) X fluctuated around 0. For cycles 3 and 4, X fluctuated around 1. Y depended on X through the transfer function model

$$
(1 - \phi B)(1 - \phi_{15} B_{15})Y(t) = [\omega X(t-1)/(1 - \delta B)] + a(t)
\tag{3.13}
$$

with $\phi = 0.3$ and $\phi_{15} = 0.3$. Residuals $a(t)$ were normally distributed with mean 0 and standard deviation 0.1. The shift $X(t-1)$ means that changes in X affect Y one sample later. I simulated time series and calculated impulse weights for 6 combinations of parameters: $\omega = -1$ or $+1$, and $\delta = -0.5, 0$, or 0.5. Note that if $\omega = 0$ then X and Y are independent.

A positive numerator (ω) term causes Y to shift upward when X is manipulated (Fig. 3.3). The first nonzero impulse weight is positive, indicating an upward shift in Y. A negative ω causes Y to shift downward when X is manipulated, and produces a negative impulse weight at lag 1. The denominator parameter δ controls the sign and magnitude of impulse weights that follow the first nonzero weight. When $\delta = 0$, only the first impulse weight is nonzero, indicating that Y shifts sharply after X is manipulated. Nonzero δ produces nonzero values for later impulse weights, indicating that Y shifts exponentially to a new level after X is manipulated. Positive δ produces impulse weights of constant sign but exponentially decreasing magnitude. Negative δ also produces weights that decline exponentially in magnitude, but their signs alternate.

In all transfer functions presented in this book, the numerator parameter ω_0 is the most interesting one. It equals the change in the response variable (e.g. the prey) that follows a unit change in the input variable (e.g. the predator).

In the case of log transformed data, the numerator parameter has a somewhat more complicated interpretation. Primary production data were transformed to common logarithms to normalize the residuals. For primary production, the meaning of ω follows from the transfer function equation

$$\log[Y(t)] = \phi \log [Y(t-1)] + [(\log \omega)/ \\ (1 - \delta B)]X(t - S) + \log[a(t)]. \tag{3.14}$$

In fitting equation 3.14, the SAS software estimates $\omega' = \log \omega$ (SAS Institute, 1988). This relatively simple model (in which ϕ_2, ϕ_C and ω_1 equal zero) was in fact the best fitting model for our primary production time series (Chapter 13). The exponential of equation 3.14 (after multiplying through by $1 - \delta B$) is

$$Y(t) = [Y(t-1)]^{\phi + \delta}[Y(t-2)]^{-\phi\delta}a(t)[a(t-1)]^{\delta}\omega^{X(t-S)}. \tag{3.15}$$

So, production changes by a factor of ω for a one unit shift in the input variable. In general, a shift in the input variable of magnitude A changes production by ω^A.

Intervention analysis

Bender, Case & Gilpin (1984) distinguished between pulse, or short-lived, manipulations and press manipulations sustained for a long time

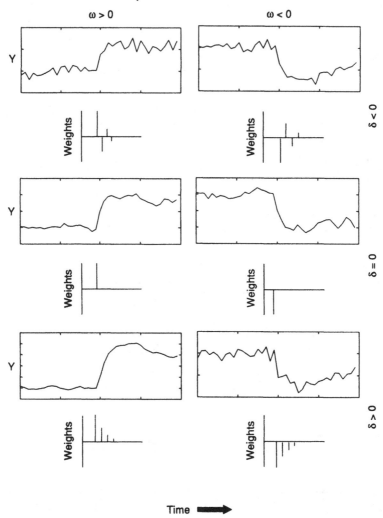

Fig. 3.3. Demonstration of how different shifts in a response variable (*Y*) are depicted by impulse response weights of a transfer function. Time series for the response variable (*Y*) were simulated using different numerator (ω) and denominator (δ) coefficients in the transfer function relating *Y* to an input variable, *x* (see text). Impulse response weights versus time after a shift in the input are shown for each transfer function. Actual numerical values are immaterial. Note how the impulse weights change depending on whether *Y* shifts upward or downward (ω positive or negative) and whether δ is positive or negative.

relative to the generation time of the target organisms. Intervention analysis (Box & Tiao, 1975) can be used to test directly for sustained responses to press manipulations.

Our intervention analyses employed the model

$$(1 - \phi B - \phi_2 B_2)(1 - \phi_C B_C)X(t) = a(t) + \alpha M_1 + \beta M_2, \qquad [3.16]$$

where $X(t)$ is the time series of the response variable and $a(t)$ is the residual time series. The left hand side of the equation is the seasonal AR(2) model introduced previously. It accounts for serial dependency and seasonality through the parameters ϕ, ϕ_2 and ϕ_C, respectively. In most cases the simpler AR(1) model was sufficient for our data, so the ϕ_2 and ϕ_C terms were omitted. The right-hand side of the equation represents two interventions through the terms $\alpha M_1 + \beta M_2$. The intervention parameters are α and β. The dummy variables M_1 and M_2 are zero prior to the intervention and one afterwards. Thus α and β represent the magnitude of the step–change shift in level of the response variable following interventions 1 and 2, respectively. If only one manipulation is studied, β and M_2 are omitted from the model.

To illustrate intervention analysis, I simulated a series using equation 3.16 with $\phi = -0.6$, $\phi_{15} = 0.6$ and only one manipulation (Fig. 3.4). Seven sampling cycles of 15 samples each were simulated, with M changing value at the beginning of the fourth cycle and $\alpha = 1$. The manipulation effect is discernible graphically, but there is doubt about its statistical significance. Analysis of the time series for X yields an estimate of $\alpha = 0.94$ with s.e. $= 0.04$ and $t = 19.22$, a highly significant result with $n = 105$. We would conclude that manipulation caused a nonrandom change of 0.94 in X, where the units of the change are the same as the units of X. The analysis does not prove that the manipulation caused the change, but does show that the noise in the series cannot explain it. The question of causality is addressed by considering the ecology of the manipulation and response, and the likelihood that the manipulation, as opposed to other factors, caused the response (Carpenter *et al.*, 1989).

In the case of primary production, which was transformed to common logarithms to normalize the residuals, the manipulation effect has a more complicated interpretation. Log primary production turns out to follow a simple autoregressive process with no seasonal parameter (Chapter 13). By equation 3.16 (using one intervention for simplicity) we have

$$\log X(t) = \phi[(\log X(t-1)] + [\log a(t)] + M[\log \alpha]. \qquad [3.17]$$

The SAS software estimates $\alpha' = \log \alpha$ when equation 3.17 is fitted to data (SAS Institute, 1988). Taking antilogarithms,

$$X(t) = [X(t-1)]^\phi [a(t)]\alpha^M. \qquad [3.18]$$

The ratio of present to past production is

$$X(t)/X(t-1) = [X(t-1)]^{\phi-1}[a(t)]\alpha^M. \qquad [3.19]$$

This ratio depends linearly on α^M. α^M is 1 when $M = 0$ (manipulation 'off') and α when $M = 1$ (manipulation 'on'). Therefore α is the factor by which X is increased or decreased by manipulation.

Randomized intervention analysis

Randomized intervention analysis (RIA) is an alternative method of testing for nonrandom change after manipulation (Carpenter *et al.*, 1989). RIA derived from the 'before–after–control–impact' experimental design of Stewart-Oaten *et al.* (1986).

RIA begins with time series for both manipulated and reference ecosystems (Fig. 3.5). A time series of differences between lakes is calculated, and the mean interlake differences before and after manipulation are determined. The change in the mean interlake difference is the test statistic. The distribution of the test statistic is estimated by randomly

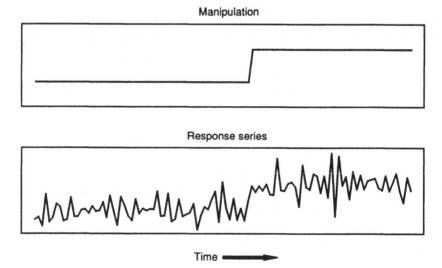

Manipulation

Response series

Time ━━▶

Fig. 3.4. Time series simulated to illustrate intervention analysis. The manipulation series $M(t)$ and response variable series $X(t)$ are shown.

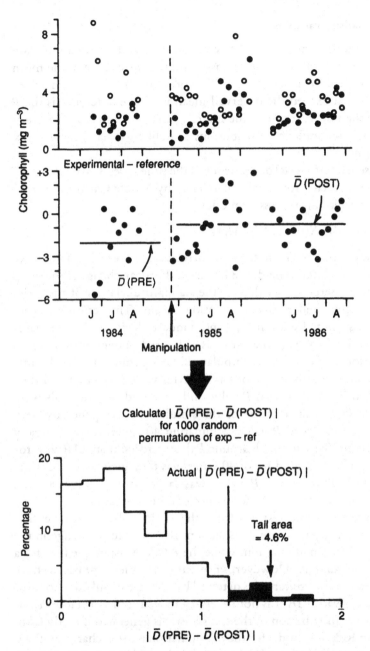

Fig. 3.5. Calculations for randomized intervention analysis (RIA). From the paired data from both experimental (filled circles) and reference (open circles) ecosystems before and after manipulation, intersystem differences are calculated. Mean intersystem differences before and after manipulation, D(PRE) and D(POST) respectively, are then calculated. Random permutation of the intersystem differences yields the distribution of the test statistic D(PRE) – D(POST). From Carpenter *et al.* (1989). Copyright © by the Ecological Society of America. Used by permission.

assigning interlake differences between the pre- and postmanipulation periods. For each of 1000 random assignments, the change in the mean interlake difference is computed. The proportion of mean interlake differences that exceed the observed mean interlake difference is the P value for the test. A low P value indicates that the difference between pre- and postmanipulation distributions is not likely to have occurred by chance alone.

RIA is sensitive to serial dependency in the time series (Carpenter *et al.*, 1989). We corrected for serial dependency by Monte Carlo simulation using the simple autoregressive model

$$(1 - \phi B)X(t) = a(t) + \alpha M(t), \qquad [3.20]$$

where ϕ was estimated from the autocorrelation of the series of interlake differences at lag 1 (Chatfield, 1984) and α is the mean change in the series of interlake differences divided by the standard deviation of the series around the pre- and post-manipulation means. The series $a(t)$ were normally distributed with mean 0 and standard deviation 1. The manipulation series $M(t)$ was the same as that used above: zero prior to manipulation, and one after manipulation. Using equation 3.20, 40 time series the same length as our observed series were generated and then subjected to RIA. The mean P value and its confidence intervals were calculated. The mean P value of the simulations was generally very similar to the observed P value. Simulating 40 experiments generally produced confidence intervals around P that were less than $0.1P$ wide for P less than 5%. A second set of 40 series was then generated, this time setting $\phi = 0$. This second P value was larger (for positive ϕ), and estimates the true P value in the absence of autocorrelation.

RIA differs in some respects from the ARIMA-based intervention analysis described in the previous section. RIA tests for a change in the distribution after manipulation, while the ARIMA technique tests for a shift in the mean only. However, since our time series appeared to have stable variances, any differences detected by RIA are mainly due to shifts in the mean. With RIA it is not necessary to assume a normal distribution, since the distribution of the test statistic is generated directly from the data. In RIA, the model for the manipulation is a step change with no lag; more complicated models are possible in ARIMA-based intervention analysis (Wei, 1990).

The main advantage of RIA appeared to be its performance when samples were too irregularly spaced in time or too infrequent to identify the noise model needed for ARIMA intervention analysis. It was difficult to fit satisfactory ARIMA models to series containing fewer than about 80 evenly spaced observations distributed equally among the seven years of the study. RIA produced credible results with as few as 40 irregularly spaced observations. Carpenter *et al.* (1989) showed that RIAs with series of about 40 observations gave the same results as *t* tests using replicate lakes. They also noted difficulties with fitting ARIMA models to such short time series.

Comment

Because this statistical approach is quite different from the norm in ecology (Matson & Carpenter, 1990), I wish to emphasize two important points about the interpretation of the results. First, these methods are conservative with regard to detecting effects. Therefore, the effects we detected are quite likely to be nonrandom. Any errors are most likely on the side of failing to detect some nonrandom patterns. Second, we have used statistics mainly to screen our data for nonrandom relationships or changes. While this screening proves very valuable, it is no substitute for a careful assessment of the ecological significance of the findings (Carpenter *et al.*, 1989).

The transfer functions are conservative because the data are filtered prior to analysis. The search for mechanistic links begins only after serial dependency and seasonality are removed. In fact, the serial dependency may be caused by dynamic relationships. The filtering may make any dynamic relationships harder to detect (Jassby & Powell, 1990).

The intervention analyses are conservative because we have used the simplest possible model for manipulation effects: a step change with no lag. Any lags in system response, or gradual responses of the system, make it more difficult to detect change under the model we used. More complex models that include lags or gradual changes can be designed (Wei, 1990). We felt that such complexities would introduce *ad hoc* assumptions that detracted from the credibility of our analysis.

We employed two kinds of questions in assessing the ecological significance of our findings. First, is the magnitude of the change meaningful ecologically? For example, an increase of 0.2 μg l^{-1} in chlorophyll concentration is detectable both analytically and statistically, but would not be important ecologically in Peter Lake. A change of 2 μg l^{-1} would double chlorophyll concentration, and is ecologically significant. Second, is the change in the direction predicted by the trophic cascade hypothesis? The hypothesis makes specific predictions about the direction of change at each trophic level under the manipulations employed. To be confirmed, the hypothesis must correctly predict the direction of a large number of changes.

Acknowledgements

I thank Tom Frost, Paul Rasmussen and John Magnuson for thorough reviews and very helpful discussions of this chapter. My exploration of time series techniques for ccosystem manipulations has been funded by the Andrew W. Mellon Foundation.

4 · *The fish populations*

James R. Hodgson, Xi He and James F. Kitchell

Introduction

Fundamental to the cascade hypothesis are the effects that fish populations can exert on species composition, biomass and productivity at other trophic levels. These may be direct or indirect (nonlethal) effects. Direct effects such as prey consumption, and indirect effects such as those influencing behavior (avoidance of predators) have been widely documented at the population level (e.g. Stroud & Clepper, 1979; Werner *et al.*, 1983) and the indirect effects expressed at the community and ecosystem levels such as those reviewed in Kerfoot & Sih (1987) & Northcote (1988). Indirect effects pertinent in the case of our studies would include behavioral responses, such as migration from or selection of specific refugia from predation (e.g. diel vertical migration of zooplankton and onshore–offshore migration of small fishes), that result in changes in foraging patterns of prey species (Carpenter *et al.*, 1987; He & Kitchell, 1990; He & Wright, 1992; Chapter 5). Another category of effects includes changes in nutrient flux due to shifts in the behavioral or structural properties of the fish populations (Carpenter *et al.*, 1992b).

Fish in our study lakes (Fig. 4.1) are common to the Great Lakes region, but some are near the limits of their geographic distributions. Largemouth bass and golden shiner are at the northern limits, while finescale and northern redbelly dace are near the southern limits (Scott & Crossman, 1973; Becker, 1983). Adult largemouth bass and rainbow trout can be keystone piscivores (Keast, 1985; Carpenter *et al.*, 1985, 1987) with an ability to limit the abundance of forage fish. Small rainbow trout, golden shiners, and both species of dace are effective zooplanktivores capable of changing the size and species composition of the zooplankton communities (Kitchell & Kitchell, 1980; Carpenter *et al.*, 1985, 1987) of our study lakes.

Currently the fish assemblages have a reduced species richness compared to other lakes of similar morphometry in the region (Tonn & Magnuson, 1982; Rahel, 1986; Tonn *et al.*, 1990). The lakes also have greater densities than limnologically comparable lakes where sport fisheries are active and fish population structure reflects exploitation effects. High densities could be reflected in growth rates, recruitment, feeding behavior, etc. of the fishes present in our study systems (Snow, 1969; Stasiak, 1978; Clady, 1974; Cochran *et al.*, 1988). For example, intense intraspecific competition can force bass to forage more as

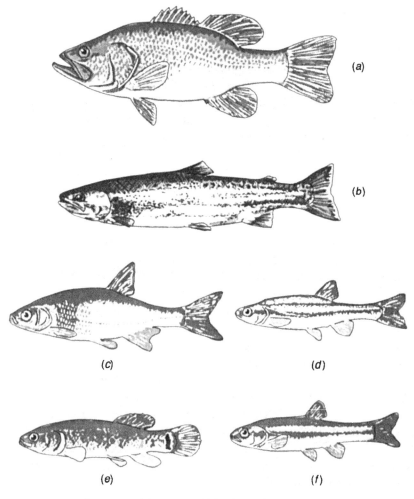

Fig. 4.1. Line drawings of the common fishes of our experimental lakes from 1984 to 1990. Drawings not to scale. (*a*) Largemouth bass; (*b*) rainbow trout; (*c*) golden shiner; (*d*) redbelly dace; (*e*) central mudminnow; (*f*) finescale dace.

generalists, including preying on *Daphnia* (Hodgson & Kitchell, 1987), than bass in lakes where fish communities are more diverse (Keast, 1978). Strong competition among dace species in Tuesday Lake could also shape diets through resource partitioning (Cochran *et al.*, 1988). Prey resources and therefore community structure of the lower trophic levels in these lakes reflect the effects of intense predation pressure by fishes.

Manipulation of the fish populations (see Fig. 4.1) in Peter and Tuesday Lakes has allowed us to test the ecosystem-level implications of predation by fishes and to evaluate both direct and indirect effects. In each lake we conducted a regular series of estimates of fish population densities, size and age structure, diet composition, and growth rates. This chapter describes the methods and primary results of those studies and serves as a background for the following two chapters. Chapter 5 provides a review of a series of experiments designed to evaluate behavioral effects involving piscivorous fishes and their potential prey. Chapter 6 elaborates the use of basic diet, growth and population data as fuel for a bioenergetics modeling approach that allows quantitative estimates of predator–prey interactions.

Methods

Capture methods

Capture and release methods (angling and boom shocker electrofishing) were employed in capturing largemouth bass and rainbow trout. When angling we used a variety of both natural and artificial baits over a range of sizes in an attempt to sample the range of fish sizes in the population. Handling mortality was minimized by short handling times and the use of antibiotic agents in live wells. Mortality was consistently low: about 3% for largemouth bass and higher but never more than 10% for rainbow trout. Other capture methods (trap netting, gill netting, or rotenone) were used only to reestablish premanipulation conditions as in the bass removal from Peter Lake in 1986 and trout removal from Peter Lake in 1990. Young-of-the-year (YOY) largemouth bass and minnows (redbelly dace, finescale dace and central mudminnow) were sampled using overnight sets of unbaited minnow traps at littoral zone stations and a cross-lake transect of traps set at two depths (Chapter 5).

All captured fish > 150 mm total length (> 195 mm in 1984–5) were marked with individually numbered floy tags. A number of fish were doubled-tagged in 1985 and 1986 in an effort to monitor tag loss, which proved to be insignificant. A total of 2657 fish were tagged, and 7535

Table 4.1. *Number of individuals, captures (tagged fish) and recaptures of largemouth bass in Paul, Peter and Tuesday Lakes during 1984–90*

Lake	Number of individuals	Number captured	Number recaptured	Percentage recapture
Paul	913	2183	1265	57.9
Peter	1302	3891	1532	39.4
Tuesday	442	1461	610	41.8

captures were recorded; hence the average fish was recaptured 2.2 times from 1984 to 1990 (Table 4.1). Each fish was measured (total length) to the nearest 1.0 mm and weighed to the nearest 1.0 g. Fish ages were determined from scale samples collected from the area dorsal to the lateral line and above the pectoral fin. Aging was usually done on fish captured during late August of each year.

Population estimates were conducted each May and August using Petersen mark-and-recapture methods (Ricker, 1975). Floy tags or caudal fin clips (fish > 150 mm or 195 mm, depending on sample year) served as marks. Both electrofishing and angling were used to establish the marked population in the first sample period. In an effort to reduce sampling biases from angling trauma, which can influence catchability, only electrofishing was used for recaptures in the second sample period.

Diet data collection

Fish used for qualitative analysis of diets were collected by angling, which proved to be the most effective means of obtaining sufficient sample sizes during periods of active foraging. Analysis of this method demonstrated that biases toward collecting fish of certain prey preferences or fish of different hunger levels appeared to be minimal (Hodgson & Kitchell, 1987; Hodgson & Cochran, 1988). Samples were generally collected between 0800 and 1130 h and 1600 and 2000 h. Each collection period from each study lake usually produced stomach contents with a wide array of prey items (e.g. zooplankton to a variety of benthic macroinvertebrates to YOY or juvenile fish) and different degrees of stomach fullness. Stomach samples were collected each year between mid-May and late August. There were 6–8 sample periods per year at approximately 2–3 week intervals, which adequately characterized intrasample (fish-to-fish) and intersample (day-to-day) variance for

Table 4.2. *Mean sample sizes for diet analysis for Paul (PA), Peter (PE) and Tuesday (TU) Lake largemouth bass (LMB) and rainbow trout (RBT) from 1984 to 1990*

Lake (spp.)	Number of samples	Mean number of fish per sample	Standard error	Range
PA (LMB)	45	29.4	2.1	10–92
PE (LMB)	35	35.5	3.9	8–96
PE (RBT)	14	22.9	0.5	18–24
TU (LMB)	13	33.4	3.7	19–72

seasonal consumption estimates computed through the use of energetics models (Hodgson, Carpenter & Gripentrog, 1989; Chapter 6). A total of 3366 largemouth bass and rainbow trout stomachs were examined in 107 sample periods representing all lakes (Table 4.2). Mean sample size was 26 (s.e. = 1.2), which proved to be statistically adequate for largemouth bass (Hodgson & Cochran, 1988) and rainbow trout.

Bass were collected from the littoral zone and trout from the metalimnion near midlake. Bass were held in a live well for < 1 hr until stomachs were flushed (usually at a shore-side processing station). To minimize temperature stress, trout were generally processed and released immediately at point of capture. Stomach flushing followed a procedure similar to that of Seaburg (1957). Water flow and pressure were maintained with a modified hand-pumped pressurized backpack sprayer. Stomach contents were washed into a 0.28 mm mesh concentrator (zooplankton collection bucket), backwashed into a specimen vial and preserved in 95% ethyl alcohol. Flushing did not appear to injure fish as evidenced by multiple recaptures during a sample period or between samples. Items too large to pass through the efferent tube of the pump (e.g. frogs, small rodents, juvenile fish, etc.) were removed from the stomach with a blunt forceps. Periodic examination of stomachs of fish that died during capture demonstrated that the stomach flushing technique removed all contents.

Stomach analyses

Individual stomach analyses were done on 3036 largemouth bass and 330 rainbow trout. All items in each stomach were enumerated ($> 3.5 \times 10^5$ individual items). Stomach contents were identified to the lowest

Table 4.3. *Prey categories and fresh masses used in diet analyses for fishes from Paul, Peter and Tuesday Lakes from 1984 to 1990*

Diet category	Fresh mass (mg)
small zooplankton[a]	0.015
large zooplankton[b]	0.206
Ephemeroptera larvae	23.60
Chaoborus spp.	1.70
Hydracarina	3.65
chironomid pupae	4.80
coleopteran larvae	52.90
odonate naiads	269.20
odonate adults	444.70
trichopteran nymphs	67.20
Notonecta	54.50
Mollusca[c]	126.40
Hirudinea	788.50
Dace[d]	999.00
YOY largemouth bass	changes with season
juvenile largemouth bass	changes with season
crayfish	4544.00
terrestrial vertebrates[e]	12000.00
other fish[f]	changes with season and species

Notes:
[a] Small zooplankton: *Polyphemus pediculus* and copepods.
[b] Large zooplankton: *Daphnia* spp. and *Holopedium gibberum*.
[c] Mollusca: gastropods and pelecypods.
[d] Dace: redbelly and finescale dace.
[e] Terrestrial vertebrates: frogs, snakes, voles, shrews, mice and birds.
[f] Other fish: central mudminnow, rainbow trout and golden shiner.

relevant taxon and were grouped into 20 prey categories (Table 4.3). Most food items were intact, permitting easy identification; however, occasional items in a fragmented state (e.g. abdominal segments, insect wings, beetle elytra, etc.) were more difficult to assign to prey categories or quantify. The number of prey assigned to these parts was the fewest possible individuals which could account for the fragments. Identification and taxonomy of macroinvertebrates followed Hilsenhoff (1975) and Pennak (1978), whereas zooplankton (i.e. cladoceran) identification and taxonomy followed Pennak (1978) and Balcer, Korda & Dodson (1984).

Diets were analyzed as frequency of occurrence (*FO*), the percentage of fish in a sample that ate a particular food item, percent number (%*N*), the percentage of each food type of the total number of food items eaten by all fish in the sample, and percent mass (%M, mg of wet biomass), the percentage of the total mass of all food items eaten. Some prey masses were determined in the laboratory while others were obtained from the literature (Cummins & Wuycheck, 1971; Wissing & Hasler, 1971; Driver, Sugden & Kovach, 1974; Dumont, Van de Velde & Dumont, 1975; Driver, 1981). These parameters were then used to generate an Index of Absolute Importance (IAI) for each prey category (George & Hadley, 1979) as follows:

$$IAI_a = \%N_a + \%M_a + FO_a.$$

A summation of all IAI values was used to calculate an Index of Relative Importance (IRI) for each dietary item (a), or:

$$IRI_a = 100 \frac{IAI_a}{\sum\limits_{a=1}^{n} IAI_a},$$

where n is the total number of prey categories. The range of IRI for any diet item is 0–100. More detailed descriptions of diet analyses are presented in Hodgson & Kitchell (1987), Hodgson & Cochran (1988), Hodgson *et al.* (1989), and Hodgson, Hodgson & Brooks (1991). All capture–recapture (tagging), growth rate and diet information, including diet data for each individual fish, were centrally archived on computer systems at St. Norbert College, De Pere, Wisconsin.

Results

Comprehensive population data for fishes in these lakes have been published previously. Detailed dietary information for largemouth bass, rainbow trout, finescale and northern redbelly dace in the lakes was provided by Knapik & Hodgson (1986), Hodgson & Kitchell (1987), Hodgson (1987), Cochran *et al.* (1988), Hodgson & Cochran (1988) and Hodgson *et al.* (1989, 1991). Here, we present fish population estimates, population size structures, and an overview of dietary data germane to the ecosystem experiments.

Paul Lake

Populations

Throughout the seven-year duration of the experiment a virtual monoculture of largemouth bass was present in Paul Lake (Table 4.4). As a

Table 4.4. *Fish species in Paul, Peter and Tuesday Lakes from 1984 to 1990*

Species include: LMB, largemouth bass; RBD, northern redbelly dace; FSD, finescale dace; CMM, central mudminnow; RBT, rainbow trout; GS, golden shiner; CC, creek chub; MIN, other species of minnows.

Year	Paul Lake	Peter Lake	Tuesday Lake
1984	LMB	LMB	RBD FSD CMM
1985	LMB	LMB RBD FSD CMM	LMB RBD FSD CMM
1986	LMB	LMB	LMB
1987	LMB	LMB RBT	RBD FSD CMM
1988	LMB	LMB RBT	RBD FSD CMM
1989	LMB	LMB RBT	RBD FSD CMM
1990	LMB	LMB RBT GS CC MIN	RBD FSD CMM LMB

reference, this bass population exhibited relatively constant densities and stable size structure from 1984 to 1990 (Fig. 4.2). August population estimates averaged about 400 adults (>150 mm) ha^{-1}. Highest estimates usually derived from relatively low catch rates for marked fish. Notable recruitment occurred in 1985 and 1987 (Fig. 4.2). The strong 1985 young-of-the-year (YOY) age class disappeared by spring 1986, presumably owing to high levels of cannibalism by the adult size classes. In contrast, a strong 1987 YOY age class survived well into the next season, even though piscivory (cannibalism) was high that year (Chapter 6). The over-winter mortality for the 1987 year class was 19%, and the over-summer mortality was 28%. These mortality rates were low compared with other largemouth bass populations in this region (Snow, 1969). The relations between biomass and density were similar from year to year, except that YOY and 1+ fish biomass increased during 1988 while densities decreased (Fig. 4.2).

The 1984 and 1985 populations were dominated by larger individuals (>200 mm), with few YOY and age-1+ fish (<80–100 mm) present (Fig. 4.3). There was relatively strong recruitment in 1985, as evidenced by a strong age-1+ cohort in 1986. The 1986 length frequency distribution was more even; by 1987, fish with a mean length of 150 mm (age-2+) became dominant and the survival of the YOY size class increased. This size distribution (a dominance of intermediate size groups) persisted through 1989, but in 1990 the population became more dominated by smaller fish (<150 mm). This shift may be more reflective of size-specific catchability than of differential mortality. From 1988 to 1990, larger fish (>250 mm) moved to deeper offshore habitats, while smaller

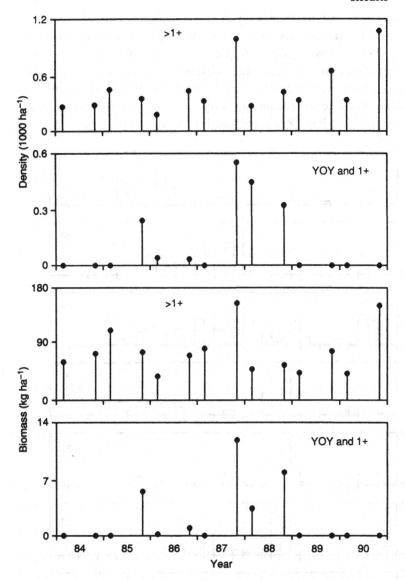

Fig. 4.2. Estimated population densities and biomass for largemouth bass in Paul Lake from 1984 to 1990.

fish continued to inhabit shallower littoral zone areas where catch per effort was much better (for both angling and electrofishing).

Diets
Largemouth bass preyed most heavily on insect larvae (trichopteran nymphs, odonate naiads and chironomid pupae) and other relatively

large prey (the 'others' diet category; Table 4.5 and Fig. 4.4). Collectively the seasonal %IRI for these two prey categories ranged from 53 to 71. Of these, odonate naiads (especially dragonfly naiads) were of particular importance (7-year mean %IRI = 15). *Chaoborus* spp. were the single most important prey item (7-year mean %IRI = 17). Zooplankton (*Daphnia*) was a consistent but less important diet component (7-year mean %IRI = 11). Fish (YOY and age-1 +) were of lesser importance in

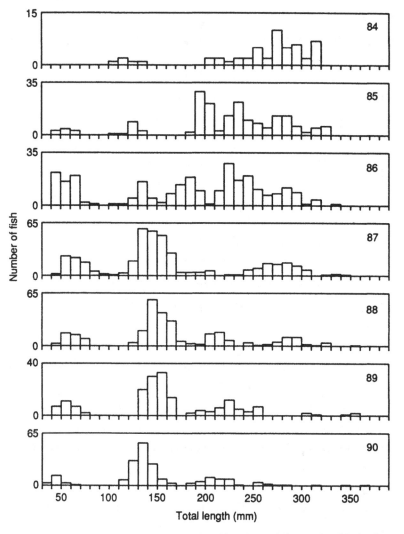

Fig. 4.3. Length frequency distribution of largemouth bass in Paul Lake from 1984 to 1990.

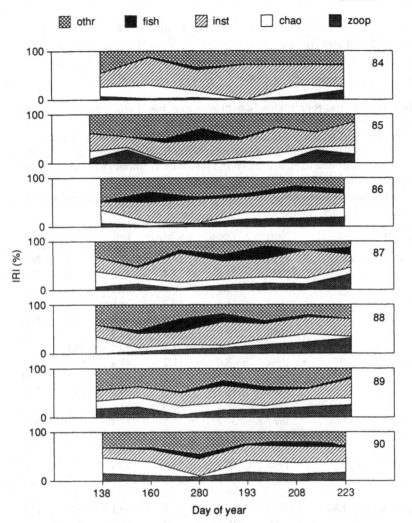

Fig. 4.4. The %IRI for five collective diet categories for largemouth bass in Paul Lake from 1984 to 1990. See text for diet category descriptions.

the diet (%IRI mean = 9) than other items, but were most important during early summer as that year's offspring became vulnerable to cannibalism.

Compared with bass from the other lakes, Paul Lake bass had less sample-to-sample and interannual variation in diet (Fig. 4.4). Although importance values for zooplankton (cladocerans) fluctuated annually (between midsummer lows and spring and late summer highs) and interannually, variations were smaller than recorded in Peter or Tuesday

Table 4.5. *Seasonal (calculated collectively on all stomach samples of each year) Index of Relative Importance (% IRI) for five diet categories of largemouth bass (LMB) and rainbow trout (RBT) in Paul, Peter and Tuesday Lakes from 1984 to 1990*

Diet categories are: zooplankton (Zoop) which includes *Polyphemus pediculus, Daphnia* spp., *Holopedium gibberum,* and copepods; *Chaoborus* spp. (Chao); insect immatures (Inst) which includes ephemeropteran nymphs, chironomid pupae, odonate naiads, trichopterans and coleopteran nymphs; fish (Fish) which includes redbelly and finescale dace, and YOY and juvenile largemouth bass; and an others category (Othr) (see Table 4.3 for those items not mentioned here).

| Lake (spp.) | Year | % IRI | | | | |
		Zoop	Chao	Inst	Fish	Othr
Paul Lake (LMB)	1984	6.3	19.2	42.8	3.4	28.3
	1985	11.8	13.4	42.8	6.0	26.0
	1986	12.1	18.0	33.6	9.8	26.5
	1987	16.5	15.7	38.0	12.5	17.3
	1988	17.7	14.6	36.6	13.6	17.5
	1989	19.1	13.0	37.4	6.1	24.4
	1990	14.3	23.4	30.4	9.0	22.9
Peter Lake (LMB)	1984	8.6	11.2	41.4	11.6	27.2
	1985	26.1	11.2	25.4	23.0	24.4
	1986	16.1	9.2	29.0	25.3	20.4
	1986[a]	38.7	6.0	17.1	0.2	38.0
	1987	31.7	3.5	17.2	0.0	47.6
	1988	17.9	11.1	36.1	11.2	23.7
	1989	10.8	5.1	48.7	4.8	30.6
(RBT)	1988	48.9	13.9	23.4	0.4	13.4
	1989	38.5	22.3	26.7	3.7	8.8
	1990[b]	15.7	6.3	31.6	14.0	32.4
Tuesday Lake (LMB)	1985	7.7	24.4	29.1	27.0	11.8
	1986	10.7	24.4	31.1	15.5	18.3

Notes:
[a] Represents age − I + fish, 1987 + age − II + fish, etc.
[b] Based on two samples (May and August).

lakes. These fluctuations were probably correlated with zooplankton dynamics. For example, the increase of %IRI for zooplankton in the 1988 bass diets corresponded to a gradual increase of *Daphnia* spp. biomass in the plankton of the lake (Chapter 8), and the 1984 mid-summer %IRI low for zooplankton occurred at the same time as a midsummer depression of *Daphnia* biomass in the plankton. However, there were some exceptions. In 1985, %IRI for zooplankton was at a seasonal low (Fig. 4.4) while the planktonic *Daphnia* biomass was at a midsummer maximum (Chapter 8). This result was probably related to an unusually high seasonal abundance of YOY bass at that time, which were probably fed on preferentially by adult largemouth bass. Similar foraging responses related to prey preferences have been reported for Peter Lake bass (Hodgson & Kitchell, 1987; Hodgson *et al.*, 1991).

Chaoborus was a relatively consistent prey item (Table 4.5), in keeping with the continuous availability of *Chaoborus* densities in the plankton (Chapter 7). However, the %IRI for *Chaoborus* varied seasonally (Fig. 4.4), usually with a midsummer low in consumption. Possible explanations for these variations include preferential foraging behavior of the bass and seasonal shifts in availability or relative costs and benefits of alternative prey items (e.g. YOY bass).

Seasonal and interannual variation was also evident for fish prey in the diets (Fig. 4.4). In 1984, when bass YOY density was the lowest, the %IRI was also the lowest (3%) of any sample year. From 1986 to 1988, when YOY densities were greater, %IRI values increased to an average of 12% (Table 4.5). Seasonal variation was correlated with abundance of YOY fish, which were usually not available until late June and quickly disappeared because of heavy predation on them. This pattern was especially evident in 1985, 1987 and 1988 (Fig. 4.4).

Peter Lake

Populations

Peter Lake underwent a series of food web manipulations (Tables 2.3 and 4.4). In 1985, 90% of the largemouth bass population was removed and 49 601 minnows were introduced (Carpenter *et al.*, 1987); they quickly disappeared as a consequence of predation by the remaining bass, emigration from the lake, and predator avoidance that forced them into a shallow-water refuge which attracted large numbers of piscivorous birds. In 1988 and 1989 rainbow trout were stocked. In 1989 and 1990 further removal of largemouth bass and rainbow trout was conducted. In early 1990 golden shiners were stocked.

The premanipulation largemouth bass population was very similar to that of Paul Lake (with respect to density and size distribution) (Figs 4.5 and 4.6). After the 1985 manipulation, only a few large adults remained ($<20\,ha^{-1}$), and a remarkably large YOY class was produced (Fig. 4.5). In August 1985, YOY density was estimated to be over $6000\,ha^{-1}$. This

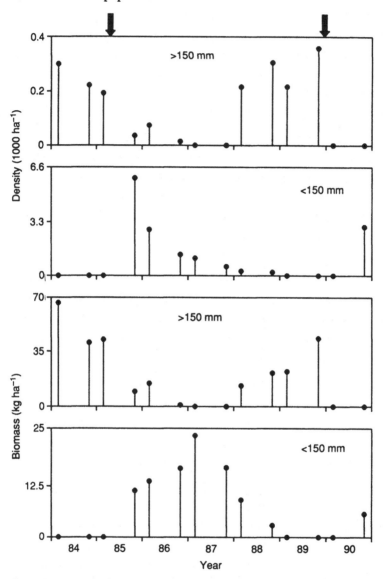

Fig. 4.5. Estimated population densities and biomass for largemouth bass in Peter Lake from 1984 to 1990. Arrows indicate dates of largemouth bass translocations in 1985 and 1989.

cohort dominated the bass population until the end of 1989 when they were removed from the lake (Fig. 4.5).

In an effort to enhance zooplanktivory, rainbow trout were stocked in 1988 ($N = 3000$; mean length = 167 mm, mean mass = 50 g) and again in 1989 ($N = 3000$; mean length = 130 mm, mean mass = 21 g) (Fig. 4.7).

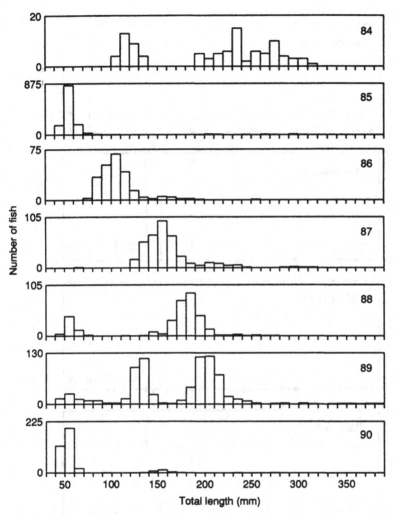

Fig. 4.6. Length frequency distribution of largemouth bass in Peter Lake from 1984 to 1990.

Survival and growth rates of the 1988 age-1 + cohort were 31.3% and 0.45 g d^{-1}, respectively. Survival of the 1989 cohort was very poor (we were unsuccessful in obtaining large enough samples to estimate population size). Piscivory by the few remaining large adult bass, the 1988 trout cohort, a resident common loon (*Gavia immer*) and daily visits by an osprey (*Pandion haliaetus*) probably accounted for the extensive mortality of this cohort. The summers of 1989 and 1990 were the only seasons of the study in which Peter Lake supported a loon. Other than Peter Lake, only Tuesday Lake in 1990, when minnows were at peak abundance, attracted a loon at any time during the study.

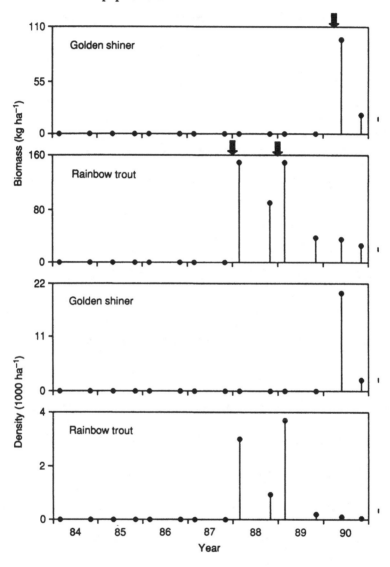

Fig. 4.7. Estimated biomass and population densities for golden shiner in 1990 and rainbow trout from 1988 to 1989 in Peter Lake. Arrows indicate dates of introduction of golden shiners in 1990 and of rainbow trout in 1988 and 1989.

In the spring of 1990, 20 000 golden shiners (mean length = 84 mm) were introduced into the lake (Fig. 4.7). Survival was 11% over the summer.

Diets

In 1984 and earlier years (Hodgson & Kitchell, 1987), diets of Peter and Paul Lake bass were very similar (Table 4.5; Figs 4.4 and 4.8). Insects and the 'others' prey category were the largest proportions of the diets. In 1985, because of strong largemouth bass recruitment and dace introduction, fish increased in importance in the diet (seasonal %IRI = 22). Dace were important diet items in the sample period which immediately followed their introduction, but they quickly disappeared from the diet record. Zooplankton also increased in importance in 1985 (seasonal %IRI increased from 9 to 26), in association with higher total zooplankton biomass in 1985 than 1984 (Chapter 8). Diet samples taken from 1986 to 1990 were mainly from the 1985 year class. *Daphnia* was the predominant prey type in 1986 (seasonal %IRI = 39, Table 4.5) for the age-1 + cohort. The 'others' prey category was also important. As this bass cohort aged and grew, mean summer %IRI for zooplankton decreased, while other prey types increased. Seasonal %IRI of benthos and other macroinvertebrates increased from 56% in 1986 to 81% in 1989 (Table 4.5). Diet change associated with increased age and size is widely known, with younger fish feeding on small prey (i.e. zooplankton) and shifting to larger prey sizes (i.e. benthos) as they become more efficient at capturing and handling larger prey. A more important cause of reduced zooplanktivory by bass in 1988 and 1989 was a resource partitioning response due to competition with rainbow trout (Hodgson *et al.*, 1991).

Zooplankton was the largest component of the rainbow trout diet (Fig. 4.9). Seasonal %IRIs were 49 in 1988 and 39 in 1989 (both trout cohorts combined) (Table 4.5). Unlike largemouth bass, *Chaoborus* was a major diet component for rainbow trout. Insect naiads and larvae (mainly chironomid pupae and trichopteran nymphs) were also important diet items (averaging 25% in 1988 and 1989). Only in 1990 did the trout become effective piscivores, preying on YOY bass (Table 4.5).

Tuesday Lake

Populations

In 1984, dense populations of small fishes (90% redbelly dace, 5% finescale dace and 5% central mudminnow) were present in Tuesday

Lake (Fig. 4.10; Table 4.4). Removal of 90% of the minnows and introduction of bass in 1985 strongly influenced trophic interactions. Nearly all the remaining dace were consumed by bass shortly after the 1985 manipulation. Removal of largemouth bass in late 1986 and reintroduction of dace and mudminnows in early 1987 reestablished the system to the planktivore-dominated condition. By 1989, biomass and species composition were similar to 1984 levels (Fig. 4.10). In 1990, age-YOY largemouth bass were reintroduced to Tuesday Lake and were

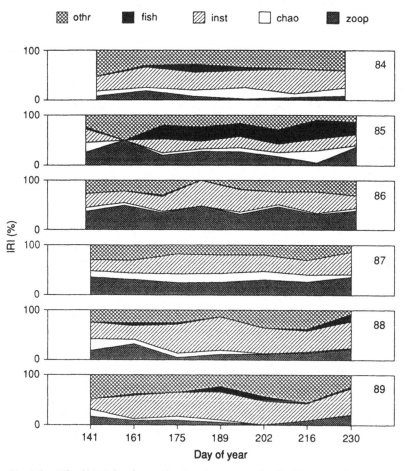

Fig. 4.8. The %IRI for five collective diet categories for largemouth bass in Peter Lake from 1984 to 1990. See text for diet category descriptions. Note that IRI values for 1984 and 1985 were from the adult largemouth bass and that IRI values from 1986 to 1989 were generally from the 1985 year class.

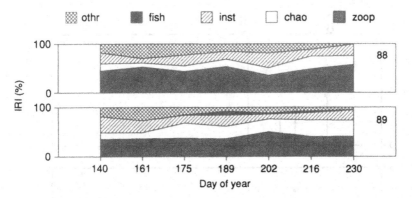

Fig. 4.9. The %IRI for five collective diet categories for rainbow trout in Peter Lake in 1988 and 1989. See text for diet category descriptions.

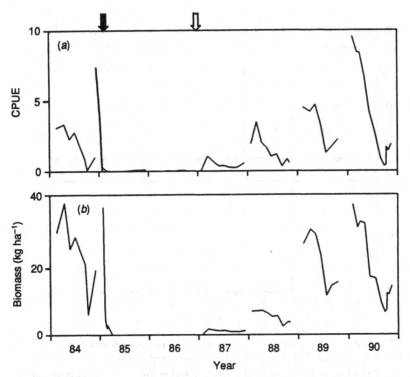

Fig. 4.10. The catch per unit effort (CPUE), expressed as number of fish trap^{-1} h^{-1} in Tuesday Lake for all minnow species from 1984 to 1990 (*a*) and estimated minnow biomass in Tuesday Lake from 1984 to 1990 (*b*). The solid arrow indicates date of removal of minnows in 1985 and the open arrow indicates date of introduction of minnows in 1987.

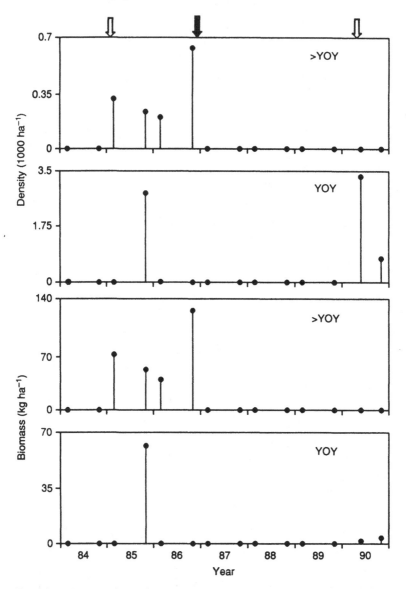

Fig. 4.11. Estimated population densities and biomass for largemouth bass in Tuesday Lake in 1985, 1986 and 1990. Open arrows indicate dates of introduction of largemouth bass in 1985 and 1990. The solid arrow indicates date of removal of largemouth bass in 1986.

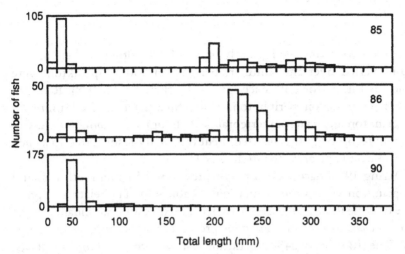

Fig. 4.12. Length frequency distribution of largemouth bass in Tuesday Lake in 1985, 1986 and 1990.

able to prey heavily on the dace by the end of the growing season (Wright & Kitchell, 1993).

Minnow catches varied considerably from year to year (Fig. 4.10). Catches were usually high in early summer, and then gradually decreased in June and July. The catches were lowest in late July and early August, and then increased again towards the end of the season. This pattern prevailed in 1984 and from 1987 to 1990. There are at least two components to this general phenomenon. First, minnows of a size vulnerable to the traps (>45 mm) suffer high mortality during midsummer. Comparisons of length frequency distributions before and after midsummer depression in catch rates indicated that mean lengths were significantly greater for the early summer population (He, 1986). Second, high epilimnion temperatures in late July and early August may have reduced dace activity, thereby reducing vulnerability to passive gear such as minnow traps.

Survival rate of bass stocked in 1985 was high (Fig. 4.11); successful spawning yielded a large year class of bass. However, few members of this cohort survived through the winter of 1985–6. Low survivorship was the result of high piscivory by adult bass (Table 4.5) and/or the result of size-selective winter mortality.

Size structure of the bass population was similar to that of Peter Lake in 1985 and 1986 (Fig. 4.12). Adults (>150 mm) were the dominant size classes in both years. Survival of the hatchery-reared YOY bass introduced in 1990 was 24%.

Diets

Dace diets were sampled regularly only in 1984. Although diet compositions varied over the summer, periphyton and zooplankton were major diet items for both dace species (Cochran *et al.*, 1988). Redbelly dace feed more on periphyton, while finescale dace feed more on zooplankton and macroinvertebrates. Although *Chaoborus* were infrequently recorded in the diet, sediment trap analysis suggested that they were also a major prey item (Chapter 15).

During 1985, dace were a major prey item of largemouth bass and zooplankton were a minor diet item (Table 4.5). The relative importance of these items in the diets paralleled their relative abundance in the lake. *Chaoborus* was an important prey category in both years (Table 4.5). The IRI value of 24% was the highest value reported for any lake or season (Fig. 4.13). The %IRI for zooplankton increased in 1986, as zooplankton became more abundant and dace and small bass became less common. As in Peter and Paul Lakes, insect naiads and larvae were important diet items (Table 4.5).

Discussion

Compared with the other study lakes, Paul Lake's consistently high rates of piscivory (Fig. 4.4; Chapter 6), usually peaking at midsummer, prevented establishment of zooplanktivores (i.e. cyprinids) and suppressed largemouth bass YOY in every year except 1985. The zooplankton assemblage was dominated by large, conspicuous species, as expected in a food web heavily regulated by piscivores (Chapter 8).

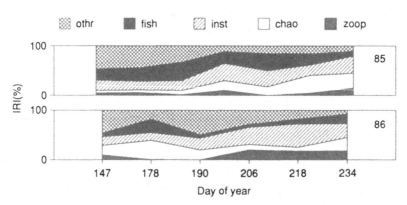

Fig. 4.13. The %IRI for five collective diet categories for largemouth bass in Tuesday Lake in 1985 and 1986. See text for diet category descriptions.

Diets of Peter Lake bass record events of the various manipulations. In 1985, for example, piscivory was the greatest recorded for any sample year (Fig. 4.8). That was due to the predominance of dace in diet samples taken during 23–31 May 1985, shortly after dace were introduced to Peter Lake. Subsequently, fish in the diet were mainly YOY bass. Even though our sampling schedule could detect routine variations in diet (Hodgson *et al.*, 1989), the dace predation event was only briefly evident (Fig. 4.8). The dace were either consumed in the period between diet samples (direct effect) and/or sought escape from bass predation by seeking refuge against the bog mat, retreating to very shallow water or emigrating through the culvert to Paul Lake (indirect effect). This response was not anticipated and was therefore poorly documented (Carpenter *et al.*, 1987). It became the basis for a series of separate experiments, which are detailed in Chapter 5.

During the sample periods that followed disappearance of the dace, YOY bass became the most important diet item (Fig. 4.8) and represented an opportunistic foraging response to prey availability (Hodgson & Kitchell, 1987). The decline in piscivory in 1986 and 1987 reflected dominance of the population by small bass. The diets (Fig. 4.8) represent those of age-1 + and age-2 + cohorts (Figs 4.5 and 4.6), respectively. In 1988 and 1989, as the bass became larger, effective piscivory on younger cohorts was possible, and fish again became an important diet component (Fig. 4.13). Indirect effects of this predator–prey interaction were expressed in the zooplankton community (Chapters 7 and 8) and ultimately in nutrient cycling and primary productivity (Chapter 13).

Rainbow trout were more effective zooplanktivores than largemouth bass in Peter Lake (Hodgson *et al.*, 1991) (Figs 4.8 and 4.9) and caused changes in the zooplankton community similar to those produced by the large YOY bass population in 1985. Predation by trout on *Chaoborus* and large daphnids led to surprising changes in the zooplankton (Chapters 7 and 8). A planktivore-dominated community (as in premanipulation Tuesday Lake) was never fully realized in Peter Lake owing to rapid growth and ontogenetic diet shifts in trout. Those included predation on both the YOY bass of 1988 and the cohort of smaller trout introduced in 1988. Additional predation by the few remaining large adult largemouth bass and piscivorous birds proved to be very effective in reducing the 1988 trout cohort and younger age classes of potentially zooplanktivorous bass.

Sustaining populations of zooplanktivorous fishes in Peter Lake proved to be surprisingly difficult. Our experience with manipulations intended to enhance zooplanktivory revealed that even small populations of piscivores can have profound effects. Predator avoidance behaviors by highly vulnerable, potential prey yielded a pronounced reduction in zooplanktivory by small fishes and the consequent, rapid cascade of interactions throughout the food web. The response of piscivorous birds was not among those we anticipated although experiences elsewhere have had a similar component (Spencer & King, 1984): egrets and herons were attracted to a minnow-dominated system but not observed in adjacent bass-dominated or fishless systems.

In Tuesday Lake, the addition of largemouth bass had a pronounced influence on the plankton communities (Chapter 8). Intense piscivory in 1985 and 1986 (Fig. 4.13) dramatically reduced vertebrate zooplanktivory. The zooplankton assemblage shifted from dominance by small-bodied animals (*Bosmina*, rotifers and small copepods) to dominance by large-bodied cladocerans (*Daphnia*). With the recovery of minnow populations during 1987–9 there was a gradual species shift to smaller-bodied cladocerans and an increased presence of rotifers and omnivorous copepods (Chapter 8). Dace densities reached premanipulation conditions by 1989–90 (Fig. 4.10).

In general, fish diets reflected the mechanisms proposed in the trophic cascade hypothesis. Diet sampling at 2–3 week intervals over the seven years was sufficient to detect the major predatory events (Hodgson *et al.*, 1989). The time lags in fish predation effects, which were due to predatory inertia and trophic ontogeny (Carpenter *et al.*, 1985), were evidenced in our samples. Nevertheless, episodes of predation that are scarcely detected by diet samples can have lasting effects on lake ecosystems and serve as evidence of the importance of 'bottlenecks', 'predation gauntlets', 'critical periods', 'survival hurdles' and the miscellany of other metaphors that attempt to represent the consequences of brief but very intense interactions. Examples include the disappearance of dace from Peter Lake in 1985, and the near-disappearance of *Daphnia pulex* from Peter Lake in 1988. In both cases, finer-grained sampling would have been helpful.

Connell (1980) reasoned that present-day community composition may derive from past interactions that are no longer apparent. We have shown that episodes of predation can also leave legacies that persist for longer than their dietary evidence. These 'ghosts' become more evident in the neolimnological observations of plankton community dynamics (Chapters 7 and 8) and in the fossil record offered by the sediments (Chapter 15).

Summary

Diets, growth, and populations of largemouth bass and rainbow trout were monitored in Paul, Peter and Tuesday Lakes during 1984–90. Population densities and size distributions were recorded for golden shiner, finescale dace, and redbelly dace.

Paul Lake, the unmanipulated reference lake, showed less interweek and interyear variation in diet than the manipulated lakes. Insect larvae were diet staples; of these, odonates were most important. The monoculture of bass had relatively constant densities and stable size structure throughout the study, though the size distribution was dominated by somewhat larger individuals in 1984–6 and somewhat smaller individuals in 1987–90. Recruitment events were common but short-lived. In general, year classes were greatly reduced within a year due to cannibalism by larger bass.

Bass diets, densities, and size distributions in Peter Lake changed as a result of the manipulations. The 1984 diets, densities, and sizes were similar to those of Paul Lake. In 1985, the year of dace introductions and a strong bass year class, fish were important prey items. After 1985, bass population density and size distribution reflected the growth and mortality of the 1985 cohort. Bass diet expanded with trophic ontogeny in 1986–7 and contracted owing to interspecific competition with rainbow trout in 1988–9.

Rainbow trout in Peter Lake were predominantly zooplanktivorous (feeding on *Daphnia* spp. and *Chaoborus*) during both years, but preyed more heavily on fish in 1989. Trout grew rapidly between 1988 and 1989.

Golden shiners introduced to Peter Lake during 1990 had 89% mortality, due, in part, to predation by the remaining trout and bass. However, golden shiners were sufficiently abundant to be significant planktivores during that year.

After their introduction to Tuesday Lake in 1985, largemouth bass preyed principally on fish. Fish were less important in diets in 1986. *Chaoborus* was a major diet item in both years. Bass grew substantially between 1985 and 1986, and recruited successfully in both years.

Density and biomass of redbelly dace, finescale dace, and mudminnows were reduced substantially by removal from Tuesday Lake in early 1985. The remaining members of those populations virtually disappeared after bass were introduced. After bass were removed in late 1986, these planktivores were reestablished in 1987 and their populations recovered to premanipulation levels by 1989.

The 49 601 planktivorous fish transplanted from Tuesday to Peter Lake in 1985 also virtually disappeared within a few weeks. Their populations were reduced by intense predation and their ecological effect as zooplanktivores diminished owing to predator avoidance and emigration behaviors.

5 · *Fish behavioral and community responses to manipulation*

Xi He, Russell Wright and James F. Kitchell

Introduction

As detailed in Chapters 2 and 4, we manipulated individual fish populations to change food web structure. In addition, we were able to follow dramatic responses in the fish community structure. Shifting a system from dominance by planktivores to one dominated by piscivores was accomplished relatively simply by introducing large numbers of piscivores. Establishing a planktivore-dominated system by eliminating piscivores first and then introducing planktivores has also worked well. These techniques were used in the manipulations in Tuesday Lake (Chapters 2 and 4). The resulting dominance by planktivores or piscivores provided dramatic contrasts in food web structure; however, most natural lake systems (such as those regularly subjected to fisheries exploitation) have mixed assemblages of piscivores and planktivores. The greater diversity and complexity of many fish communities encourages a search for general ecological principles (Werner & Gilliam, 1984). Fish communities are often further influenced by stocking and/or species- and size-selective harvest regulations. Thus, the interests of managers in facilitating or diminishing the relative abundance of selected species suggests that we must better understand the interactions among piscivorous and planktivorous fishes, and the resulting effects on other trophic levels.

In Peter Lake, attempts to establish dominance by planktivorous fishes in 1985 and 1989 failed (Chapters 2 and 4). The minnow population introduced in 1985 declined much more rapidly than could be accounted for by direct predation by known numbers of largemouth bass in the lake. Using a bioenergetics model (Kitchell, Stewart & Weininger, 1977; Rice & Cochran, 1984; Bevelhimer, Stein & Carline, 1985), we estimated that consumption by piscivorous bass could have accounted for only 11% of the minnow biomass that disappeared from the lake (Chapter 6). Large rainbow trout succeeded when stocked in Peter Lake in 1988 but quickly outgrew their intended role as zooplanktivores and switched to larger prey. A similar number of smaller rainbow trout was introduced in 1989 but quickly disappeared as predators responded to their presence. These observations suggest that: (1) short-term interactions among introduced fishes and resident fishes are important in determining the final outcome of manipulations; (2) establishment of a fish assemblage requires a better understanding of community responses to manipulations.

We argue that predator avoidance behavior by prey fishes is important in short-term interactions between piscivorous and planktivorous fishes. The nonlethal effects of predators on prey fishes, such as an increase of emigration or the use of littoral refuges (Werner & Gilliam, 1984), can effectively reduce prey-fish populations that affect the outcome of the manipulations intended for the pelagic zone. This effect has been demonstrated experimentally for piscivore–planktivore systems where the presence of the piscivore did not cause increased mortality in the planktivore but did produce reduced mortality in the zooplankton populations and altered zooplankton community dynamics (Turner & Mittlebach, 1990). We also argue that responses of prey fishes to manipulations must be evaluated at the scale of the fish community, because responses of prey fishes can be size- and species-specific.

In this chapter, we present the results from a series of experiments designed to elucidate the effects of a piscivore (northern pike, *Esox lucius*) on behavior and community characteristics of an assemblage of prey fishes. These experiments were conceived in response to our observations of anti-predator behaviors exhibited by minnows introduced to Peter Lake in 1985 (Carpenter et al., 1987). They were conducted over the course of four years in a lake (Bolger Bog) near those employed for the cascade manipulations.

In the first portion of this chapter, we focus on species-specific and behavioral responses of prey fish populations to the introduction of northern pike. We evaluate the hypothesis that predator avoidance behavior (e.g. increase in emigration) can account for a large proportion of decline in prey fish populations. In the second part of the chapter, we focus on the community responses to predation. Based on the results of multivariate analysis (Gauch, 1982; Zimmerman, Goetz & Meilke, 1985), we develop evidence that long-term effects on prey fish assemblages are strongly related to the history of predation effects.

Methods

Study site

The study site, Bolger Bog, is a small lake located within the University of Notre Dame Environmental Research Center, about two kilometers southeast of Peter Lake. The lake is surrounded by bog mat and has an open water area of about 1 hectare (Fig. 5.1). It is shallow ($Z_m = 4.0$ m), acidic (pH = 6.0), and has low transparency (Secchi depth = 0.2–0.7 m). During summer stratification, the depth of the thermocline is 0.5–1.0 m and the hypolimnion is anoxic between May and late September. The lake is largely anoxic during the months of ice cover, November–March.

A small tributary extending about 200 m from the lake gathers groundwater and flows into one end of the lake. At the opposite end, the lake is connected to a nearby stream via a 1–2 m wide outflow channel. Fish can migrate between the lake and the stream through this outflow channel. The channel, however, is a potentially severe habitat for fish because oxygen in the channel can fall below 1 mg $O_2 l^{-1}$ at night.

The fish community is diverse, consisting mainly of small planktivores and omnivores (He & Kitchell, 1990). The principal species include: northern redbelly dace (*Phoxinus eos*), central mudminnow (*Umbra limi*), finescale dace (*Phoxinus neogaeus*), fathead minnow (*Pimephales promelas*), brook stickleback (*Culaea inconstans*), brassy minnow (*Hybognathus hankinsoni*), golden shiner (*Notemigonus crysoleucas*), blacknose shiner (*Notropis heterolepis*), creek chub (*Semotilus atromaculatus*) and juvenile sunfish (*Lepomis* spp.). These fishes have a size range of 30–140 mm in total length. This assemblage, which has no large piscivores, is common in lakes that are subject to periodic winterkill conditions (Tonn & Magnuson, 1982). To avoid winter hypoxia, most fish species emigrate from the lake in late autumn and early winter, then return in spring (Magnuson *et al.*, 1985). The zooplankton community is dominated by predatory rotifers (*Asplanchna* spp.), other rotifers, and small copepods. Cladocera are rare although a few *Chydorus* spp. have been found in the lake.

Sampling and experimental design

The experiments were conducted from late May to late October during 1986 through 1989. Planktivorous fishes were sampled with minnow traps (height = 42 cm, diameter = 22.5 cm, mesh = 6 mm). Each sample was derived from 6 traps set in each channel, 12–26 traps set along the

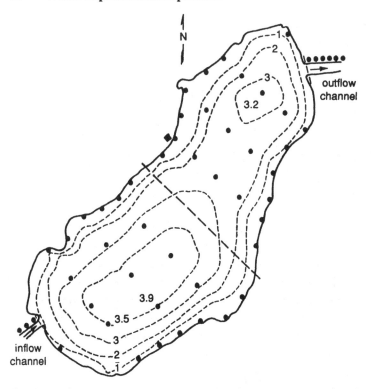

Fig. 5.1. The sampling stations in Bolger Bog. The division where the metal fence was set across the lake is shown by a straight dashed line. Numbers indicate the depth contours (m) of the lake. Minnow trapping stations are shown by circles; the boat landing is indicated by a diamond.

perimeter of the lake, and 9–18 traps set on transect lines in the open-water habitat (Fig. 5.1). All traps were set just below the surface for a duration of 22–24 h. Fishes caught in each trap were identified to species, counted and released. In 1986, about 30 fish of each species were taken for measurements of mass on each sampling date. Seasonal mean masses for each species were then calculated. During summer months (June to August), samples were taken every 2–3 weeks in 1986 and every week in 1987–9. During fall months (September and October), samples were taken every 2–3 weeks. In 1987, each sample was replicated the following day and the average of two day samples was used in the analysis.

Densities of prey fishes were estimated by calibrating catch per unit effort (CPUE, number of fish caught per trap per hour) from minnow traps with mark-and-recapture methods (Petersen method; Ricker, 1975). The mark-and-recapture was conducted in June 1986 and June

1988 to estimate population densities of redbelly dace, the most abundant species. Relative densities of other species were estimated based on the CPUE ratio of redbelly dace to individual species. Total prey fish biomass was then estimated by density and mean mass of each species.

The experiments, designed as unreplicated paired-system experiments, were conducted in two phases (1986–7 and 1988–9). The first year (1986) served as a baseline for experimental manipulations in the following three years (1987–9). No northern pike were introduced into the lake in 1986. On May 23, 1987, 15 adult northern pike (mean total length = 507 mm, s.d. = 30 mm) were removed from a nearby lake by hook-and-line and introduced to Bolger Bog. The outflow channels were blocked by wire fences (mesh size 2.5 cm × 2.5 cm) that retained northern pike in the lake but allowed prey fishes to pass through. Each northern pike was measured, weighed and tagged before release. Their diets and growth rates were determined on June 29 and again on July 29 using overnight gillnet sets to capture nine individuals each time. Pike removed in this fashion were replaced with animals of similar size. On August 21 of 1987, all northern pike were removed from the lake using gillnets.

The second phase of the experiment began in June 1988, when the lake was divided into halves by a metal fence (mesh size 5 cm × 2.5 cm) that would allow planktivorous fishes to pass through but not northern pike. The division of the lake was originally designed to provide the prey fishes with half of the lake as a refuge area. Later analyses suggested that responses of prey fish to predators were at the scale of the whole lake (He, 1990) either because the fence failed to retain all of the northern pike or because minnows did not perceive the relative safety of the pike-free side of the lake. In this chapter, we assume that the effect of northern pike was uniform across the entire lake. On June 6, 1988, eight adult northern pike (mean total length = 470 mm, s.d. = 15 mm) were released into the east side of the lake. From July 16 to 22, two gillnets were set on the east side of the lake and a total of four northern pike were caught. We assume that the other four northern pike had died. After July 22, the lake was left pike-free until 26 August when 12 adult northern pike (mean total length = 474 mm, s.d. = 48 mm) were released into the west side of the lake. Those northern pike remained in the lake until the last sampling date (October 15). On 21 June 1989, 12 northern pike (mean total length = 480 mm, s.d. = 44 mm) were released into the lake. All remaining northern pike were removed from the lake during the week of August 16.

Data analysis

Year-to-year variance of prey densities was analyzed using the catch per unit effort (CPUE) of individual species compared among the four years. In the comparison, the whole lake was treated as a sampling unit and catches from all traps set in the whole lake were included. The comparison between the baseline year (1986) and the manipulated years (1987–9) was emphasized. Ratios of CPUE in the outflow channel to the lake perimeter were used to indicate habitat preference by prey species. Because the outflow channel is shallower and has 2–4 mg l^{-1} lower oxygen content than the lake, we expected that fish should prefer the lake to the channel habitat. Therefore, higher catches of prey fishes in the channel would indicate either that fish emigrate through the channel or that fish gather in the channel as a predator avoidance behavior.

Because the experiment was conducted on a single lake, no replicated data can be obtained and traditional statistical techniques (t-tests, ANOVAs, etc.) cannot be used to test whether there is change due to the manipulation. Randomized intervention analysis (RIA, Chapter 3) was used to test whether nonrandom change occurred in the CPUE following the introduction of northern pike into the lake. In the analysis of 1986 and 1987 data, the lake was considered the manipulated system and the outflow channel was the reference system. Premanipulation data were from the periods before the addition of northern pike and after the removal of northern pike. Postmanipulation data were for the period when northern pike were present in the lake. Similar approaches were used in the analysis of 1988 and 1989 data.

Although individual species abundances might vary from year to year, there is strong evidence of rapid compensatory responses in minnow-dominated assemblages such as that of Bolger Bog (Tonn, 1985). Accordingly, we assumed that prey fish biomass in the manipulated years (1987–9) would have been the same as that in 1986 if no northern pike had been introduced to the lake. Therefore, the biomass lost to emigration in the manipulated years was estimated from the difference between the prey biomass in 1986 and the prey biomass in the manipulated year plus the direct consumption by northern pike in that year. The direct consumption of prey biomass by northern pike was estimated using a bioenergetics model for northern pike (Kitchell et al., 1977; Bevelhimer et al., 1985) which provided an estimate of prey eaten based on the observed growth of stocked pike.

The data sets used in the community analysis are mean CPUE

(number of fish caught per trap per hour) of individual species from all traps set in the lake from 1986 to 1989. In the data set, a sampling unit (SU) was mean CPUE from each sampling date. An indirect gradient analysis, Detrended Correspondence Analysis (DCA) (Hill & Gauch, 1980; Gauch, 1982), was used in the analysis. The percentage CPUE data, transformed by arcsine-square-root (Sokal & Rohlf, 1981), were used in the analysis, because this normalization allows a better representation of community structure (He, 1990). Eleven species were included in the data set.

The multiresponse permutation procedure (MRPP) was used to determine whether significant differences existed between sample groups on the ordination space separated by presence/absence of northern pike (Zimmerman *et al.*, 1985).

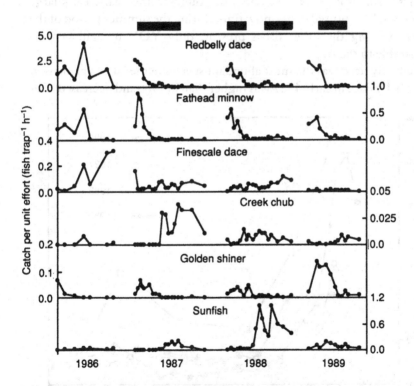

Fig. 5.2. Mean catch per unit effort (CPUE) of six prey species for Bolger Bog from 1986 to 1989. Means were calculated from all traps set in the lake and both channels. Periods of northern pike presence in the lake are indicated on the top of upper panel by bars. Note that different scales were used on Y-axes.

Results

Introduction of northern pike significantly reduced total prey fish densities. Reduction of CPUE of prey fish differed among species (Fig. 5.2). In comparison of the unmanipulated year (1986) with the three manipulated years (1987–9), catch reduction occurred in the three most abundant species: redbelly dace, fathead minnow and finescale dace (Fig. 5.2). Increased catches occurred as creek chub, golden shiner and sunfish, which were regularly present, became more abundant (Fig. 5.2). Comparisons for other species were not made either because their abundances were generally low and exhibited high year-to-year variability in catches (brassy minnow, brook stickleback and mudminnow) or because they were present only in the manipulated years (blacknose shiner and yellow perch). Numerical responses due to reproduction by the minnow species would not be perceived because young-of-the-year are not large enough to be captured in minnow traps during the summer period of the study. Among the other fishes present, only YOY bluegill became vulnerable to the traps.

Redbelly dace was the most abundant species in the lake. The density of redbelly dace in early 1987 was higher than in 1986 and decreased after

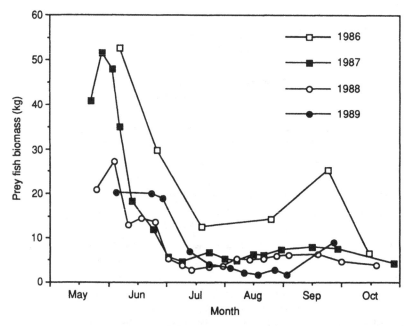

Fig. 5.3. Estimated total prey fish biomass in Bolger Bog from 1986 to 1989.

the northern pike addition (Fig. 5.2). Only one month after the introduction, redbelly dace were rarely found in the lake, and density
remained very low throughout summer. Responses of redbelly dace in
1988 and in 1989 were similar to those in 1987, except that density
slightly increased during the pike-free period from late July to late
August in 1988. Responses of fathead minnow were similar to those of
redbelly dace (Fig. 5.2). Catches of finescale dace declined after the
northern pike introduction in 1987, and catches remained low in 1988
and 1989.

Densities of creek chub and sunfish were higher in the manipulated
years than in the baseline year (Fig. 5.2). Highest catches were found in
1988 for sunfish and in 1987 for creek chub. Catches of golden shiner
were higher in 1989 than the first three years.

Removals of northern pike in the middle of summer were originally
designed to relieve predation pressure on prey fishes. It was expected that
catch of all species should increase by immigration after the removal of
the predator. Catches did increase during the pike-free period in 1988
(July 23 – August 26) but not during the pike-free periods in 1987 and
1989 (late August – October) (Fig. 5.2). This suggested that immigration

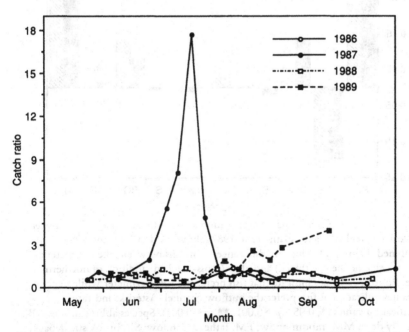

Fig. 5.4. CPUE ratios of outflow channel to lake perimeter of Bolger Bog for all
species combined, from 1986 to 1989.

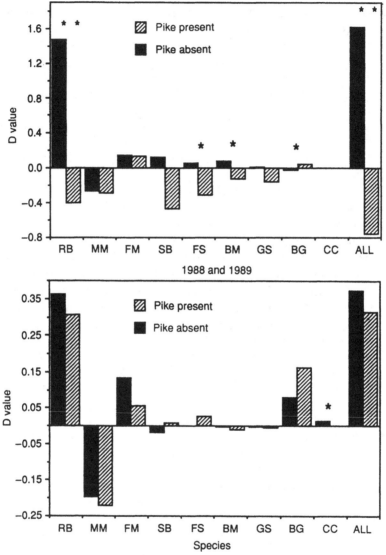

Fig. 5.5. Results of randomized intervention analysis (RIA) for CPUE between outflow channel and lake perimeter in 1987 (above) and in 1988 and 1989 combined (below). *D* values are differences of means between lake perimeter and outflow channel before (black) and after (cross-hatched) addition of northern pike. Positive *D* values indicate that fish prefer the lake perimeter while negative *D* values indicate that fish prefer the outflow channel. Asterisks indicate significant *p* values (*, $0.05 > p > 0.001$; **, $p < 0.001$). Species abbreviations: RB, redbelly dace; MM, mudminnow; FM, fathead minnow; SB, brook stickleback; FS, finescale dace; BM, brassy minnow; GS, golden shiner; BG, sunfish; CC, creek chub; ALL, all species combined.

occurred primarily during July and August and stopped after August. The high catch in July 1986 also indicated active immigration in July.

The annual mean biomass for all species observed during 1987, 1988 and 1989 was reduced to 60%, 41% and 39% of that recorded in 1986. Total biomass was always the highest at the beginning of each sampling season (Fig. 5.3). However, total biomass in the beginnings of 1988 and 1989 was much lower than in 1986 and 1987 (Fig. 5.3). We do not know whether this difference resulted from the carry-over effects of predation from 1987, or from weak recruitment of these year classes of prey fishes.

Many fish moved to the outflow channel after the northern pike addition in late May 1987, as indicated by higher catch ratios of the outflow channel to the lake perimeter (Fig. 5.4). Beginning in June 1987, catch ratios increased to as high as 17.7 before decreasing in July. In the same period of 1986, the ratios tended to decrease, with a low point of 0.17. However, catch ratios in 1988 were similar to those in 1986 and did not increase after the northern pike addition (Fig. 5.4). In 1989, catch ratios were similar to those in 1986 until August and then began to increase although northern pike had been removed from the lake after August. Reasons for the increase in catch ratio with absence of northern pike are unclear. It might be due to the few fish caught during that period and therefore higher variances of catches among traps.

The behavioral responses observed in 1987 were different among species. The RIA of 1986 and 1987 data showed that significant habitat shifts from the lake perimeter to the outflow channel occurred in four species (Fig. 5.5). More redbelly dace, finescale dace and brassy minnow shifted from the lake to the outflow channel after the addition of northern pike. More sunfish shifted from the outflow channel to the lake. All other species showed no significant shifts into any particular habitats. The RIA of 1988 and 1989 data indicated that only creek chub showed a significant shift between the outflow channel and the lake perimeter. Creek chub showed preference for the outflow channel after the northern pike introduction although the catch difference between the lake perimeter and the outflow channel was very small. Lack of a shift between the outflow channel and the lake perimeter in 1988 and 1989 suggests that the pike-free side of the lake provided a partial refuge for prey fishes.

In 1987, the estimated biomass of prey fish plus the estimated biomass consumed by northern pike were much lower than the estimated biomass of prey fish in 1986 (Fig. 5.6). If we assume that the prey fish biomass in 1987 would have been the same as in 1986 without the presence of northern pike in the lake, the decline of the biomass of prey fish in 1987 was largely due to emigration by prey fishes. In fact, biomass of prey fish lost to emigration was about 2–9 times higher than biomass of prey fish lost to direct consumption by northern pike during early June to mid-July. However, in 1988 and 1989, the same analytic method indicated that biomass of prey fish lost to emigration was negligible. Predation accounted for the majority of apparent change in abundance.

DCA ordinations for mean CPUE (percentage catch) are presented in Figure 5.7. The first two axes extracted more than 90% of variance in the data, the first axis (DCA1) accounting for about 60% and the second axis (DCA2) accounting for about 30%. Redbelly dace was negatively correlated with DCA1 ($r = -0.95$, $p < 0.001$), while sunfish was positively correlated with DCA1 ($r = 0.90$, $p < 0.001$). Thus, samples with relatively high abundances of redbelly dace and relatively low abundances of sunfish will have low scores on DCA1; conversely, samples with relatively low abundances of redbelly dace and relatively high abundances of sunfish will have high scores on DCA1. As DCA1 accounts for the majority of variance in the data, we infer that the major difference among samples is correlated with variances in abundance of redbelly dace and sunfish.

The MRPP test shows that sample groups defined by presence/absence of northern pike were significantly separated in ordination space ($p < 0.005$). Thus presence/absence of northern pike was related to changes in community structure of prey fishes. The effects of northern

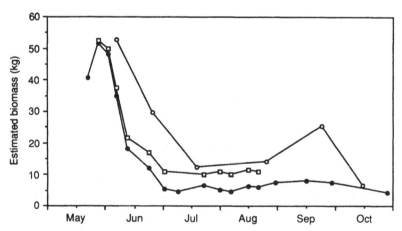

Fig. 5.6.   Estimated biomass of prey fish in 1986 (open circles) and 1987 (solid circles), and 1987 biomass plus biomass consumed by northern pike (open squares). The difference between the open circle line and open square line indicates the biomass lost to emigration.

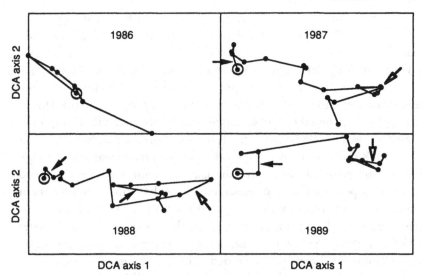

Fig. 5.7. DCA ordination of percentage CPUE of prey species from 1986 to 1989. Scales of both axes are the same for all four panels. Lines join sample points according to the sequence of sampling. Circled points are samples from the first sampling date of each year. Solid arrows indicate the dates of northern pike introduction while open arrows indicate the dates of northern pike removals.

pike predation can also be seen by tracing the sample trajectories on the ordination space along with the manipulation schedules of northern pike (Fig. 5.7). For example, although DCA1 is highly correlated with day of year ($r = 0.72$, $p < 0.01$), samples from the later periods of each year (September–October) do not have highest scores on DCA1. Samples with high scores on DCA1 are those from late summer when northern pike were present in the lake. In late August 1987 and late July 1988, removals of northern pike from the lake caused sample points to move to the left along DCA1, reversing the positive correlation between day of year and DCA1 score. In late August 1988, reintroduction of northern pike into the lake caused sample points to move back to the right along DCA1. These shifts of sample points along DCA1, associated with removal and reintroduction of northern pike, indicate that northern pike predation had immediate and pronounced effects on the prey fish community.

Discussion

This study has demonstrated that predator avoidance behavior exhibited by prey species, which is expected from theory (Werner & Gilliam,

1984) and demonstrated in experiments (Werner, 1986), is an important determinant of both population density and community structure when evaluated at the scale of whole systems. Rapid increases in the catch ratios of the outflow channel to the lake demonstrated increased emigration as the major cause of reduction in prey fish biomass. We also found that behavioral responses of prey fish to predation may depend on the density of prey fish in the lake. For example, significant behavioral responses were observed in 1987 when total biomass of prey fishes was high, but not in 1988 and 1989 when total biomass of prey fishes was low. At high density, the prey fish may leave the lake to avoid both predators and intense competition for food resources. Those that appear in subsequent catches may have recently entered the lake or remained there because the benefit due to reduced competition offsets the risk of predation (Werner & Gilliam, 1984). At low prey fish density, the direct effect of predation may account for the greatest proportion of prey biomass losses.

Bolger Bog is a winterkill lake and its fish assemblage is reestablished through the migratory process each spring. Therefore, the effects of northern pike on the prey fish community in this experiment are more important in the short-term dynamics (i.e. greater within-year variance) than in the long-term dynamics. Prediction of long-term effects of northern pike predation based on the observations on Bolger Bog could be misleading because of differences in scales of observation. The predation effects could be very different if Bolger Bog were a closed and non-winterkill lake. However, with four years of evidence, repeated experimental treatment, and a whole-lake scale, some consistent responses were apparent and provide some insight for long-term prediction. For example, during the pike-present periods of all three manipulated years (1987–9), abundance of redbelly dace always decreased while abundance of sunfish increased. The variances of abundances of these two species also correspond to the strongest patterns in the community ordination. Thus, it is reasonable to predict that among all prey fish species, redbelly dace might be the first species to disappear from the community while sunfish could be the species that would most likely coexist with northern pike. Fathead minnow and other small-bodied, soft-rayed fishes show a similar response to redbelly dace while golden shiner and creek chub exhibit responses similar to those of sunfish. Both the shiner and the chub may benefit from larger body size, which reduces vulnerability to predation. By using similar analysis, but based on the presence/absence of species, He and Wright (1992) found that complete changes of species composition in Bolger Bog may also be possible as the rare species (yellow perch, golden shiner, blacknose shiner and creek chub) become important species in determining the community structure.

Our conclusions are similar to those based on species-specific responses to predation (He & Kitchell, 1990) when adult body size and morphology were used to rank fish in terms of vulnerability to pike predation. This suggests that other interactions among prey species (i.e. competition) are relatively weak compared with the effects of the predator and the behavioral responses that altered the fish community.

This study cannot provide direct evidence of competition among prey species. If competition for food or space were intense among prey species, replacement of species would be more likely to be due to density compensation so that the total prey densities in Bolger Bog would remain relatively constant with or without predation (Tonn, 1985). Alternatively, competition may exist between prey species, such as between redbelly dace and sunfish, and its effects might be amplified by the predator effect. We can only affirm that abundances for these two species were negatively correlated when the predator population was manipulated.

The community analyses performed in this study take advantage of the fact that species responses are not independent; this relation reduces the dimensionality of the data. An accounting of each species is not required. This simplification permits interpretation of important trends and greater understanding of trophic 'events' in the system although it does not elucidate the relative importance of causative factors. As a result, the general characteristics of a fish assemblage may be used to estimate their functional effect on lower trophic levels and may be used to anticipate highly nonlinear responses to changes in piscivore abundance. As portrayed in the previous and following chapters, this generality can be extrapolated to the ecosystem scale.

Summary

Population changes documented in Chapter 4 suggest that behavioral responses of planktivores in Peter Lake significantly affected predation rates in the pelagic zone. Here we compare the direct effects of piscivory with the indirect effects of predator avoidance on planktivore density. This experiment used northern pike as a piscivore and was conducted in Bolger Lake separately from the other whole lake experiments.

The first year (1986) was a baseline year in piscivore-free Bolger Lake.

Northern pike were introduced in each of the subsequent three years. Presence of relatively few pike caused rapid changes in the prey fish community. Predator avoidance behavior by prey species accounted for a large proportion of these changes. Magnitude of prey response probably depended on predation pressure and intensity of competition among prey species. Prey fish responses to predation were evaluated from both population and community perspectives. Community analysis techniques (Detrended Correspondence Analysis and multi-response permutation procedure) provided simple, useful, integrated indications of prey fish response.

Acknowledgements

We thank Dave Benkowski and Angela Dunst for help with fish sampling. Steve Carpenter, Daniel Schindler and Wayne Wurtsbaugh provided many helpful comments on the early draft.

6 · *Roles of fish predation: piscivory and planktivory*

Xi He, James F. Kitchell, James R. Hodgson,
Russell A. Wright, Patricia A. Soranno, David M. Lodge,
Phillip A. Cochran, David A. Benkowski and
Nicolaas W. Bouwes

Introduction

Understanding the impacts of fish predation on lower trophic levels is a generally important goal (Wootton, 1990). In the special case of our studies, fishes are the reagents of whole-lake experiments. Because many fishes are opportunistic predators capable of complex behavior (Chapters 4 and 5; Hodgson & Kitchell, 1987), manipulation of fish populations may change predation pressure on lower trophic levels in unexpected ways. Therefore, it was essential to measure rates of predation on key food web components during the course of our experiments.

In piscivore-dominated systems, some species of planktivorous fishes may not persist or may be maintained at very low population densities (Tonn & Magnuson, 1982). Juvenile fishes are typically planktivorous and may be very abundant after hatching. Although a cohort of juveniles may be dramatically reduced owing to intense, continuous predation by adult piscivores, their effect as predators of zooplankton may be intense for very short periods. The prospect for a pulse of zooplanktivory followed by a pulse of piscivory heightened our interest in providing quantitative measures of intensity and duration of such short-term dynamics in predator–prey interactions revolving around fishes.

Habitat heterogeneity and habitat selection also influence predator–prey interactions (Werner & Gilliam, 1984). The relatively simple habitats in our study lakes provide only a modest amount of refuge where prey fishes may escape piscivores. Lack of refugia in Peter Lake explains the quick disappearance of the minnows introduced in 1985 and the rapid decline of rainbow trout in 1989 (Chapter 4). Intense cannibalism is often another distinct feature of piscivore-dominated systems (Smith &

Reay, 1991), where low recruitment success has been regularly observed. As evident in Paul and Peter Lakes, cannibalism constrains survivorship and recruitment of young-of-the-year (YOY) and juvenile bass (Chapter 4).

Many fishes progress through a trophic ontogeny: diets change as habitat shifts occur and the relative sizes of predator and prey change with growth dynamics (Werner & Gilliam, 1984). Adult piscivorous fishes can alter their diets and eat a wide range of other prey when small fishes are scarce (Hodgson & Kitchell, 1987; Hodgson et al., 1991). The shift of diets between fish and non-fish prey directly affects growth rates of larger fishes and may regulate survivorship of juvenile bass that eventually become piscivorous. Diet shifts also have implications for alternative prey populations, such as those of benthic or pelagic invertebrates.

Planktivorous fishes exhibit much of the same trophic ontogeny that is found in the early life history of piscivores and exhibit shifting diets as a function of prey availability as well. For example, redbelly dace switch both ontogenetically and seasonally between eating zooplankton and periphyton.

Small-bodied organisms like the small planktivorous fishes (primarily minnows) found in our lakes tend to have shorter generation times and higher intrinsic rates of population growth than large organisms such as large piscivorous fishes. Cannibalism in the planktivore populations does not appear to be a strong influence in limiting the cohort strength, as can be the case in piscivore populations. These features of planktivore populations suggest populations that would not show cycles of strong cohorts. Populations of small planktivores should respond rapidly to changes in resource abundance, recovering quickly after disturbance. One would predict that these small planktivores, because of their higher intrinsic rate of increase, would have relatively more resilient and stable abundance than populations of piscivores (Chapter 16). This apparent stability breaks down rapidly and state change occurs when piscivores are introduced to a planktivore system, owing to the strong behavioral shifts and the high vulnerability of the minnow species in our lakes (Chapter 5).

The goal of this chapter is to present predation rates by fishes on their primary prey, which include other fishes, zooplankton, benthos, and predatory invertebrates as components of a diverse diet. The importance of fishes, zooplankton and *Chaoborus* is emphasized because these prey populations are critical actors in the trophic cascade. Growth rates of piscivorous fish (largemouth bass) as a function of their densities are compared among lakes as an indicator of the density-dependent components of predator–prey interactions.

Methods

Predation rates were estimated for the period from late May to late August of each year using the bioenergetics model developed by Kitchell *et al.* (1977). This model is generally employed to estimate rates of feeding based on the observed average growth rates, diets and energy densities of predator and prey, and thermal history (Kitchell, 1983). It uses the size structure and dynamics of the predator population as a basis for calculating total predation rates. Modeling analyses were conducted using the computer software developed by Hewett & Johnson (1987, 1992). The energetics modeling approach is a preferred means for estimating feeding rates because it is less likely to be biased by the artifacts that derive from laboratory studies of digestion rates, respiration rates, excretion rates, etc. (Adams & Breck,1990; Wootton, 1990). Rigorous and direct comparisons of the daily ration method and the bioenergetics modeling approach reveal that the latter is a fully satisfactory substitute and offers the benefit of integrating short-term variability in feeding–growth dynamics (Kitchell, 1983; Rice & Cochran, 1984). Our previous analyses of sampling methods (Hodgson & Cochran, 1988) and sampling frequency (Hodgson *et al.*, 1989) assured a most appropriate application in this study.

The model is based on calculation of a daily energy budget, which takes the form:

$$dB/(Bdt) = C - (R + F + U),$$

where B = mass of fish, C = food consumption (predation rate), R = respiration, F = egestion and U = excretion.

Specific growth rates of fish ($dB/(Bdt)$) were estimated from the fish growth data (Chapter 4). Respiration was estimated from temperatures occupied in the lake, and fish mass. Egestion (F) and excretion (U) were calculated from temperature, fish mass, food consumption rates and food quality. We estimated daily consumption of each of the diet categories for all cohorts of fish from the observed diet composition (Chapter 4). Total consumption by a population of fish (i.e. for the whole lake and expressed on an areal basis) was the sum of all of the cohorts as calculated from estimates of population densities and mortality rates. Total predation estimates therefore integrate changes in diet, dynamics of the populations, and changes in the species composition of the fish assemblage.

Water temperatures used in the model were taken from weekly temperature profiles. Mean epilimnion temperature was used for large-mouth bass and minnows (Chapter 13). Owing to their restriction to the metalimnion during the period of summer stratification, a constant temperature of 15 °C was assumed for rainbow trout.

Diets for largemouth bass and rainbow trout were obtained from biweekly sampling in all three lakes (Chapter 4). Percentage mass of each diet item was used in the model. Six diet categories were used: fish (minnows, YOY and juvenile largemouth bass); zooplankton (all zoo-plankton, which was dominated by *Daphnia*, but excluded *Polyphemus*); *Chaoborus* spp.; benthos (benthic invertebrates other than those inverte-brates included in other categories); and other prey, which includes amphibians and terrestrial vertebrates. Energy densities of predator and prey were taken from the compilation offered in Hewett & Johnson (1992).

Growth rates of largemouth bass and rainbow trout were calculated from the equation:

$$G = (M_t - M_0)/t,$$

where G = mean growth rate (g d^{-1}), M_t = mass at time t (g), M_0 = mass at time 0 (g) and t = days between sampling (d).

For largemouth bass, estimates of mass were obtained from individu-ally tagged fish caught both in the late May sampling and in the late August sampling. Owing to small sample sizes in some of the sampling periods, growth rates for some of the age groups were estimated from the most recent year containing that age group from the same lake.

Scale samples of largemouth bass were taken during the late August sampling of each year, and were used to determine ages of fish. Age distributions were used to calculate the numbers of fish in each age group. Mortality rates for each age group were assumed to be constant for the period from late May to late August.

Only single cohorts of rainbow trout were introduced into Peter Lake in 1988 and again in 1989. Individuals were not tagged. Therefore, mean growth rates were obtained from the difference between the averages of the measurements of mass taken in late May and again in late August.

The minnow population in Tuesday Lake was divided into two cohorts, juvenile and adult, because no age distributions were obtained.

We assumed an operational definition of minnow cohorts; juvenile minnows were those too small to be caught by minnow traps while adult minnows were those that could be caught. Only population sizes of adult minnows were estimated (Chapter 4). In 1988, after minnows were reestablished in Tuesday Lake during 1987, we observed rapid recruitment to the adult size class. Accordingly, we assumed that juvenile minnows would mature in one year. Annual mortality was assumed to be 90%, evenly distributed through the whole year (He, 1986). Growth rates for juvenile minnows for 1984 and 1985 were estimated from otoliths (He, 1986) and were assumed to be comparable for the similar periods of 1987–90. There may have been a strong but unrecorded density effect on growth in 1987 when the adult minnow population was very low and large zooplankton prey were abundant. By 1988, however, minnow density had increased to higher levels and was causing rapid changes in composition of the zooplankton community (Chapter 8). Growth rates for adult minnows were estimated from length frequency distributions of fish caught in minnow traps. This procedure is fully described by He (1986).

Diet information for the minnows in Tuesday Lake in 1984 was obtained from Cochran *et al.* (1988). Because no diet data were taken after 1984, we assumed that all minnows' diets were similar to those previously determined: 27% zooplankton, 40% periphyton and 33% small insects. This assumption does not give accurate seasonal dynamics of predation, but is useful for comparison of interannual total predation rates on plankton. Occasional examination of gut samples indicated that the same assumptions were applicable in estimating diets of golden shiners in Peter Lake during 1990.

Results

Paul Lake

Predation on YOY largemouth bass by adult largemouth bass was correlated with the densities of YOY in the lake (Fig. 6.1; see also Fig. 4.3). The lowest predation rates occurred in 1984 when YOY densities were low. Likewise, the greatest predation rates occurred in 1987 when both YOY and adult densities were highest (Fig. 4.3). The highest predation rates in any season were observed in July and August, when temperatures were highest and YOY bass had grown to vulnerable sizes. Cannibalism by adult largemouth bass on the YOY largemouth bass can eliminate a large proportion of the YOY biomass and is the major

Table 6.1. *Annual average daily consumption (mg m⁻² d⁻¹) for each diet category estimated using the bioenergetics model*

Abbreviations: Fish, YOY largemouth bass and minnows; Chao, *Chaoborus* spp.; Zoop, zooplankton (mainly *Daphnia* spp.); Bent, benthos (mainly insects); Othr, vertebrates; Poly, *Polyphemus*; Peri, periphyton.

Lake/year	Fish	Chao	Zoop	Bent	Othr	Poly	Peri
Paul							
1984	0.83	1.42	0.03	31.88	12.85	0	0
1985	3.81	0.89	0.08	34.52	10.72	0	0
1986	10.70	1.17	0.20	29.78	0	0.00	0
1987	20.78	2.53	0.97	55.87	0	0.02	0
1988	4.64	1.60	1.34	18.45	0.86	0.03	0
1989	8.29	1.92	0.48	47.03	10.49	0	0
1990	12.71	4.55	0.16	31.37	18.78	0	0
Mean %	16.18	3.69	0.85	65.20	14.07	0.01	0
Peter							
1984	5.63	1.10	0.06	47.60	17.09	0	0
1985	18.67	0.39	19.35	31.09	0.09	2.61	0
1986	20.05	0.20	0.15	45.62	7.46	0	0
1987	0	1.55	3.70	47.50	0	0.00	0
1988	7.80	29.81	50.63	152.04	0	0.35	0
1989	28.49	15.95	7.93	197.65	21.63	0.01	0
1990	2.54	9.64	42.20	90.36	0	0.02	0
Mean %	8.98	6.33	13.39	66.05	4.99	0.26	0
Tuesday							
1984	0	0	177.7	0	0	0	212.90
1985	21.84	0.79	11.98	34.13	0	0.10	8.24
1986	7.15	0.50	0.03	12.55	3.30	0	0
1987	0	0	63.32	0	0	0	83.93
1988	0	0	342.74	0	0	0	454.35
1989	0	0	344.52	0	0	0	456.71
1990	2.63	1.06	376.51	4.66	0	0.00	485.86
Mean %	1.02	0.08	42.37	1.65	0.11	0.00	54.77

control of recruitment success in any year (Table 6.1). In Paul Lake, recruitment is usually poor, except when an unusually large year class is formed and a greater proportion of juveniles escape from cannibalism.

Predation on *Chaoborus* was low in Paul Lake compared with predation rates on fish (Fig. 6.1). Generally, predation rates were relatively high during both the beginning and the end of the summer and low in midsummer, corresponding to the availability of larger instar *Chaoborus*

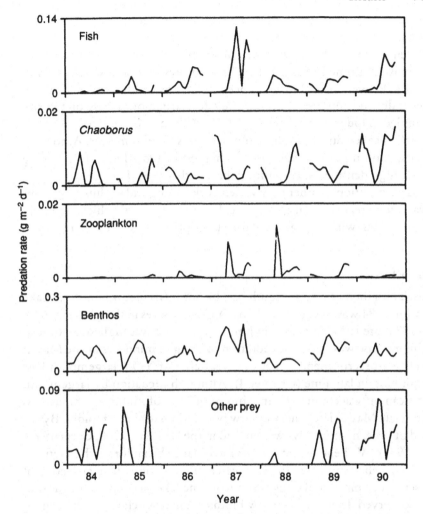

Fig. 6.1. Estimated predation rates on five main prey items by all fish in Paul Lake from 1984 to 1990. See Table 4.3 for details of each prey category. Note different vertical scales for each prey.

(Chapter 7). This pattern generally corresponds with high predation rates on fish and other invertebrate prey in midsummer (Fig. 6.1).

Predation on zooplankton (mainly *Daphnia* spp.) was low and relatively insignificant compared with that on other prey items (Table 6.1 and Fig. 6.1). The total biomass of zooplankton consumed by fish was a small proportion of the total biomass in the lake (Chapter 10). The relatively high predation rates on zooplankton observed in 1987 and 1988 resulted from the high densities of YOY and young largemouth bass in the those years.

Benthic invertebrates were the most important prey items in terms of biomass consumed (Table 6.1). This indicates that largemouth bass in Paul Lake were food-limited, since the sizes of benthos in the diets are generally not among preferred prey for largemouth bass of the size sampled (Hodgson & Kitchell, 1987). Within a season, consumption usually peaked during the high temperatures of midsummer. Among all years, consumption of benthos was highest in 1987 (Fig. 6.1), resulting from high densities of largemouth bass in that year (Fig. 4.1).

Consumption of other prey varied both among and within years. This was due to relatively large prey (i.e. birds and rodents) which are rarely captured but which can account for a large proportion of total prey mass.

Peter Lake

Predation on YOY largemouth bass by adult largemouth in Peter Lake during 1984 was low, owing to low YOY densities in the lake (Fig. 6.2). In 1985, predation increased in June and July and was highest in August. This increase was due to our addition of 55 kg of minnows in late May as well as to recruitment of a strong year class of YOY largemouth bass beginning in late June (Chapter 4). Although predation by largemouth bass can only account for a maximum of 11% of minnow biomass loss from late May to late June, minnow populations declined rapidly. By the end of June, no minnows were found in the lake (He, 1986; Carpenter *et al.*, 1987). We believe that predator avoidance behavior, such as refuge-seeking behavior in very shallow water (and the consequent attraction of piscivorous birds) and emigration from the lake, accounted for most of the observed decline in minnow biomass (He & Kitchell, 1990; Chapter 5).

Cannibalism by adult largemouth bass continued in 1986, mainly on the survivors of the 1985 year class which were age 1 + (Fig. 6.2). In 1987, however, predation on fish was negligible because surviving members of the 1985 year class had grown and there were very few fish small enough to be vulnerable as prey. Predation on fish in 1988 was low until late summer when a weak recruitment of YOY largemouth bass occurred. In 1989, predation on fish was the highest of all years owing to the relatively large size of the 1985 cohort of largemouth bass, which could prey on the 1 + rainbow trout that were smaller than those introduced in 1988 (Chapter 4). In 1990, only a few adult largemouth

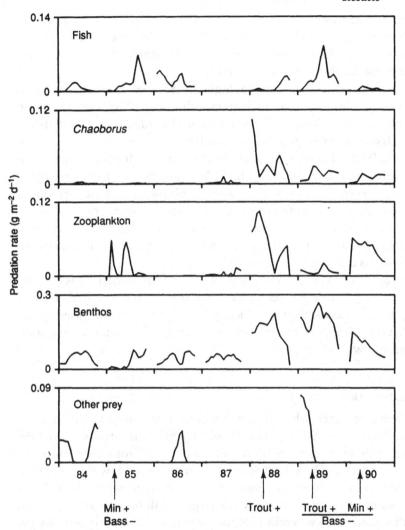

Fig. 6.2. Estimated predation rates on five main prey items by all fish in Peter Lake from 1984 to 1990. See Table 4.3 for details of each prey category. Note different vertical scales for each prey.

bass and rainbow trout remained in the lake; predation on fish was very low.

Predation on *Chaoborus* was low until the introduction of rainbow trout in 1988. Predation rates were highest immediately following the introduction, and gradually declined from 1988 to 1990 as *Chaoborus* densities decreased (Chapter 7). Predation on *Chaoborus* in 1990 was mainly by the 1990 year class of largemouth bass. The effects of high predation rates on *Chaoborus* from 1988 to 1990 are discussed in detail in Chapter 7 and had profound effects on the distribution of nutrients among the trophic levels (Chapter 17).

Predation on zooplankton was high in 1985, 1988 and 1990, owing to the strong year class of YOY largemouth bass in 1985, the introduced rainbow trout in 1988, plus the introduced golden shiners and strong year class of YOY largemouth bass in 1990.

As in Paul Lake, benthic invertebrates were the dominant prey items for fishes in Peter Lake (Table 6.1). A major increase in the predation rates on benthos began in 1988 when rainbow trout were introduced into the lake. These trout grew rapidly, switching from zooplankton to benthos as their size increased. Lake-wide predation rates on benthos were highest in 1989 but decreased in 1990 as a result of the removal of piscivorous fishes at the end of 1989.

As in Paul Lake, consumption of other prey varied between and within years. Large fluctuations can result from the presence of a few large vertebrates in the diets, which drives the estimates for consumption of that diet category to a very large percentage of the total consumption compared with other prey.

Tuesday Lake

Piscivory occurred when largemouth bass were present in the lake in 1985, 1986 and 1990 (Fig. 6.3). Highest predation rates were in 1985 when some minnows remained and there was a strong recruitment of YOY largemouth bass. In 1986, predation rates on fish were lower than in 1985 because minnows were absent and there were fewer YOY largemouth bass. In 1990, the YOY largemouth bass began to prey on small minnows a few weeks after their introduction in early July. By the end of the summer, bass that preyed heavily on minnows earlier in the summer grew large enough to prey on all sizes of minnows as well as the smaller members of the stocked bass cohort (Wright & Kitchell, 1993).

Predation on *Chaoborus* and invertebrates by largemouth bass followed the same pattern as predation on fish (Fig. 6.3). Predation rates were highest in 1985, 1986 and 1990 and low in other years. It is still unclear how much predation on *Chaoborus* was due to minnows. The limited diet data available from 1984 studies showed few *Chaoborus* in minnow diets (Cochran *et al.*, 1988). High *Chaoborus* densities were observed in Tuesday Lake during 1984 when minnow populations were

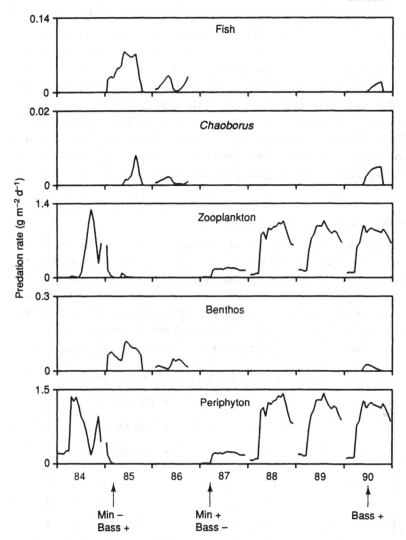

Fig. 6.3. Estimated predation rates on five main prey items by all fish in Tuesday Lake from 1984 to 1990. See Table 4.3 for details of each prey category. Note different vertical scales for each prey.

dense. However, *Chaoborus* did not recover during 1987–90 during and after the time when minnows were reestablished. The causes of this lag in recovery remain the subject of speculation and interest (Chapters 7 and 17).

Minnows were much more important than largemouth bass as predators of zooplankton in Tuesday Lake. Predation rates corresponded to minnow densities and growth rates (Fig. 6.3 and Fig. 4.9) and were highest in 1984 and 1988–90 (Table 6.1). Predation rates were relatively consistent from 1988 to 1990 because the zooplankton component of the diet was assumed to be constant for all three years (27% of total consumption). The low predation rates in the beginning of each summer from 1988 to 1990 were due to the late appearances of the juvenile minnows, which we assumed did not become zooplanktivorous until late June. Consumption of periphyton followed the same pattern as that for zooplankton because we also assumed constant dietary composition (40%) of periphyton during the period from 1985 to 1990 (Fig. 6.3).

Growth rates of largemouth bass

We compared growth rates for largemouth bass among different lakes and years as a basis for gaining insights about density dependence of growth, prey selection, and the effects of largemouth bass on their prey. Because growth rates can depend in part on food availability, faster growth rates generally indicate greater prey abundance and/or selection of higher-quality prey (Rice & Cochran, 1984; Hodgson & Kitchell, 1987). Predation effects on prey populations can also depend on the

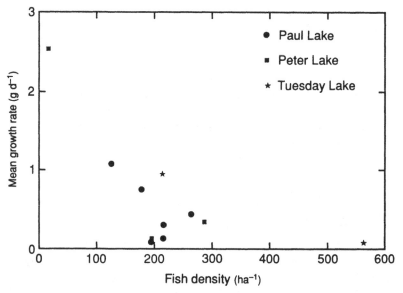

Fig. 6.4. Mean summer growth rates for adult largemouth bass (> 195 mm) versus densities in all three lakes. Rates were calculated from only those years in which adequate data were available.

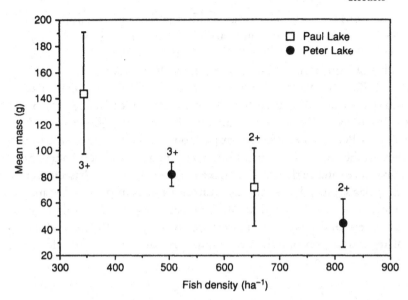

Fig. 6.5. Mean and standard deviations of summer growth rates for age 2 + (1987) and age 3 + (1988) largemouth bass in Paul and Peter Lake compared with densities of largemouth bass (excluding YOY).

population density of the predator because estimates of total consumption are a function of prey selection (as evidenced in diets), growth rate and population density.

Mean growth rates for adult largemouth bass for each year in all three lakes are presented in Figure 6.4. The highest growth rate in the adults occurred in Peter Lake in 1985 after the removal of 90% of the largemouth bass (Chapter 4). This time corresponds to the lowest density of adults and highest YOY recruitment for largemouth bass in Peter Lake. As a result, the most preferred prey – small fishes – were very abundant. Nevertheless, growth remained well below the maximum possible rates estimated by the bass bioenergetics model (Rice & Cochran, 1984). The lowest growth rates were observed in Tuesday Lake during 1986 when bass densities were the highest. Highest and lowest growth rates were observed in the manipulated lakes where densities of largemouth bass were at both extremes.

Clearly, food for largemouth bass in all three lakes was limiting and growth rates depended primarily on densities of predators. In the unmanipulated system, Paul Lake, growth rates were variable and also negatively correlated to densities (Fig. 6.4), although within a smaller range compared with Peter and Tuesday Lakes.

Comparisons of mean mass for a specific age class of largemouth bass in Peter and Paul Lakes also indicate that growth rates are density-dependent (Fig. 6.5). Mean weight for 2+ largemouth bass was higher, although not significantly (t-test, $p > 0.2$), in Paul Lake than in Peter Lake in 1987. By 1988, the 3+ largemouth bass in Paul Lake had significantly greater mean mass than those in Peter Lake (t-test, $p < 0.05$). These differences reflect intense intraspecific competition for food, especially in Peter Lake, where competition was most intense among members of the same cohort. In addition, competition in Peter Lake was probably more intense than in Paul Lake because Peter Lake has a lower littoral : pelagic ratio than Paul Lake. Littoral area is important because fish densities were calculated as number-per-area of the whole lake. Benthic invertebrates were the main prey for largemouth bass and the role of terrestrial prey would be roughly proportional to the relative extent of littoral habitat.

Discussion

Over time, densities and size structures of largemouth bass in Paul Lake were more consistent than in Peter and Tuesday Lakes (Chapter 4). It is not surprising, therefore, that effects of largemouth bass on the lower trophic levels of Paul Lake also remained relatively consistent for all years (Chapters 7, 8 and 10). In this oligotrophic and unexploited system, competition between largemouth bass is intense, especially for fish prey. Cannibalism plays an important role and is a key factor in determining the density and size structure of the largemouth bass population in Paul Lake.

As the result of a series of manipulations, predation rates in Peter Lake changed dramatically over the seven years with respect to both prey composition and absolute predation rates. The 1985 manipulation failed to establish planktivorous minnow populations, yet the remaining adult bass reproduced very successfully and offered low levels of cannibalistic constraint. As a result, survivorship of YOY bass was high and food web interactions became dominated by the 1985 year class. This year class preyed on zooplankton during its first summer of life and gradually switched to benthic prey. Growth rates were sufficient to allow an escape from predation by the remaining adults. By 1987, benthos comprised nearly 90% of total prey mass consumed by the bass population. By 1989, members of the 1985 cohort were large enough to impose heavy cannibalism on younger age classes (Table 6.1).

The ontogenetic shift in diet exhibited by the 1985 cohort of bass in Peter Lake has an important implication in terms of the trophic cascade. As a cohort matures, the strength of interactions within the food web shifts from dominance by planktivory to benthivory and then piscivory. Strong or weak year classes (Ricker, 1963) produce a wave of predation effects that pass through food webs over the course of years.

Introduction of rainbow trout in 1988 increased total predation rates and changed the composition of the prey community in three ways. First, predation rates on *Chaoborus* increased dramatically because rainbow trout are more pelagic than largemouth bass and inhabited waters near the thermocline throughout the summer. This increase in predation rates on *Chaoborus* had cascading effects on the zooplankton community in Peter Lake because it decreased invertebrate predation on small zooplankton (Chapter 8). Second, predation rates on large *Daphnia* also increased and caused zooplankton biomass to initially decline then recover and increase dramatically which was due, in part, to the indirect effect of trout predation on *Chaoborus* (Chapter 8). Third, introduced rainbow trout may have caused largemouth bass to shift their diet to include more benthos prey because trout were more efficient predators on zooplankton and *Chaoborus* than largemouth bass (Hodgson et al., 1991). The compensatory changes in size and species composition of the zooplankton community yielded lower total predation rates on zooplankton and *Chaoborus* in 1989 than in 1988, even though biomass of rainbow trout and largemouth bass were about the same in both years. At the same time, increased predation rates on small, zooplanktivorous fishes were apparent in the 1989 diets of both trout and bass. The net result was that overall predation rates on zooplankton changed dramatically from 1988 to 1989 and were an important factor in the regulation of the plankton community in Peter Lake (Chapter 8).

In general, Peter Lake was more variable and less predictable than the other lakes. Many unexpected events occurred over the seven years of intensive study. Among those were the failure to establish minnow populations in 1985, strong YOY classes of largemouth bass in 1985 and 1990, disappearances of introduced rainbow trout in 1989 (Chapter 4), and unusually high zooplankton biomass in 1989 and 1990 (Chapter 8). These events were strongly related to the tenor of fish predation in the lake, and further suggest the importance of predator–prey interactions in controlling food web dynamics.

Important lessons to learn from the Peter Lake experiment, from the point of view of piscivory, are that behavioral responses by prey fish have to be more effectively considered in predator–prey interactions (see also Chapter 5) and as a highly responsive component of the trophic cascade. In unexploited systems, potential for piscivory is usually high and cannibalism can strongly regulate recruitment success. Although piscivore densities might be reduced through exploitation or stochastic events, the behavioral response of vulnerable planktivorous fishes can effectively eliminate them from the epilimnion. Thus, low densities of piscivorous fish can effect the ecological equivalent of high piscivory. A successful recruitment event can restart the ontogeny of these interactions and will persist for the duration of the dominance by that cohort.

Over time, invasion by spiny-rayed and less vulnerable fishes such as yellow perch and bluegill might eventually yield a planktivore assemblage less subject to the effects of a modest piscivore population (Hambright, 1991). As a result, fish communities in lakes subject to regular exploitation will be dominated by planktivores that are less efficient predators than the competitively superior cyprinids, which can exploit the food refuge of periphyton when preferred prey are strongly depressed (Persson & Greenberg, 1990). The zooplankton communities in these kinds of systems will include a diversity of small to intermediate-sized forms such as *Bosmina*, *Chydorus* spp., *Diaphanosoma*, small *Daphnia* spp. and omnivorous copepods (e.g. *Diacyclops bicuspidatus*). In keeping with the ubiquitous distribution of anglers, the plankton and its consequent epilimnetic trophic system described above is among those most commonly observed from surveys of lakes of the temperate zone.

Compared with Peter Lake, the manipulation responses of Tuesday Lake were more predictable. Predation by largemouth bass introduced in 1985 successfully eliminated the remaining minnow populations and the lake became piscivore-dominated within just one year. The sequence of replacement in the zooplankton was *Bosmina–Diaphanosoma*–small *Daphnia*–large *Daphnia* and occurred within a single summer. Reversal of the process along much the same path required two years (Chapter 8). The system was fully returned to being planktivore-dominated by late 1988, within two years after removal of piscivores and reintroduction of a small number of minnows (Chapter 4). Predation effects on both prey fishes and zooplankton are well reflected in the responses of prey abundances (Chapters 3 and 8).

We were also able to relate the diet shifts of fish to their habitat use. For example, based on previous work (Hall & Ehlinger, 1989) the golden shiners introduced into Peter Lake during 1990 were expected to distribute evenly in both pelagic and littoral zones and feed primarily on zooplankton. Instead, we found that they aggregated in the littoral zone during the daytime and that their diets consisted of a large proportion of invertebrate prey (X. He and J. R. Hodgson, unpublished data). At night, shiners migrated offshore to feed on zooplankton and *Chaoborus* which had ascended into the epilimnion. As a result, predation on zooplankton by golden shiners was not as effective as we had expected.

Although our emphasis was on predation effects in the pelagic zone, both the behavioral responses evident in predator avoidance and the preponderance of benthic invertebrates evident in diet data for bass demonstrate a strong fish-mediated linkage between the pelagic and the littoral. Clearly, fishes and their effects serve to integrate interactions at the whole system level. Another linkage is also apparent in the sporadic but quantitatively important evidence of terrestrial vertebrates in the diets of these opportunistic predators. One jumping mouse is the biomass equivalent of 10 000 *Daphnia*. It also represents an allochthonous source of nutrients recruited from the terrestrial environment and recycled through the immediate effects of excretion and the long–term effects of accumulated fish biomass.

How fish predation affects the food web dynamics in lake ecosystems depends on the strength of linkage between fish and other food web components (Paine, 1980; Carpenter *et al.*, 1987). This study has shown that the strength of these linkages is not only determined by the abundances and size structure of fish populations, but can also depend on the vulnerability and relative abundance of prey populations. Large-mouth bass, an effective piscivore, can become non–piscivorous and maintain sufficient growth rates on alternate prey when small fishes are not available. Studies have shown that juvenile bass and minnows are the preferred prey for largemouth bass, followed by benthic insects, molluscs and large-bodied *Daphnia* (Werner, 1979; Hodgson & Kitchell, 1987). In all three lakes, we found that high piscivory occurred when prey fish were temporarily abundant, as in Paul Lake in 1987, in Peter Lake in 1985 and 1989, and in Tuesday Lake in 1985. As a result, relative predation rates on alternative prey, such as benthos, were reduced in those years. This suggests that studies on the alternate prey, such as the food refuge offered in benthos, are needed to effectively forecast predation effects. A piscivore response capacity waits in reserve and may be poorly anticipated in food webs described by contemporary diet surveys.

Interannual differences in the strength of trophic cascade effects derive from trophic ontogeny. Following the 1985 manipulation, the density and size structure of the Peter Lake bass population more nearly resembled that of lakes where fishery exploitation is common. The resultant wave of predation effects as a strong cohort passed through trophic ontogeny suggests that similar food web effects would be periodically demonstrated in lakes where fisheries are regularly practiced. Unlike Peter, Paul and Tuesday, most lakes are fished. Thus, the effects of trophic ontogeny are probably being expressed to some degree in most of the lakes in this part of the world.

Summary

Analysis based on bioenergetics models revealed that predation rates varied with time, but were more consistent in Paul Lake, the reference lake, than in Peter and Tuesday Lakes, the manipulated lakes. Piscivory rates were mainly determined by the availability and abundance of prey fishes. Even at low densities, piscivores were able to reduce planktivory as predator avoidance behaviors caused planktivores to remain in near-shore habitats.

Among all prey items, predation rates were the greatest on benthos. In Paul Lake, predation rates on benthos were relatively similar from year to year. In Peter Lake, a step increase in predation on benthos occurred when rainbow trout were introduced in 1988. In Tuesday Lake, significant predation on benthos occurred only when largemouth bass were present during 1985, 1986 and 1990.

Predation rates on *Chaoborus* were generally low. However, a substantial increase in predation on *Chaoborus* occurred in Peter Lake after rainbow trout were introduced in 1988. Predation rates on large, rare prey items (amphibians, birds, reptiles) fluctuated considerably within and among years and contributed to high variability in the bass diets.

Predation rates on zooplankton were low when largemouth bass were present in all three lakes. Predation rates on zooplankton in Tuesday Lake were high in 1984 and 1988–90 when dace were abundant. Introductions of rainbow trout into Peter Lake in 1988 and 1989 and the introduction of golden shiner into Peter Lake in 1990 also increased zooplanktivory rates.

The trophic ontogeny of a strong cohort of predators can play a major role in regulating long-term (i.e. interannual) responses in food webs and the variability of the trophic cascade.

7 · Dynamics of the phantom midge: implications for zooplankton

Patricia A. Soranno, Stephen R. Carpenter and
Susan M. Moegenburg

Introduction

Invertebrate planktivores occupy an intriguing position in the pelagic food web. Like other zooplankton, they are vulnerable to planktivory by fishes, yet they can prey heavily on certain zooplankton (Dodson, 1972; Neill, 1981; Black & Hairston, 1988; Hanazato & Yasuno, 1989). In Paul, Peter, and Tuesday Lakes, the most important invertebrate planktivores are larvae of three species of *Chaoborus*, the phantom midge. Owing, in part, to behavioral responses and ontogenetic diet shifts of both fishes and *Chaoborus*, the interaction between the two and consequences for zooplankton communities can be difficult to predict (Luecke, 1986; Elser *et al.*, 1987*b*; Neill, 1988; Hanazato & Yasuno, 1989). The trophic cascade hypothesis states that planktivory by invertebrates is inversely related to planktivory by fishes (Carpenter *et al.*, 1985). Nevertheless, we anticipated little change in *Chaoborus* during our experiments, because of their cryptic morphology and pronounced diel vertical migration behavior (Carpenter & Kitchell, 1987). Here we examine how *Chaoborus* populations responded to the fish manipulations, and consider the consequences for the trophic cascade.

Chaoborus develop through four instars as planktonic larvae before pupating. Their body is relatively transparent, except for well-developed hydrostatic organs, hence the common name phantom midge (Fig. 8.1, Chapter 8). Nevertheless, they are susceptible to predation by fishes. In lakes with fish, third- and fourth-instar *Chaoborus* will spend the days in either the hypolimnion, metalimnion, or bottom sediments and come to the surface at night to feed (Roth, 1968; von

Ende, 1979; Luecke, 1986). In lakes without fish, *Chaoborus* species often show no migratory behavior (Northcote, 1964; Carter & Kwik, 1977). The three species found in our experimental lakes are *C. punctipennis*, *C. flavicans* and *C. trivittatus*. One species migrates into the sediments during the day (*C. punctipennis*), while the others migrate to the hypolimnion (von Ende, 1982; X. He, unpublished data). Studies performed on *Chaoborus* from this region found that *C. flavicans* emerges in late May, *C. punctipennis* in early July, and *C. trivittatus* at the end of August (von Ende, 1982). These populations overwinter as either second or third instars, except for *C. trivittatus*, which overwinters as fourth instar (von Ende, 1982). The larger species (*C. trivittatus* and *C. flavicans*) are most abundant in the presence of piscivorous fish, such as largemouth bass. In the presence of planktivorous fish, such as minnows, the smaller *C. punctipennis* is most abundant.

Head capsule length of individuals is correlated with the maximum size of prey eaten (Fedorenko, 1975). Measured as head capsule length of fourth-instar larvae, *C. trivittatus* are the largest (1.8 mm); *C. flavicans* are intermediate (1.3 mm); and *C. punctipennis* are the smallest (1.0 mm) (von Ende, 1982). *Chaoborus* feed on most rotifers, small to medium-sized crustaceans, and large algal cells such as dinoflagellates. Predation by *Chaoborus* on zooplankton populations can be substantial (Dodson, 1972; Pastorok, 1980; Zaret, 1980; Elser *et al.*, 1987b; Yan *et al.*, 1991). Prey most vulnerable to invertebrate predators are those small enough to be handled and ingested, and detectable through mechanoreception (Gliwicz & Pijanowska, 1989).

The diet of *Chaoborus* can vary seasonally depending on its development stage, the numerical availability of different food items, and the amount of overlap in the spatial and temporal distributions of predator and prey populations (Fedorenko, 1975; Lewis, 1977; Pastorok, 1980). In our lakes, third- and fourth-instar *Chaoborus* prey heavily on prereproductive daphnids (length less than 1.4 mm), juvenile and adult copepods, and solitary rotifers (Elser *et al.*, 1987b). Many studies have shown how *Chaoborus* impacts populations of certain vulnerable prey (Dodson, 1972; Moore & Gilbert, 1987; Black & Hairston, 1988; Yan *et al.*, 1991). However, few studies have addressed how the zooplankton community may be affected by invertebrate planktivores at the whole-lake scale (Neill, 1988; Hanazato & Yasuno, 1989).

Many species of zooplankton have evolved adaptations to invertebrate predators (Havel, 1987). The morphological and life history responses of zooplankton to *Chaoborus* are distinctive. These include: the elongation of spines; the production of neck teeth (small spines on the dorsal crest of the helmet of *Daphnia*) (Havel, 1987); and reproduction at a larger size (Lynch, 1980). In our lakes, *Daphnia* are especially vulnerable to predation by late-instar *Chaoborus* (Elser *et al.*, 1987*b*; MacKay *et al.*, 1990). In response, juveniles produce neck teeth that serve as a defense by increasing the escape efficiency from invertebrate predators (Krueger & Dodson, 1981; Havel & Dodson, 1984). Laboratory experiments have shown that animals with neck teeth suffer less from predation by *Chaoborus* than animals lacking neck teeth (Havel & Dodson, 1984). However, daphnids only sometimes produce neck teeth in the presence of *Chaoborus*, but never in their absence, suggesting an energetic cost to production of these defenses (Krueger & Dodson, 1981; Havel & Dodson, 1985; Black & Dodson, 1990; Reissen & Sprules, 1990). In addition, the costs of these defenses are higher at lower resource levels (Walls, Caswell & Ketola, 1991). Therefore, the presence of these defensive spines indicates elevated predation pressure by *Chaoborus* when sufficient food is available.

In later chapters we discuss the response of the zooplankton community to changes in both vertebrate and invertebrate planktivory (Chapter 8). Here, we address how *Chaoborus* populations responded to different levels of fish predation by examining the biomass, individual mass and species shifts of the *Chaoborus* populations. We consider the potential impact of *Chaoborus* on the zooplankton communities based on studies of invertebrate planktivory by *Chaoborus* (Elser *et al.*, 1987*b*; MacKay *et al.*, 1990) and the morphological responses of prey species of these lakes. Finally, we consider the implications of changing prey abundance for *Chaoborus* biomass.

Methods

Chaoborus populations

Procedures for the collection and processing of *Chaoborus* samples have been presented elsewhere (Elser *et al.*, 1987*b*). Here we provide a brief summary of previously published procedures, and more detail on methods we have not published before. *Chaoborus* were sampled after dusk at biweekly intervals from the end of May to mid-September in 1985–90, and from 1 July to late September in 1984. Three to nine vertical tows were taken from 3 m using a 202 μm mesh net (30 cm diameter) in 1984 and 1985. Although head capsules of instar I larvae are 140–150 μm across, we believe that instar I larvae generally encountered the net lengthwise. Comparisons with plankton trap samples indicate that instar I were adequately represented in the samples. In subsequent years, three tows from 3 m were taken using a 147 μm mesh net (30 cm diameter). All samples were preserved in a 4% sugared, buffered formaldehyde solution. Net efficiencies were estimated in 1989 three times during the summer by calibrating two vertical tows against a vertical profile at 1 m intervals obtained with a Schindler–Patalas trap. The three trials were averaged to obtain an efficiency of 58% (s.e. = 3.8), which was applied to all years.

The density of *Chaoborus* was averaged from the three vertical tows. Measurements of head capsule length, body width at the third segment, and body length were taken on approximately 100 individuals. From 1986 to 1990 individual biomass was estimated from the following length–mass regression ($N = 12$, $R^2 = 0.86$, m.s.e. = 0.06):

$$\log (B) = 1.189 \log (\text{Vol}) - 8.644,$$

where B is individual biomass in micrograms and Vol is volume in cubic micrometers obtained from length and width measurements assuming cylindrical body shape. This regression was fitted in 1988 using individuals from the third and fourth instar of all species (*C. flavicans*, *C. punctipennis* and *C. trivittatus*) and applied to all samples from all years. To obtain the species composition of the populations, larvae were enumerated by species at the beginning and the end of the summer season. In 1984 and 1985, *Chaoborus* were enumerated by species and instar by C. von Ende (Carpenter *et al.*, 1987). To estimate the total biomass from these years, we multiplied the density of each species and instar present on a given day by the average individual mass that was measured for each species and instar from 1986 to 1989.

Daphnia morphology and induction experiments

In 1989, reciprocal transfer experiments of *Daphnia pulex* were performed in Peter, Paul, and Long Lakes to examine the induction of neck teeth in *Daphnia* by *Chaoborus*. Long Lake is located within 2 km of Paul and Peter Lakes and has zooplankton and *Chaoborus* communities similar to those of Paul Lake (S. R. Carpenter, unpublished data). The transfer experiments took place between a lake in which *D. pulex* were found to have neck teeth (either Paul or Long Lake) and a lake where neck teeth were not present (Peter Lake). For each experiment, in each lake, microcosms were set up in two triplicate sets: one with *Daphnia* from the other lake, and another set with *Daphnia* from the host lake. *Daphnia* were collected on the day of the experiment using a 80 μm mesh Nitex net and brought back to the laboratory. Ten to fifteen gravid females were selected for each microcosm using a dissecting microscope. The animals were placed in the microcosms, which were one-gallon plastic jugs with the four sides cut out and replaced with 355 μm mesh net that allowed passage of algae but retained newborn daphnids. Microcosms were suspended in the lake at approximately 2 m depth and incubated for 3–4 days to allow adequate time for egg development to the neonate stage (Lei & Armitage, 1980). At the end of the experiment, the water from the microcosms was filtered and the animals were preserved in 4% sugared buffered formaldehyde solution and the percentage of juveniles with neck teeth was determined.

The presence or absence of neck teeth on *Daphnia* from the weekly zooplankton samples were observed in each lake in 1987 and 1988, but were not quantified. In 1989 and 1990, we quantified the percentage of daphnids containing neck teeth, based on 20–47 animals per sample from the weekly zooplankton samples.

Results

Paul Lake

Chaoborus biomass was relatively consistent between 1985 and 1990 (Fig. 7.1*a*). Seasonal trends are not known in 1984, because only one sample was taken (Carpenter *et al.*, 1987). Total biomass declined towards the end of each summer owing to declining individual mass (Fig. 7.1*b*) as the dominant cohort pupated in early to midsummer and also to recruitment of newly hatched larvae. The species composition was variable, but was dominated by either *C. flavicans* (1985–7) or *C. trivittatus* (1988–90).

Peter Lake

During the years that bass dominated the fish assemblage, *Chaoborus* populations changed little, except for a pulse in total biomass in early 1987 (Fig. 7.2*a*). This pulse was due to an increase in individual body mass (Fig. 7.2*b*) rather than to density. At this time, the species composition was similar to that of Paul Lake. Upon introduction of rainbow trout in 1988, predation on *Chaoborus* by fishes increased substantially (Chapter 6) and *Chaoborus* biomass declined to barely detectable levels by midsummer. At the same time, individual size declined, especially in

1989 and 1990 (Fig. 7.2*b*). During the years that trout and golden shiners were present in Peter Lake, the relative importance of *C. flavicans* increased, in contrast to Paul Lake where the larger *C. trivittatus* dominated (Fig. 7.2*c*).

Tuesday Lake

In 1985, after the removal of minnows and addition of bass, *Chaoborus* biomass increased substantially (Fig. 7.3*a*). In 1986, biomass decreased to levels similar to those of 1984, and continued to decline for the rest of the

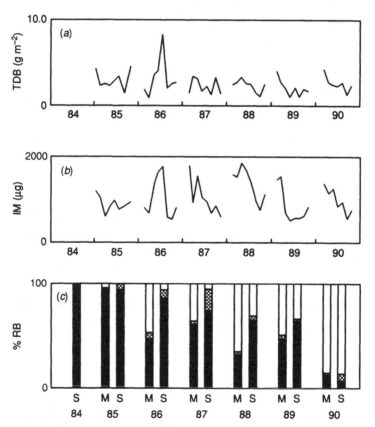

Fig. 7.1. Paul Lake: (*a*) total dry biomass (TDB) of *Chaoborus* populations from 1985 to 1990 during the summer stratified season; (*b*) the average individual mass (IM) of all instars of *Chaoborus*; (*c*) the percentage relative biomass (%RB) of species at the beginning and end of each summer (end of May (M) and early September (S)). Black is *C. flavicans*; hatched is *C. punctipennis*; white is *C. trivittatus*.

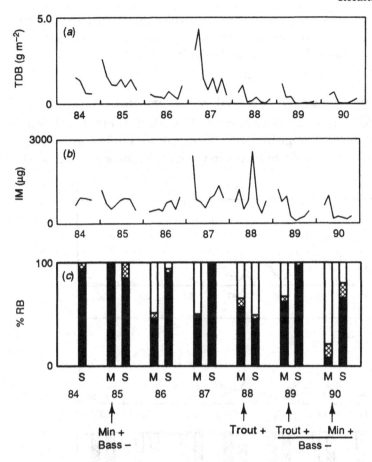

Fig. 7.2.   Peter Lake: (a) total dry biomass (TDB) of *Chaoborus* populations from 1984 to 1990 during the summer stratified season; (b) the average individual mass (IM) of all instars of *Chaoborus*; (c) the percentage relative biomass (%RB) of species at the beginning and end of each summer (end of May (M) and early September (S)). Black is *C. flavicans*; hatched is *C. punctipennis*; white is *C. trivittatus*.

study. Individual mass was consistent from 1984 to 1986 owing to the dominance of one species, *C. punctipennis* (Fig. 7.3b,c). During the years that planktivory by fishes was either low or gradually increasing (1986–9), the two larger species of *Chaoborus* (*C. flavicans* and *C. trivittatus*) were present, especially in 1987 when there were few fish in the lake. In 1988 and 1989, even though the populations of minnows were similar to premanipulation levels (Chapter 4), the larger chaoborids made up a large portion of the biomass. By 1990, the species composition was once again dominated by *C. punctipennis*, the smaller species that was present prior to manipulation. However, total *Chaoborus* biomass was still low and had not returned to premanipulation levels.

Daphnia morphology and reciprocal transfer experiments

In Paul Lake, neck teeth were present on *Daphnia pulex* individuals from 1984 to 1988. The average frequency of neck teeth was 27% in 1989 and 22% in 1990. However, from week to week the frequency of neck teeth

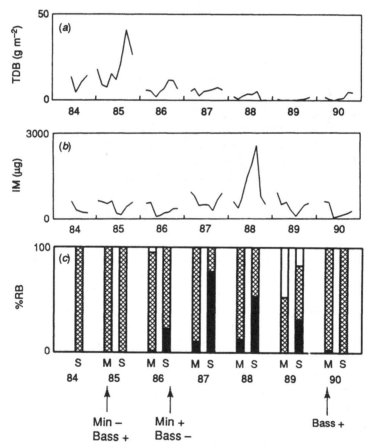

Fig. 7.3. Tuesday Lake: (*a*) total dry biomass (TDB) of *Chaoborus* populations from 1984 to 1990 during the summer stratified season; (*b*) the average individual mass (IM) of all instars of *Chaoborus*; (*c*) the relative biomass of species at the beginning and end of each summer (end of May (M) and early September (S)). Black is *C. flavicans*; hatched is *C. punctipennis*; white is *C. trivittatus*.

could be highly variable. In 1989 and 1990, from 5% to 78% of the *D. pulex* adults and juveniles that were examined bore neck teeth.

In Peter Lake from 1984 to 1987, neck teeth were observed on *D. pulex* individuals. However, in 1988 and 1989 no individuals were found with neck teeth. On the last three sampling dates of 1990, a few *D. pulex* were found with neck teeth. These represented 6% of the animals examined.

In 1989, reciprocal transfer experiments were performed to assess whether populations of daphnids not presently producing juveniles with neck teeth could be induced to do so. In experiments exchanging gravid females between Peter and Paul lakes, all of the juveniles born to *Daphnia* from Peter Lake that were transferred to Paul Lake produced neck teeth (Table 7.1), while ones in Peter Lake did not produce any regardless of their lake of origin. Therefore *Daphnia* in Peter Lake retained the ability to produce neck teeth in the presence of appropriate stimuli. Similar responses, but with a much lower induction rate, occurred in reciprocal transfers between daphnids in Long Lake and Peter Lake. Predation by *Chaoborus* may have been low in Long Lake at the time of the experiment, since no neck teeth were observed on daphnids from a sample taken from the lake prior to the experiment.

Discussion

Despite their crypsis and migratory behavior, *Chaoborus* were sensitive to changes in planktivory by fishes. In addition, the species of planktivore and the availability of food items were important in determining the response of *Chaoborus*.

In Peter Lake, *Chaoborus* did not respond to the changes in the bass populations from 1984 to 1987 (Chapter 4). However, upon introduction of rainbow trout, consumption of *Chaoborus* increased substantially (Chapter 6) and led to significant declines in the total biomass of *Chaoborus*. Declines in mean individual mass in 1989 and 1990, associated with elevated consumption by fishes, suggest that the larger instars and larger species suffered the highest mortality rates. Prior to the trout addition, *Chaoborus* species composition in Peter Lake was similar to that of Paul Lake. However, in 1989, the intermediate-sized *C. flavicans* made up a greater percentage of the population in Peter Lake compared with Paul Lake, where the larger *C. trivittatus* dominated. *C. flavicans* can coexist with rainbow trout (Luecke & Litt, 1987), whereas in another study *C. trivittatus* was eliminated from a lake two years after trout were added (Northcote, Walters & Hume, 1978). The high densities of

Table 7.1. *Results from* Daphnia *reciprocal transfer experiments*

Gravid *D. pulex* from two lakes (one where neck teeth were evident, +NT, and one where neck teeth were not produced, −NT) were exchanged and incubated for 3–4 days to induce the production of neck teeth on the juveniles born during the experiment. % Neck teeth is the percentage of the juveniles born from the original females that exhibit neck teeth.

Experiment	Origin of Daphnia	Transferred to	% Neck teeth
1	Peter Lake (−NT)	Paul Lake (+NT)	100
	Paul Lake (+NT)	Paul Lake (+NT)	64
	Paul Lake (+NT)	Peter Lake (−NT)	0
	Peter Lake (−NT)	Peter Lake (−NT)	0
2	Peter Lake (−NT)	Long Lake (+NT)	18
	Long Lake (+NT)	Long Lake (+NT)	5
	Long Lake (+NT)	Peter Lake (−NT)	0
	Peter Lake (−NT)	Peter Lake (−NT)	0

zooplankton that are the main food items of *Chaoborus* (small cladocerans, solitary rotifers and juvenile copepods) suggest that these declines were not due to food limitation (Chapter 8).

In Tuesday Lake, *Chaoborus* biomass increased in response to the reduced minnow density in 1985. This increase was accompanied by increased survivorship of all instars of larvae, facilitated by persistent populations of the zooplankton that were the main foods of these *Chaoborus*: rotifers and juvenile and adult copepods (Elser *et al.*, 1987*b*). Although consumption of *Chaoborus* by fishes was low in 1986 and 1987, *Chaoborus* populations declined. This decline was most likely to be due to food limitation of early instars, since rotifer and juvenile copepod populations were very low during this time (Chapter 8). Also, dinoflagellates make up almost 50% of the diet of *Chaoborus* at times in Tuesday Lake (Elser *et al.*, 1987*b*). The virtual disappearance of large dinoflagellates in Tuesday Lake after 1985 (Chapter 11) may have contributed to declines in *Chaoborus* (Yan *et al.*, 1991). Individual chaoborid size increased in 1987 and 1988, because of the shift to larger species. This shift to larger species may have been delayed by food limitation, as early instars may have suffered high mortality by starvation (Neill & Peacock, 1980). By 1990, however, *C. punctipennis* regained dominance, although the total biomass of the population was substantially lower than prior to manipulation.

In 1985, the effects of *Chaoborus* predation on the zooplankton communities of Paul, Peter and Tuesday Lakes were analyzed by Elser *et al.* (1987*b*). They estimated *Chaoborus* predation on zooplankton communities by calculating consumption rates of zooplankton from bioenergetics and diet analysis. The potential impact of *Chaoborus* predation was great in Tuesday Lake in 1985, owing to the dramatic increases in *Chaoborus* biomass. High predation rates on rotifers and copepods during June and July contributed to declining densities of these species. These declines corresponded to maximal densities of fourth instar *C. punctipennis*, which can exert strong predation pressure on rotifers (Moore & Gilbert, 1987). In addition, especially high predation rates and positive selection for *Bosmina* in early summer of 1985 may explain its low densities at this time (Chapter 8).

The most striking implication of predation rates estimated for *Chaoborus* populations in Peter and Paul Lakes in 1985 was the potentially strong impact on *Daphnia pulex* and *D. rosea*. In both lakes, third- and fourth-instar *Chaoborus* larvae had positive electivities for *Daphnia*, and consumed up to 20% of the standing stock on some days (Elser *et al.*, 1987*b*). The suggestion that *C. flavicans* could suppress *Daphnia* populations was directly tested in a mesocosm experiment done in Peter Lake in 1987 (MacKay *et al.*, 1990). In bags where *C. flavicans* larvae were added at densities found in the lake, populations of *D. pulex* declined significantly compared with controls where no *Chaoborus* were added. The results show that predation by *Chaoborus* can directly control populations of *D. pulex*, and raises the possibility that *Chaoborus* can limit the success of *D. pulex* when planktivory by fish is absent or low (MacKay *et al.*, 1990).

Further support of the potential importance of *Chaoborus* predation for *Daphnia* populations lies in morphological responses of the prey. In Peter and Paul Lakes, *D. pulex* juveniles are often found with neck teeth (small spines on the dorsal crest of the helmet), an adaptation that increases daphnid survivorship in encounters with *Chaoborus*, but incurs metabolic costs (Krueger & Dodson, 1981; Havel & Dodson, 1984, 1985; Black & Dodson, 1990; Reissen & Sprules, 1990). The negative effects of predation can be intense enough to offset the energetic costs of producing neck teeth (Havel & Dodson, 1985). In both Paul and Peter Lakes, neck teeth were found on *D. pulex* from 1984 to 1987. However, in 1988, 1989 and almost all of 1990, neck teeth were not found on any daphnids in Peter Lake, although an average of 27% and 22% of the *D. pulex* populations contained neck teeth in Paul Lake in 1989 and 1990. This suggests that predation by *Chaoborus* on *Daphnia*, and possibly on the rest of the zooplankton community, was reduced from 1988 to 1990. This inference is consistent with the low biomass of *Chaoborus* during these years. These morphological responses of the *D. pulex* populations in both Paul and Peter Lake suggest that planktivory by *Chaoborus* is an important regulatory factor. These defenses may have been historically important in the coexistence of *D. pulex* with *Chaoborus*.

The changes in the zooplankton following fish manipulations in both Peter and Tuesday Lakes may have proximal causes in shifting predation by *Chaoborus*. In Tuesday Lake, predation by *Chaoborus* contributed to declines of *Bosmina* in early summer and of rotifers and copepods in late 1985 (Elser *et al.*, 1987*b*). In Peter Lake, decreases in the *Chaoborus* populations in 1989 and 1990 probably contributed to the community changes that occurred. Decreased invertebrate predation would have favored the increases in smaller taxa of zooplankton that occurred, and contributed to the replacement of the dominant daphnid by the small-bodied *D. dubia*. Our data cannot separate this mechanism from the possibility that the removal of *D. pulex* by predation released smaller zooplankton taxa from competition.

Implications for the trophic cascade

The trophic cascade hypothesis states that, in general, planktivory by fishes and by invertebrates are inversely related (Carpenter *et al.*, 1985). In the specific case of *Chaoborus*, we expected that crypsis and diel vertical migration would weaken this inverse relationship (Carpenter & Kitchell, 1987; Chapter 16). This specific hypothesis proved incorrect. The *Chaoborus* populations responded strongly to changes in planktivory by fishes. Planktivory by *Chaoborus* interacted with that by fishes to influence zooplankton communities in these lakes. This inference is consistent with the whole-lake time series, as well as bioenergetic calculations (Elser *et al.*, 1987*b*), enclosure experiments (MacKay *et al.*, 1990) and expression of morphological defenses of prey species.

Our general hypothesis was only partly supported. In Tuesday Lake, the decrease in planktivory that occurred in 1985 resulted in only a transitory increase in populations of *Chaoborus*. Owing to food limitation in 1986, the biomass of *Chaoborus* returned to levels found prior to manipulation, and actually decreased the following year despite low levels of planktivory by fishes. In Peter Lake, on the other hand, increases in planktivory by rainbow trout did lead to significant declines in the populations of *Chaoborus*, with apparent repercussions for the zooplankton community (Chapter 8).

Summary

Like herbivorous zooplankton, invertebrate predators such as *Chaoborus* are sensitive to changes in planktivory by fishes. The dominant invertebrate planktivore in our lakes, *Chaoborus*, can prey heavily on certain species of herbivorous zooplankton. Therefore interactions between *Chaoborus* and fish may have complex effects on the zooplankton community. Manipulations of the fish populations allowed us to assess effects on *Chaoborus* in relation to changes in the zooplankton community as a whole.

In Tuesday Lake, a decrease in planktivory by fishes prompted transitory increases in *Chaoborus*. High *Chaoborus* biomass was not sustained in the following years, apparently owing to food limitation. In Peter Lake, *Chaoborus* populations were not significantly altered until the introduction of rainbow trout in 1988. From 1988 to 1990, total *Chaoborus* biomass was substantially lower than in earlier years. This reduction caused major changes in the zooplankton community. No comparable changes occurred in the reference lake.

The importance of invertebrate planktivory was demonstrated by the changes in the zooplankton community that followed severe changes in *Chaoborus* biomass, and by bioenergetic and enclosure studies performed to assess the importance of *Chaoborus* predation (Elser *et al.*, 1987*b*; MacKay *et al.*, 1990). In Tuesday Lake in 1985, decreases of *Bosmina*, rotifers and copepods corresponded in time to elevated predation rates by *Chaoborus* on these species (Elser *et al.*, 1987*b*). The potential regulation of *Daphnia* populations by *Chaoborus* was suggested by the high predation rates on *D. pulex* and *D. rosea* in Peter and Paul Lakes, the presence of defensive neck teeth on daphnids, and the suppression of *D. pulex* populations by *C. flavicans* in a mesocosm experiment performed in Peter Lake (Elser *et al.*, 1987*b*; MacKay *et al.*, 1990). Neck teeth were induced by *Chaoborus* in populations of *D. pulex* that had not produced such defenses for at least two years. Other work has shown that neck teeth reduce mortality by *Chaoborus* predation (Havel & Dodson, 1984). The persistent ability of daphnids to produce neck teeth may have been historically important in the coexistence of *Daphnia* and *Chaoborus* in these lakes.

Acknowledgements

We thank Bart DeStasio, Jim Kitchell and Peter Leavitt for helpful reviews of earlier drafts, and Carl von Ende for his contributions to *Chaoborus* work in 1984 and 1985.

8 · *Zooplankton community dynamics*

Patricia A. Soranno, Stephen R. Carpenter and
Monica M. Elser

Introduction

Whole-lake manipulations of the fish assemblages in Peter and Tuesday Lakes provided an excellent opportunity to ask how vertebrate and invertebrate predators affect zooplankton communities. A central element of the cascade hypothesis is the regulation of large herbivores by visually feeding planktivores. In the light of the prominent position that large herbivores can occupy in zooplankton communities, repercussions in populations of less conspicuous taxa are expected (Lewis, 1979; Zaret, 1980; Kerfoot & Sih, 1987). These include compensatory shifts in the dominant species and increases in previously rare species.

Planktivory by fishes and invertebrates (such as *Chaoborus*) has both direct and indirect effects on zooplankton populations (Zaret, 1980; Neill, 1981; Vanni, 1986). Fish predation can constrain the maximum adult body size of prey, while invertebrate predation can restrict the minimum size. Either can interact with effects of food limitation to alter zooplankton populations (Hall *et al.*, 1976). At the community level, however, the significance of predation is less clear. Changes in density alone may not be sufficient to alter competitive exclusion, diversity, or their consequences for the community (Thorp, 1986). Interpretations of zooplankton community structure must consider competitive interactions among species, especially among the dominant herbivores (Lynch, 1979).

In this chapter, we focus on the changes in the zooplankton communities of Peter and Tuesday Lakes as a result of fish manipulations and contrast them to the natural variation exhibited by Paul Lake. We also consider the possible effects of food limitation (Chapter 11) and invertebrate predation (Chapter 7). We explore how vertebrate and invertebrate predators in these lakes regulate zooplankton community structure, indexed by the relative biomass of key zooplankton groups and the dominant species. We also evaluate how *Daphnia*, the dominant herbivore in Peter Lake, responds to changes in predation pressure by examining the population's demographic characteristics and by comparing them to *Daphnia* populations in Paul Lake.

Related chapters discuss invertebrate predation (Chapter 7), the behavioral responses of the dominant herbivores to fishes (Chapter 9), total zooplankton biomass, body size and implications for ecosystem processes (Chapter 10), and the effects of herbivory on algal dynamics (Chapters 11–13). Here, we concentrate on the dynamics of the zooplankton community as a whole, focusing on key species and selected groups of species.

Methods

Detailed methods for zooplankton sampling and enumeration have been presented elsewhere (Elser *et al.*, 1986a; Soranno, 1990). We provide a brief summary here.

Zooplankton were collected weekly from the end of May to early September in each lake from 1984 to 1990. Pooled duplicate vertical hauls were taken using an 80 μm mesh Nitex net (30 cm diameter) and preserved in 4% sugared, buffered formaldehyde solution. Additional duplicate tows were taken for *Conochilus* colonies, which were counted live since they dissociated in formaldehyde. Net efficiencies were estimated yearly for each lake by comparing a set of duplicate tows to an evenly spaced vertical profile (1 m intervals) obtained with a 12 l Schindler–Patalas trap. Zooplankton were counted and measured using a binocular dissecting scope. Lengths were converted to dry mass using equations from Downing & Rigler (1984).

Demographic parameters of *Daphnia* populations

Demographic parameters have proven useful for inferences about the relative effects of resource limitation and mortality on *Daphnia* populations (Hall, 1964; Wright, 1965). We therefore attempted to compile time series of *Daphnia* birth and death rates for this study. In the early years of our study, funding limitations precluded analyses of *Daphnia* demography. In 1986, an intensive study of *Daphnia* brood sizes and reproductive ratios was performed by P. Soranno for her undergraduate research project at the University of Notre Dame. From 1988 through

1990, we were able to collect and process a separate series of samples to determine demographic parameters. Birth and death rates for 1984, 1985 and 1987 were estimated by analysis of archived samples (see below). The accuracy of estimates from the archived samples was checked by analyzing archived samples from 1986 and comparing the results with those of Soranno's intensive study in that year. We have not attempted species-specific estimates of birth and death rates. *D. pulex* was the dominant species in most of the samples analyzed. In most cases, we have insufficient data to make separate analyses for subdominant species. Despite their limitations, these demographic data provide some clues about the causes of fluctuations in *Daphnia*, the most important herbivore in our ecosystem study.

From 1988 to 1990, demographic parameters of *Daphnia* populations in Paul and Peter Lakes were obtained from additional weekly samples. Pooled duplicate tows were taken from 8 and 12 m for Paul and Peter Lakes at three stations over the deepest portion of the lake, and preserved in cold 4% sugar-buffered formaldehyde solution. Samples were stored on ice until they were brought back to the laboratory and processed on the same day to prevent animals from 'ballooning' and releasing their eggs.

Animals were selected haphazardly from each sample to measure body length (exclusive of tail spine and helmet) and determine the brood size. When possible, eggs were counted from a minimum of 25 animals from each station. Data from all stations on each date were pooled to obtain adequate sample sizes for accurate estimates of the demographic parameters of the population. Instantaneous birth rates were calculated using the egg ratio method (Edmondson, 1960; Paloheimo, 1974). The egg ratio, the number of eggs per adult female, was calculated by multiplying the mean brood size by the reproductive ratio of the population (total number of females with broods/total number of adult females). Birth rates (*b*) were calculated using the following formula:

$$b = \ln (E_a + 1)/D,$$

where E_a is the egg ratio (eggs per adult female) and D is the egg development time (days) calculated from the temperature that the animals experience (Lei & Armitage, 1980). We assumed that the *Daphnia* populations spent 17 h at 6 °C in deep waters, and 7 h in the epilimnion at its temperature (Chapter 9). The instantaneous rate of population increase (*r*) was estimated by:

$$r = (\ln N_t - \ln N_0)/t,$$

where N_t and N_0 are the population densities at times t and 0, respectively, and t is the interval between sampling. The instantaneous death rate was calculated as the difference between the birth rate and the growth rate. Growth and death rates are presented as three-point running averages.

Estimates of the demographic parameters of the daphnid populations prior to 1988 (except for 1986) were obtained from archived samples of the weekly zooplankton hauls. We excluded from the calculations any animals that had 'ballooned', since they might have lost their eggs. Since the likelihood of an animal having an everted carapace did not appear to be size-dependent we probably obtained an accurate representation of the population by only examining intact animals. We included in the analysis only those samples with a standard error of the mean brood size of approximately 10% of the mean or less.

For an intensive study of *Daphnia* in these lakes in 1986 we measured the brood sizes and reproductive ratios by randomly choosing 100 individuals from weekly vertical tows. Mean lengths and densities of *Daphnia* were taken from the routine weekly zooplankton tows.

As a measure of the accuracy of the method of using archived samples (for 1984, 1985 and 1987), we compared mean brood sizes and reproductive ratios from archived samples taken in 1986 with measurements made on fresh samples that year. We calculated the ratio of the two methods (method using archived samples/method from daphnid study). The ratios of the two methods averaged over the season for mean brood size (0.82, 95% c.i. = 0.15), and for reproductive ratio (0.72, 95% c.i. = 0.23) indicated that estimates from archived samples were close to, but generally lower than, estimates from fresh samples. We did not attempt to correct values from archived samples for the following reasons. First, the changes in brood size and reproductive ratio that occurred in the lakes were larger than the differences between archived and fresh samples. Second, we usually compared the manipulated and reference lake in a given year when all were analyzed by the same method, so any intermethod differences are moot. Third, time series for Paul Lake (see below) show no apparent changes in brood size and reproductive ratio associated with changes in methods.

For samples from 1984, it was necessary to use a different counting procedure since almost all animals had lost their eggs. To estimate the egg ratio, we counted the total number of loose eggs in the sample and the total number of adult females. The calculations for birth, growth and death rates were identical for all years.

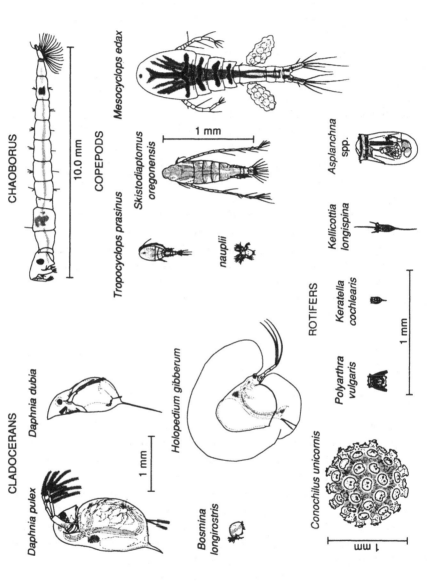

Fig. 8.1. The dominant zooplankton taxa of the lakes, grouped as described in the text.

The size at first reproduction, determined in 1986 and 1988–90, was the smallest reproductive female found in all of the triplicate samples for each date.

Results

The zooplankton communities found in Paul, Peter and Tuesday Lakes in 1984, the year before our manipulations began, were very similar taxonomically to those found in 1981–3 (Bergquist, 1985). Paleolimnological data suggest that the 1981–4 communities were also similar to those of the 1960s and 1970s (Chapter 15). Shifts in zooplankton community structure lasting up to several years occurred as a result of ecosystem experiments between 1951 and 1976 (Leavitt et al., 1989; Chapter 15).

The zooplankton of Paul, Peter and Tuesday Lakes were divided into five main groups (Fig. 8.1): (1) cladocerans, including the largest herbivores (*Daphnia*); (2) cyclopoid copepods, including omnivorous and predaceous species; (3) calanoid copepods, including herbivorous and omnivorous species; (4) juvenile copepods, including the naupliar and copepodid stages of all species; and (5) rotifers, including small herbivores, one genus (*Asplanchna*) predaceous on other rotifers, and a large colonial species (*Conochilus unicornis*). While these groupings are in some respects arbitrary, they did allow us to effectively compare and contrast the changes that occur in the communities. We will also discuss the species in each group that dominate the dynamics of the community as a whole.

Table 8.1 presents a list of common species found in Paul, Peter and Tuesday Lakes, grouped as described above. The primary herbivores in these lakes are the cladocerans, dominated by *Daphnia* species. Daphnids are effective grazers and filter-feed on a wide spectrum of algal sizes ranging from 1 to > 70 μm in diameter (Burns, 1968). Adults of the larger species, *D. pulex*, are susceptible to fish predation while juveniles are also vulnerable to invertebrate planktivores such as *Chaoborus* and predaceous copepods (Lane, 1978; Elser et al., 1987b). The large-bodied *Holopedium gibberum* is also a dominant species at times in these lakes. *Holopedium* generally peaks in early summer before daphnid populations, partly as a result of its preference for colder water temperatures (Walters, Robinson & Northcote, 1990). However, in contrast to *Daphnia*, it is relatively resistant to planktivory because of the large gelatinous mantle covering its carapace, which not only increases the animal's

Table 8.1. *Zooplankton species in each of the main groups*

Length ranges, in millimeters, are for all three lakes combined. See text for further details.

Group	Species	Length range (mm)
cladocerans	*Daphnia pulex*	0.70 – 2.00
	Daphnia dubia	0.70 – 1.20
	Daphnia rosea	0.50 – 1.40
	Daphnia parvula	0.40 – 1.00
	Holopedium gibberum	0.60 – 1.60
	Diaphanosoma birgei	0.40 – 0.80
	Bosmina longirostris	0.20 – 0.60
cyclopoid copepods	*Mesocyclops edax*	1.10 – 1.80
	Orthocyclops modestus	0.70 – 1.40
	Cyclops varicans rubellus	0.50 – 0.90
	Tropocyclops prasinus	0.40 – 0.65
calanoid copepods	*Epischura lacustris*	1.50 – 2.20
	Diaptomus spp.	0.70 – 1.30
rotifers	*Asplanchna* sp.	0.30 – 0.70
	Conochilus unicornis	0.10 – 0.18
	Conochilus dossaurius	0.11 – 0.16
	Trichocerca cylindrica	0.25 – 0.40
	Trichocerca multicrinis	0.14 – 0.20
	Polyarthra major	0.16 – 0.20
	Polyarthra vulgaris	0.10 – 0.17
	Synchaeta sp.	0.12 – 0.25
	Ascomorpha ecaudis	0.10 – 0.17
	Filinia terminalis	0.12 – 0.18
	Kellicottia longispina	0.12 – 0.13
	Kellicottia bostoniensis	0.10 – 0.16
	Ploesoma sp.	0.11 – 0.15
	Keratella testudo	0.10 – 0.15
	Keratella taurocephala	0.10 – 0.12
	Keratella cochlearis	0.09 – 0.13
	Gastropus hyptopus	0.10 – 0.13
	Gastropus stylifer	0.08 – 0.13

volume but may make it distasteful to fishes (McNaught, 1978; Balcer *et al.*, 1984). However, juveniles and recently molted animals have been shown to be vulnerable to young fishes and invertebrate predators (Tessier, 1986).

Although *Polyphemus* are abundant in diets of fish in these lakes

(Chapter 6), they are rarely found in samples taken from the pelagic zone. *Polyphemus* are concentrated in the littoral zone.

Bosmina longirostris and *Diaphanosoma birgei* (previously identified as *D. leuchtenbergianum* in Bergquist *et al.*, 1985; Carpenter *et al.*, 1987; Elser *et al.*, 1987*b*) are small-bodied cladoceran species which dominate the cladoceran populations in these and many other lakes in the presence of vertebrate planktivores. They are smaller than daphnids and *Holopedium*, and consequently feed on a smaller and narrower range of food particles, ranging from 1 to 15 μm (Gliwicz, 1969; Burns, 1968; Makarewicz & Likens, 1975). These small cladocerans may be less competitive than larger cladocerans as well as less effective grazers on algal standing stocks (DeMott & Kerfoot, 1982; Vanni 1986). In addition, they are extremely vulnerable to invertebrate planktivores, such as *Chaoborus* and predaceous copepods (Lane, 1978; Elser *et al.*, 1987*b*).

The common calanoid copepods in these lakes are the herbivorous diaptomids, which include *Diaptomus oregonensis*, *D. pallidus* and *Leptodiaptomus siciloides*. These feed in a similar size range and prefer algae from 12 to 22 μm (Richman, Bohon & Robbins, 1980). Because of their relatively large body size, adults may be susceptible to predation by fishes (Balcer *et al.*, 1984), while juveniles are susceptible to invertebrate predators (Elser *et al.*, 1987*b*).

The cyclopoid copepods include the predaceous *Mesocyclops edax*, which feeds on small daphnids, rotifers (including *Asplanchna*) and copepodids and nauplii (Gilbert & Williamson, 1978; Lane, 1978; Williamson 1980). *M. edax* has also been shown to filter nanoplankton (Balcer *et al.*, 1984). Owing to its large body size, it is vulnerable to predation by fishes. The smaller *Tropocyclops prasinus* is omnivorous, but can prey heavily on rotifers, *Bosmina*, *Diaphanosoma* and juvenile copepods (Lane, 1978). Other omnivorous species include *Orthocyclops modestus* and *Cyclops varicans rubellus*.

The 18 species of rotifers that we found in these lakes (Table 8.1) generally feed on particles in the range of 1–5 μm diameter (Gliwicz, 1969; Bogdan & Gilbert, 1984). However, *Polyarthra vulgaris* can feed selectively on particles as large as 17 μm in diameter (Bogdan & Gilbert, 1984). One of the more prominent rotifers, in terms of biomass and size, is the predaceous *Asplanchna* sp., which prefers rotifers less than 0.3 mm, but itself is vulnerable to predaceous copepods, *Chaoborus* and fish larvae (Stemberger & Gilbert, 1987; Moore & Gilbert, 1987). Another prominent rotifer was the colonial *Conochilus unicornis*, which includes 80–300 individuals per colony. These colonies can make up a large portion of the total rotifer biomass, and at times, the total zooplankton biomass. Colonies protect individuals from invertebrate predation, and may even confer an advantage on the individuals by increasing feeding efficiency (Wallace, 1987; Edmondson & Litt, 1987). However, colonies are vulnerable to predation by larval and YOY fishes, and especially minnow species (Carpenter *et al.*, 1987; Stemberger & Gilbert, 1987). The remaining rotifers include: (1) spined taxa, *Keratella* and *Kellicottia* spp.; (2) soft-bodied taxa, *Conochilus dossuarius* (identified as *Conochiloides dossuarius* in Carpenter *et al.*, 1987 and Elser *et al.*, 1987*b*), *Synchaeta* sp., *Ascomorpha ecaudis*, and *Gastropus* spp.; (3) loricate taxa, *Trichocerca* spp. and *Ploesoma* sp.; and (4) species that exhibit escape responses to invertebrate predators, *Polyarthra* spp. and *Filinia terminalis* (Stemberger & Gilbert, 1987).

Three species of *Chaoborus* are present in Paul and Peter Lakes (Fig. 8.1; Chapter 7). The larger species *C. flavicans* and *C. trivittatus* dominate in Paul and Peter Lakes. In Tuesday Lake, the smaller species *C. punctipennis* dominates.

We assessed the responses of the zooplankton community to the manipulations by examining the relative contributions of the major groups to the total biomass of zooplankton, the shifts in the dominant taxa in each group and the abundances of all relevant taxa. The changes observed in the zooplankton compared to consumption estimates by fishes on zooplankton (Chapter 6), predation rates by *Chaoborus* on zooplankton (Elser *et al.*, 1987*b*) and available food resources (Chapter 11). Food limitation will be assessed indirectly through changes in biomass of zooplankton and edible algae (< 20 μm greatest axial linear dimension, GALD) and, for daphnids only, birth rates.

Paul Lake

Cladocerans dominate the zooplankton community in Paul Lake. During the summer seasons, 30–65% of the total zooplankton biomass was composed of large-bodied cladocerans (Fig. 8.2*a*). Juvenile copepods and nauplii also made up a large portion of the biomass (20–50%), and at certain times during the summer made up as much as 75% of the total (Fig. 8.2*d*). Adult cyclopoids and calanoids, however, made up a relatively small percentage of the total zooplankton biomass, each never exceeding 20% (Fig. 8.2*b,c*). Similarly, rotifers made up a relatively small proportion of the biomass: from 5 to 15% (Fig. 8.2*e*).

The zooplankton community was clearly dominated by relatively large-bodied species. *Daphnia pulex* and *Holopedium gibberum* were the most numerous cladocerans, with *D. rosea* attaining lesser densities. Although *H. gibberum* occurred in densities comparable to those of *D. pulex*, the major peaks of these species appeared to be offset (Fig. 8.3*a–c*). *Holopedium* is sensitive to both low food levels and competition with large-bodied daphnids (Tessier, 1986). In fact, except for 1990, the abundances of *Holopedium* were lowest when edible algae in Paul Lake were lowest: in 1985, 1986 and early 1987 (Chapter 11).

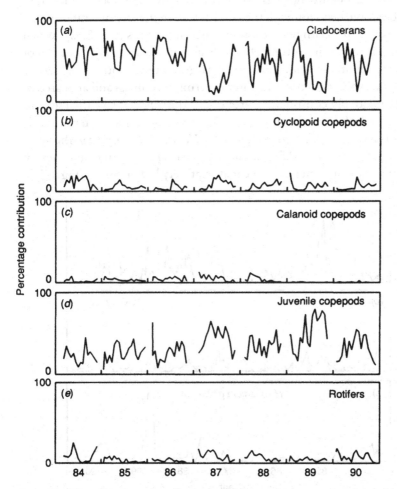

Fig. 8.2. The percentage contribution of each zooplankton group (see text) to the total zooplankton biomass in Paul Lake versus time during the summer stratified season for: cladocerans (*a*), cyclopoid copepods (*b*), calanoid copepods (*c*), juvenile copepods of all species (*d*) and rotifers (*e*).

The major copepod species were the cyclopoids *Orthocyclops modestus* and *Tropocyclops prasinus*, and sometimes the calanoids *Diaptomus* spp. (Fig. 8.4a–c). The predaceous *Mesocyclops edax* appeared in relatively high levels in certain years (1984, 1987 and 1989), which coincided with relatively high densities of their prey: rotifers and juvenile copepods (Gilbert & Williamson, 1978; Williamson, 1980). Clearly, a major portion of the biomass of copepods was in copepodid and naupliar stages (Fig. 8.4e,f).

The rotifer assemblage was also dominated by a few species, although 15 species have been found in the lake. The total density of rotifers in the lake was variable with no consistent seasonal patterns (Fig. 8.5a). When present, the predatory rotifer, *Asplanchna* sp. made up a major portion of the rotifer biomass (Fig. 8.5b). The large colonial rotifer *Conochilus unicornis* was also a dominant rotifer in terms of biomass and appeared in the lake relatively consistently throughout the summer stratified season, tending to decline by August (Fig. 8.5c). The dominant rotifer numerically was *Keratella cochlearis*, followed by *Kellicottia longispina* in the years that it was present (Fig. 8.5d, e). *Polyarthra vulgaris* appeared consistently throughout the summer every year at relatively low densities (Fig. 8.5f).

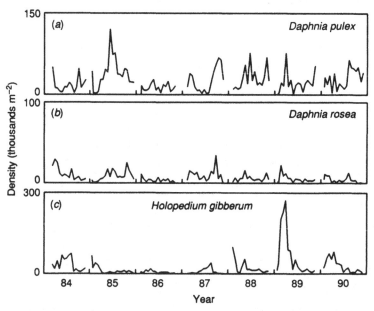

Fig. 8.3. Paul Lake: density of cladoceran species as thousands of individuals m⁻² versus time during the summer stratified season for *Daphnia pulex* (a), *D. rosa* (b) and *Holopedium gibberum* (c).

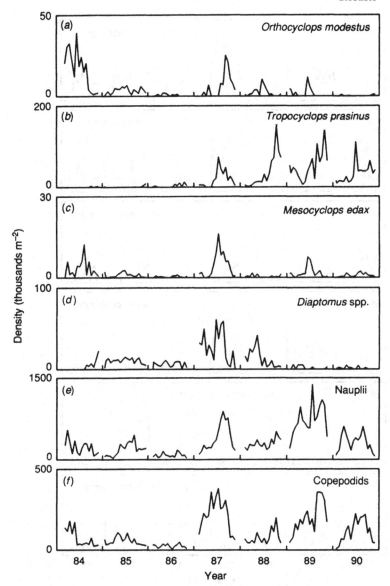

Fig. 8.4. Paul Lake: density of copepod species versus time during the summer stratified season for *Orthocyclops modestus* (*a*), *Tropocyclops prasinus* (*b*), *Mesocyclops edax* (*c*), *Diaptomus* spp. (*D. oregonensis*, *D. pallidus*, *Leptodiaptomus siciloides*) (*d*), nauplii of all species combined (*e*) and copepodids of all species combined (*f*).

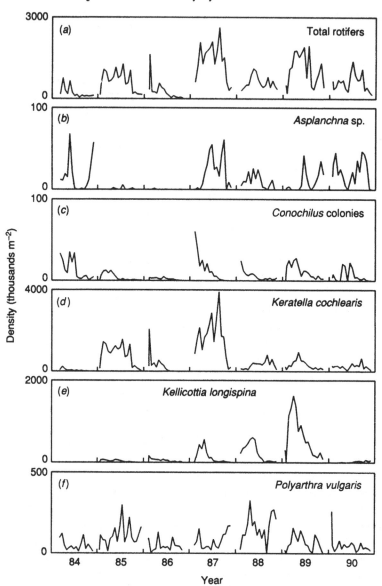

Fig. 8.5. Paul Lake: density versus time during the summer stratified season for total rotifers (*a*) and for selected species: *Asplanchna* sp. (*b*), *Conochilus unicornis* colonies (*c*), *Keratella cochleraris* (*d*), *Kellicottia longispina* (*e*) and *Polyarthra vulgaris* (*f*).

Other species that appeared sporadically in some of the years were *Gastropus stylifer*, *G. hyptopus*, *Trichocerca cylindrica*, *Filinia terminalis*, *Synchaeta* sp. and *Conochilus dossuarius*.

Peter Lake

Zooplankton in Peter Lake responded to major shifts in planktivory that occurred as a result of manipulation of the top piscivores and planktivores. Key events were (1) the strong year class of YOY bass in the latter half of 1985 and their transition to piscivory in 1986 and 1987; (2) the addition of rainbow trout in 1988 and 1989; and (3) the piscivore removal and minnow addition in 1990.

Prior to manipulation (1984), the zooplankton community resembled that of Paul Lake. Total biomass was dominated by large-bodied cladocerans, *Daphnia pulex*, *D. rosea* and *Holopedium gibberum* (Figures 8.6 and 8.7). Copepod populations were made up of juveniles (copepodids and nauplii) and a few adult cyclopoids and calanoids (Fig. 8.8). Adults of both groups made up as much as 20% of the total biomass at times, and included *Diaptomus* spp. (*D. oregonensis*, *D. pallidus*, and *Leptodiaptomus siciloides*), *Orthocyclops modestus* and *Mesocyclops edax*. Rotifers contributed very little to the total biomass (<5%) (Fig. 8.6e). The most common species were *Conochilus unicornis* colonies, *Keratella cochlearis* and *Polyarthra vulgaris* (Fig. 8.9).

Initially, the response of the zooplankton to the removal of piscivores and addition of planktivores in 1985 was contrary to expectation (Carpenter *et al.*, 1987). Densities of large-bodied zooplankton were slightly higher in the first half of the summer than in 1984 (Fig. 8.7a,b) since the introduced minnows moved inshore to escape predation by the remaining piscivores. However, as recruitment of YOY bass was unusually high that year, planktivory increased in midsummer (Chapter 6) and densities of *Daphnia* declined by August. At the same time, a small-bodied cladoceran, *Diaphanosoma birgei*, appeared in August. *D. birgei* was not found in Peter Lake in 1981–3 (Bergquist, 1985; Bergquist *et al.*, 1985) or in 1984. *H. gibberum*, which usually peaks in early summer in Paul and Peter Lakes, exhibited only a minor peak in early summer, but increased again and at higher densities in late summer.

Copepods, on the other hand, declined in response to the increase in planktivory that occurred immediately after the manipulation, and did not recover through the rest of the summer (Fig. 8.8). All categories of copepods made smaller contributions to the biomass in 1985 than in 1984

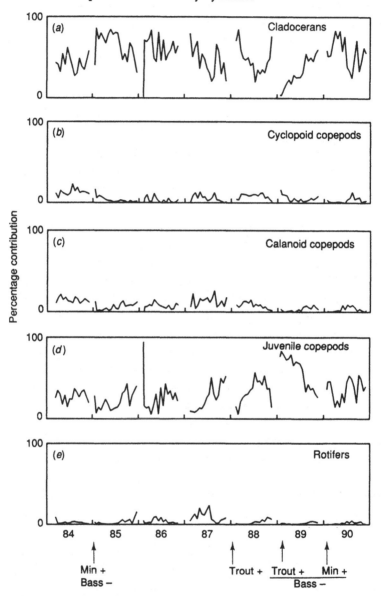

Fig. 8.6. The percentage contribution of each zooplankton group (see text) to the total zooplankton biomass in Peter Lake versus time during the summer stratified season for: cladocerans (a), cyclopoid copepods (b), calanoid copepods (c), juvenile copepods of all species (d) and rotifers (e).

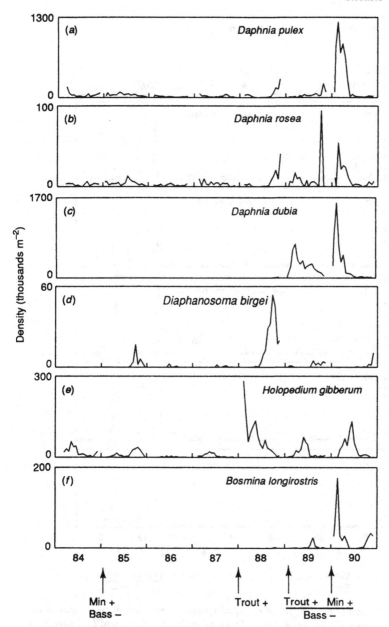

Fig. 8.7. Peter Lake: density of cladoceran species versus time during the summer stratified season for *Daphnia pulex* (*a*); *D. rosea* (*b*); *D. dubia* (*c*); *Diaphanosoma birgei* (*d*); *Holopedium gibberum* (*e*) and *Bosmina longirostris* (*f*).

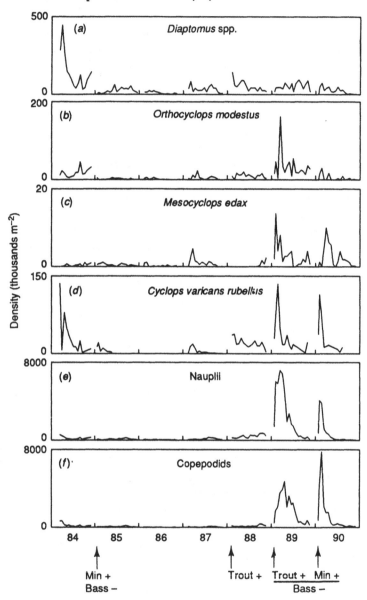

Fig. 8.8. Peter Lake: density of copepod species versus time during the summer stratified season for *Diaptomus* spp. (*D. oregonensis, D. pallidus* and *Leptodiaptomus siciloides*) (*a*), *Orthocyclops modestus* (*b*), *Mesocyclops edax* (*c*), *Cyclops varicans rubellus* (*d*), nauplii of all copepod species (*e*) and copepodids of all species (*f*).

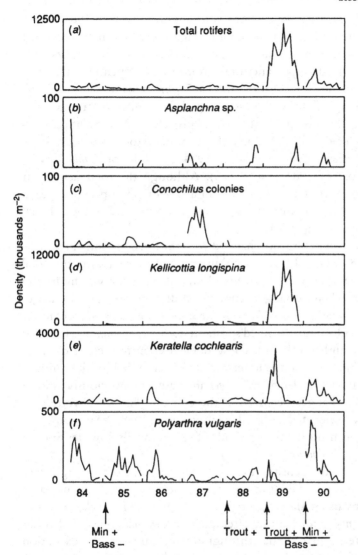

Fig. 8.9. Peter Lake: density versus time during the summer stratified season for total rotifers (*a*) and selected species: *Asplanchna* sp. (*b*), *Conochilus unicornis* colonies (*c*), *Kellicottia longispina* (*d*), *Keratella cochlearis* (*e*) and *Polyarthra vulgaris* (*f*).

(Fig. 8.6). Also, rotifer populations were slightly lower in 1985 except for the colonial *Conochilus unicornis* (Fig. 8.9*a,c*), which decreased initially while it was found in large numbers in minnow guts, but increased after the disappearance of the minnows (Carpenter *et al.*, 1987; Chapter 6). The summers of 1986 and 1987 were transitional as the 1985 cohort of young-of-the-year bass changed from planktivory to piscivory, causing the bass population to approach premanipulation conditions (Chapter 4). By 1987, many of the zooplankton populations were found at densities similar to those of 1984. Densities of *Daphnia* were similar to those of 1984 (Fig. 8.7*a,b*). Although the total density of rotifers in 1987 was closer to that found in 1984 than in previous years, biomass increased substantially owing to the dominance of the colonial rotifer *Conochilus* (Fig. 8.9*c*).

Zooplankton in Peter Lake changed with the introduction of rainbow trout in 1988. The relative contribution of cladocerans to the total biomass dropped steadily when vertebrate planktivory was highest in mid-1988 (Fig. 8.6*a*), owing to the virtual disappearance of the large-bodied *Daphnia pulex* (Fig. 8.7*a*). At the same time, the species composition of the cladocerans changed substantially. *H. gibberum* was found at exceptionally high levels just a few days after introduction of trout (Chapter 4; Fig. 8.7*e*), and as it declined in the latter half of the summer the small-bodied *Diaphanosoma birgei* increased to previously unseen levels. Similar changes occurred at the end of 1985 when vertebrate planktivory was elevated in the lake. However, in 1988, the response was much stronger, most likely as a result of exceptionally low densities of the predatory *Chaoborus* (Chapter 7).

Copepods and the density of total rotifers responded only slightly to the trout addition in 1988. On the other hand, because of the increased planktivory by fishes, the large predaceous copepod *Mesocyclops edax* did not appear until late August, after planktivory had relaxed. Rotifer densities were slightly higher throughout the summer. An exception was the colonial rotifer *Conochilus unicornis*, which disappeared after the second week of sampling and did not reappear for the rest of the study (Fig. 8.9*c*). The colonies are vulnerable to predation by *Leptodora* (Edmondson & Litt, 1987) and some fishes (Carpenter *et al.*, 1987), and were presumably eliminated by trout predation.

Responses to the trout addition were short-lived. Planktivory declined toward the end of the summer in 1988 (Chapter 6). The relative contribution of cladocerans increased in the end of August, owing to an increase in all daphnid species as well as the appearance of a new species,

Daphnia dubia (Fig. 8.7*a–c*). High densities of *D. dubia* and the returns of *Mesocyclops edax* and the predatory rotifer *Asplanchna* were all facilitated by reduced densities of *Chaoborus* (Chapter 7).

The dynamics of the fish community in 1989 were more complex and led to continued changes in the zooplankton. Overall, planktivory by fishes was intermediate between 1988 and 1990 (Chapter 6). Nevertheless, consumption by fishes was probably sufficient to suppress both *Chaoborus* populations and large *Daphnia*, enabling smaller invertebrate predators such as *M. edax* and *Asplanchna* to respond to the elevated densities of prey species that also occurred (Kerfoot, 1987*a,b*). The contribution of cladocerans to total biomass was low (between 5 and 35%) for the first two thirds of the summer, despite the replacement of *D. pulex* by *D. dubia*. Other cladocerans benefitted from fish changes in 1989. Cladocerans that appeared in 1989 were *H. gibberum* in lower densities than 1988, and small pulses of *Diaphanosoma birgei* and *Bosmina longirostris* (Fig. 8.7*d–f*). Increases also occurred in the copepods: *Orthocyclops modestus*, *Mesocyclops edax*, *Cyclops varicans rubellus*, and especially the juveniles (Fig. 8.8*a–f*). Although rotifers continued to contribute a small percentage to the total biomass, they exhibited major shifts in 1989. Rotifer densities increased tenfold, primarily because of an increase in *Kellicottia longispina* and a smaller increase in *Keratella cochlearis* (Fig. 8.9*d,e*). The predaceous *Asplanchna* again appeared in the end of the summer. On the other hand, *Polyarthra vulgaris*, a prominent species in the lake in most years, disappeared by the end of June and remained rare for the rest of the season.

Manipulations in 1990 included the removal of many of the remaining piscivores and additions of golden shiners in June and other minnows throughout the summer (Chapter 4). Planktivory by fishes was variable but generally quite high (Chapter 6). As might be expected, the zooplankton community was quite different from that of 1989. Cladocerans regained their dominance of the total zooplankton biomass, owing in part to the resurgence of the large-bodied *Daphnia pulex* in early summer coincident with a large pulse of *D. dubia* and a smaller one of *D. rosea* (Fig. 8.7*a–c*). These high densities were sustained only in early summer, and declined rather sharply by July. *Bosmina longirostris* reached unprecedented densities in early summer, possibly as a result of spatial segregation from *D. pulex*, which migrated strongly at the time and rarely rose above 5 m (Chapter 9). The larger *Eubosmina tubicen* appeared for the first time in the study in early summer. As in 1989, many of the zooplankton groups had high biomass in early summer, but declined rapidly by

August. However, in 1990 the declines occurred earlier. *Holopedium gibberum* again appeared in large numbers and peaked as the daphnids declined in July. The copepods *Mesocyclops edax, Cyclops varicans rubellus*, and all juveniles were abundant in early summer, but declined in July to the lowest levels since the trout introduction in 1988 (Fig. 8.8*a–f*).

Although rotifer densities were substantially lower in 1990 than in 1989, they were still higher than in 1984–8. The decline from 1989 to 1990 was due to the return of *Kellicottia longispina* to typical densities. At the same time, *Keratella cochlearis* and especially *Polyarthra vulgaris* were found in very high densities until the end of the summer (Fig. 8.9*a–f*). In addition, a few species also increased to levels not seen before in our study: *Conochilus dossuarius, Synchaeta* sp. and *Ascomorpha ecaudis*. These are soft-bodied rotifers that are especially vulnerable to invertebrate predation.

Demographic responses of *Daphnia* in Paul and Peter Lakes

We calculated birth, death and growth rates, and the size at first reproduction (SFR) of *Daphnia* in Peter and Paul Lakes. Prior to manipulation (1984), the densities, mean lengths and demographic parameters of the *Daphnia* populations were similar in the two lakes (Figures 8.3, 8.7 and 8.10; Chapter 10). Substantial increases in both birth and death rates in Peter Lake occurred in late summer 1985 following elevated levels of planktivory by YOY bass (Chapter 6). Increases in birth and death rates also occurred in late 1986. In 1987, planktivory by fishes appeared to be similar in Peter and Paul Lakes (Chapter 6). SFR was similar in both lakes in 1986 (Fig. 8.10*d*).

In response to the first trout addition in 1988, birth and death rates were only slightly elevated for a short time in midsummer, well after the peak in planktivory (Chapter 6). However, SFR of the *Daphnia* in Peter Lake declined soon after introduction of the trout and deviated substantially from SFR in Paul Lake and from SFR prior to trout addition (1986) for the remainder of the year. Since the species composition at this time was similar to that prior to trout addition, the SFR shift was probably a population response to size-selective predation by fishes. By August, the death rates of *Daphnia* in Peter Lake were the lowest of the study, and relatively high (though declining) birth rates were allowing the population to recover. With the introduction of trout in 1989, birth and death rates of the *Daphnia* population declined to their lowest levels, as did SFR, while densities increased to unusually high levels. The shift of dominance to the smaller-bodied *D. dubia* partly accounts for these

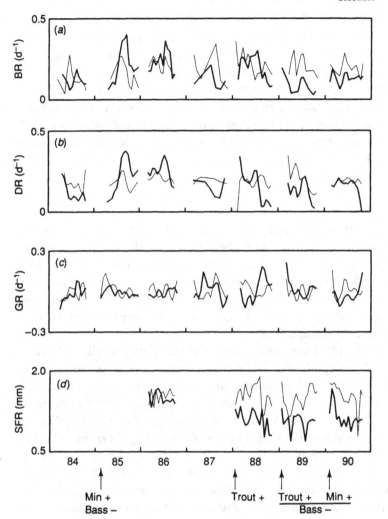

Fig. 8.10. Demographic parameters of *Daphnia* spp. populations in Paul Lake (thin line) and Peter Lake (thick line) versus time during the summer stratified season. (*a*) Birth rates (BR); (*b*) death rates (DR) presented as weighted three-point running averages; (*c*) growth rates (GR) presented as weighted three-point running averages; and (*d*) size at first reproduction (SFR).

trends, since mean brood sizes, and consequently birth rates, are sensitive to body size. However, there were only subtle changes in the mean length of daphnids (Chapter 10).

In 1990, following the introduction of golden shiners, planktivory by fishes increased substantially compared with 1989 (Chapter 6). Birth and death rates of *Daphnia* increased slightly from levels found in 1989. Birth rates were still lower than those in Paul Lake, while death rates were very similar. SFR of the *Daphnia* populations continued to be low since the smaller *D. dubia* constituted a large percentage of the population. However, the other *Daphnia* species (*D. pulex* and *D. rosea*) in Peter Lake were still reproducing at a smaller size compared with Paul Lake and 1986 in Peter and Paul Lakes. For ten sampling dates from July through August, the average size at first reproduction for those *Daphnia* was 1.27 mm ($n = 10$, s.e. $= 0.027$) in Peter Lake and 1.44 mm ($n = 10$, s.e. $= 0.080$) in Paul Lake.

Tuesday Lake

The zooplankton of Tuesday Lake responded readily to the decrease in planktivory that occurred with the removal of the resident minnow population and addition of largemouth bass in 1985. The gradual increase in planktivory that occurred after piscivore removal (fall 1986) and minnow reintroduction (spring 1987) also altered the zooplankton community, although these changes did not completely or rapidly reverse the 1985 manipulation.

In 1984, zooplankton biomass was dominated by juvenile and naupliar stages of copepods (Fig. 8.11*d*) that constituted between 50 and 75% of the total. Adult cyclopoids (the carnivorous *Tropocyclops prasinus* and the omnivorous *Orthocyclops modestus*) only made up 5–30% of the total biomass. Calanoid copepods were rarely present (Fig. 8.11*b,c*). Cladocerans, mainly the small-bodied species *Bosmina longirostris* and *Diaphanosoma birgei*, made up approximately 1–20% of the total biomass. Rotifers contributed 4–20% of the total biomass (Fig. 8.11*a,e*). This small-bodied zooplankton assemblage was similar to that found in the lake in 1981–3 (Bergquist, 1985) and in earlier decades (Chapter 15).

The most obvious response of the zooplankton community to the decrease in planktivory in 1985 was the increase in cladocerans from approximately 7% to 95% of total biomass by late summer (Fig. 8.11*a*). The initial increase was due mainly to *Holopedium gibberum*, although *Daphnia pulex* and *D. rosea* increased as well (Fig. 8.12*a,b,d*). Small-bodied cladocerans that dominated prior to manipulation disappeared early in the summer (Fig. 8.12*e,f*). Both adult and juvenile copepods decreased near the end of 1985 (Fig. 8.13*a,d,e*). *Tropocyclops prasinus*, the dominant copepod prior to manipulation, declined at the end of 1985 and did not recover until 1988. Elser *et al.* (1987*b*) attributed the decline in 1985 to intense predation by *Chaoborus*. In addition, likely prey of *T.*

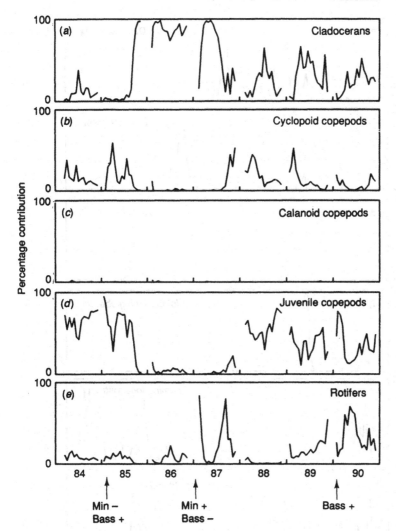

Fig. 8.11. The percentage contribution of each zooplankton group (see text) to the total zooplankton biomass in Tuesday Lake versus time during the summer stratified season for: cladocerans (*a*), cyclopoid copepods (*b*), calanoid copepods (*c*), juvenile copepods of all species (*d*) and rotifers (*e*).

prasinus (rotifers, nauplii and some algae) were uncommon until 1988. The contributions of the copepods to the total biomass dropped precipitously and stayed low until *D. pulex* was practically eliminated from the system in 1988 (Fig. 8.11*d* and 8.12*a*).

Rotifer populations responded to the fish manipulations as well. The density of all rotifer species decreased steadily by the end of the summer

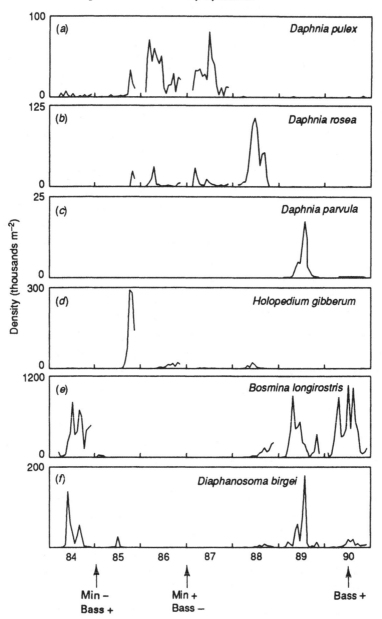

Fig. 8.12. Tuesday Lake: density of cladoceran species versus time during the summer stratified season for *Daphnia pulex* (*a*), *D. rosea* (*b*), *D. parvula* (*c*), *Holopedium gibberum* (*d*), *Bosmina longirostris* (*e*) and *Diaphanosoma birgei* (*f*).

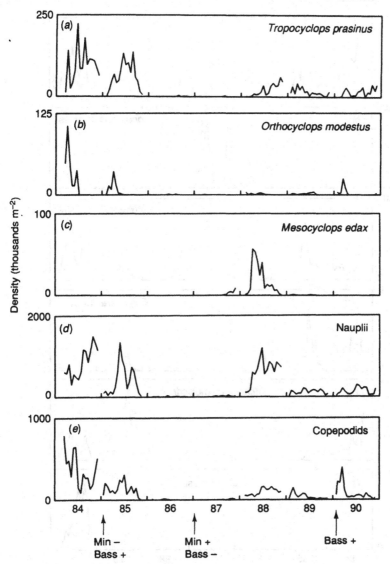

Fig. 8.13. Tuesday Lake: density of copepod species versus time during the summer stratified season for *Tropocyclops prasinus* (*a*), *Orthocyclops modestus* (*b*), *Mesocyclops edax* (*c*), nauplii of all species (*d*) and copepodids of all species (*e*).

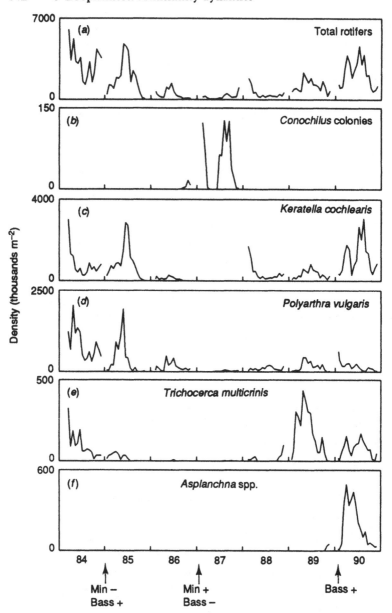

Fig. 8.14. Tuesday Lake: density versus time during the summer stratified season for total rotifers (a) and selected species: *Conochilus unicornis* colonies (b), *Keratella cochlearis* (c), *Polyarthra vulgaris* (d), *Trichocerca multicrinis* (e) and *Asplanchna* spp. (f).

in 1985 (Fig. 8.14*a–f*). This decline included the more common species (*Keratella cochlearis, Polyarthra vulgaris, Ploesoma* sp. and *Trichocerca multicrinis*), and coincided with increases in the *Daphnia* populations and increases in *Chaoborus* predation (Elser *et al.*, 1987*b*). *D. pulex* interferes with rotifers, and exploits limiting food resources more effectively than smaller *Daphnia* spp. (Gilbert, 1988; Neill, 1984; Vanni, 1986). Two of the species (*T. multicrinis*, Fig. 8.14*e*; and *Ploesoma* sp., not shown) were absent from the system except when the biomass of *Daphnia* was very low.

The changes in crustaceans and rotifers that occurred in 1985 were sustained in 1986, as well as most of 1987. The zooplankton community resembled that of Paul Lake, and continued to be dominated by large-bodied cladocerans to the exclusion of smaller species. In addition, densities of juvenile and adult copepods stayed very low. Although rotifer densities were low, the colonial *Conochilus unicornis* appeared for the first time in Tuesday Lake in August 1986 (Fig. 8.14*b*).

From August 1987 to 1990, planktivory in Tuesday Lake gradually increased to premanipulation levels (Chapter 6). One of the earliest indications of the increase in planktivory in 1987 was the decrease in the contribution of cladocerans to the total biomass from over 90% in early summer to about 20% by August. This decline was due to a decrease in density of *D. pulex* and the unprecedented increase in the colonial rotifer *Conochilus* (Fig. 8.14). The major peak in *Conochilus* occurred as *Daphnia* declined in the latter half of the summer, while planktivory by fishes was still relatively low (Chapter 6).

Although planktivory by fishes reached premanipulation levels by 1988, the zooplankton community still differed in important ways from the premanipulation state. Although *D. pulex* declined in 1987 and did not recover the following year, it was replaced in 1988 by a smaller species, *D. rosea*, which was found in very high numbers (Fig. 8.12*a,b*). At the same time, there was a slight resurgence of *Holopedium* in midsummer, a gradual increase in *Bosmina* densities throughout the summer and an increase in density of *Diaphanosoma*. A similar response occurred in the copepods. *Tropocyclops* gradually increased throughout 1988. *Mesocyclops edax*, which first appeared in the lake in August 1987, increased markedly in 1988 along with juvenile and naupliar copepods (Fig. 8.13). The late appearance of *Mesocyclops* could be due to the increase in prey species that occurred in 1988, coupled with low levels of vertebrate planktivory. Total rotifer densities did not increase substantially in 1988, although certain species that had dominated before manipulation did begin to increase, particularly *Keratella cochlearis* and *Polyarthra vulgaris* (Fig. 8.14c,d).

The changes in the zooplankton community that occurred in 1989 and 1990 reflect a continued increase in planktivory by fishes, coincident with a decrease in planktivory by *Chaoborus* (Chapter 7). *Daphnia* species virtually disappeared in 1989 except for minor pulses of *D. dubia* in early summer and of *D. parvula* in midsummer. *D. parvula*, the smallest daphnid found in these lakes, was not found in any of the lakes prior to its appearance in Tuesday Lake in 1989. At the same time, densities of *Bosmina* and *Diaphanosoma* returned to premanipulation levels through 1990. Copepod densities, however, did not respond as predicted. The species that was dominant before manipulation, *Tropocyclops prasinus*, did not increase to premanipulation levels, and actually decreased after 1988. The same response was evident in the juvenile and naupliar stages of all species. Rotifers, on the other hand, did return to premanipulation levels. By 1990, densities of all rotifers (including the dominant species *K. cochlearis*, *Ploesoma* spp. and *Trichocerca multicrinis*) were similar to those of 1984. In addition, rotifers made up a greater portion of the total biomass of the zooplankton community. Although densities of rotifers were similar to those of 1984, rotifer biomass was substantially higher in 1990 compared to 1984, owing to the dominance of *Asplanchna*. Its dominance may result from the decrease in predation by *Chaoborus*, as well as increasing levels of prey species. There were some rotifers (such as *Trichocerca cylindrica* and *Synchaeta* sp.) that were found in relatively low numbers throughout the study and did not appear to be influenced by changes that were occurring with the rest of the community.

Discussion

Manipulations of the fish assemblages in Peter and Tuesday Lakes substantially altered the structure of the zooplankton community. The changes included responses of the dominant species of the community and of the major groups of zooplankton; changes in the demographic parameters of the dominant herbivore (*Daphnia*); changes in the relative biomasses of the major groups; and the appearances of species not known from premanipulation studies or the reference lake. The significance of these changes depended on the magnitude, nature and timing of the manipulations of the fish populations. While direct effects of predation on community structure were clearly demonstrated, the indirect effects of interspecific competition and resource limitation were strongly suggested and probably contributed to the often surprising responses of the community. However, we were not able to disentangle completely this complex of factors.

Zooplankton community change in Peter Lake

The manipulations of the fish populations in Peter Lake resulted in a series of pulses of planktivory that were short (< 1 summer) and quite different from year to year. While the community was relatively similar to that of Paul Lake before manipulation (1984), it deviated substantially after some manipulations. Peter Lake's zooplankton were most different from those of Paul Lake while planktivory deviated most from the premanipulation state (1988–90). However, the responses of the zooplankton to a milder manipulation (1985) were also informative.

The manipulations of 1985 yielded an exceptionally large year class of the piscivore that was already present in the system (largemouth bass). We tracked the impact on zooplankton dynamics as the cohort developed from planktivores as YOY in 1985 to piscivores by 1987 (Chapters 4 and 6). Planktivory by YOY bass intensified in midsummer 1985 (Chapter 6) and large-bodied *Daphnia* declined. Birth rates of *Daphnia* were unusually high prior to the decline, suggesting that food was not limiting. Elevated death rates occurred during the decline. Large individuals of *Daphnia pulex* disappeared from the population (Carpenter *et al.*, 1987). These changes are all consistent with heavy mortality due to fish predation. *Holopedium* populations, which usually peak in early summer in these lakes, increased after *Daphnia* were declining. The late appearance of *Holopedium* may be explainable by decreased competition with *Daphnia* and/or its low susceptibility to planktivory by fishes (McNaught, 1978; Tessier, 1986). At the same time, more subtle changes in other zooplankton occurred, such as the appearance of the small cladoceran *Diaphanosoma* and decreases in total rotifer densities.

By 1986, daphnid densities were low. Death rates remained higher in Peter Lake than in Paul Lake, although total losses to planktivory by fishes were similar in the two lakes (Chapter 6). Birth rates were lower in Paul Lake than in Peter Lake, suggesting that the two populations were experiencing different constraints. By 1987, *Daphnia* populations in Peter Lake approached premanipulation conditions. The effects of the unusually large 1985 year class of piscivorous fish passed as a wave of predation through the *Daphnia* population within three years.

Nevertheless, there were no major changes in the zooplankton species composition in Peter Lake from 1984 through 1987. The contributions of the major groups to the total biomass did not deviate appreciably during the four years, and exhibited similar variability to that of Paul Lake. With the exception of the appearance of *Diaphanosoma* in 1985, there were no appearances and disappearances of species, or changes in the dominant species of the functional groups. Apparently, zooplankton community structure can remain relatively stable in the face of highly variable recruitment of largemouth bass.

Introduction of a new species of planktivore (rainbow trout), however, did alter the zooplankton community structure of Peter Lake. Vulnerable populations of zooplankton and *Chaoborus* suffered, while resistant species flourished. Rainbow trout feed deeper in the water column than other planktivores in Peter Lake. The deep-water refuge of *Daphnia pulex* and *Chaoborus* from fish predation was reduced or eliminated. Predation on *Chaoborus* increased and held chaoborids to unusually low densities in 1988 through 1990 (Chapters 6 and 7).

The rainbow trout introduction in 1988 had substantial effects on cladocerans in particular. Their biomass declined relative to other groups, previous years and Paul Lake. *Daphnia* decreased to their lowest densities of the study. Further evidence that size-selective predation caused these declines lies in the changes in the demographic parameters of the *Daphnia* populations. Birth rates were elevated in June and July and were comparable to those in Paul Lake, suggesting that food was not limiting. Death rates, however, were high. Second, the size at first reproduction of Daphnia declined soon after the addition of trout in 1988 and deviated substantially from that of Paul Lake. This decrease in SFR was not due to a species shift, since the species composition of *Daphnia* in Peter Lake did not change at the time. *Daphnia* initiate reproduction at a smaller size to reduce the impact of size-selective predation by fishes (Culver, 1980). Small-bodied cladocerans responded as well. *Diaphanosoma* increased to unusually high densities. A new daphnid species, *Daphnia dubia*, appeared for the first time in the lake. These shifts in cladoceran composition, along with the increases in *Holopedium* and rotifer density, may have been facilitated by decreased competition with large *Daphnia* species and/or reduced predation by *Chaoborus* (Elser *et al.*, 1987*b*; Gilbert, 1988; MacKay *et al.*, 1990; Chapters 6 and 7). However, it is not possible to distinguish the relative importance of these processes. Many responses of the zooplankton were short-lived, and were reversed by the end of summer 1988 as planktivory by fishes declined (Chapters 4 and 6).

Although planktivory by fishes was lower in 1989 than in 1988 in Peter Lake, it was sufficient to cause additional changes in the community structure. Small-bodied species that previously were held at low densities in Peter Lake (small cyclopoids, juvenile copepods and rotifers) increased in 1989. Further shifts in the cladoceran populations took place as the mid-sized *D. dubia* replaced *D. pulex* as the dominant species and attained remarkable densities. Cladocerans as a whole made up a very small percentage of the total biomass compared to previous years. Birth rates of all *Daphnia* species were very low throughout the summer. Low birth rates are partly a function of the small body size of *D. dubia*, since brood sizes are strongly size-dependent, but may also reflect food limitation. Death rates were also low, however, allowing *D. dubia* populations to increase. The dominance of *D. dubia* was probably facilitated by the low density of *Chaoborus*, which is a very effective predator on small cladocerans in Peter Lake (Elser *et al.*, 1987b; MacKay *et al.*, 1990). It is curious that *D. dubia* benefitted from decreased levels of *Chaoborus* planktivory instead of *D. rosea*, another small daphnid species that has been present in the lake at relatively low levels for decades (Chapter 15; Leavitt *et al.*, 1989). Although the body sizes of the two are similar, the helmet of *D. dubia* may confer resistance against small invertebrate predators such as *Mesocyclops* (Kerfoot, 1980). Planktivory by fishes may facilitate increases in small *Mesocyclops* by suppressing more effective competitors such as *Chaoborus* (Kerfoot, 1987a,b). *Chaoborus* can also consume large numbers of juvenile copepods, including those of *Mesocyclops* (Elser *et al.*, 1987b). The populations of *Mesocyclops* and juvenile copepods were substantially higher in 1989 and 1990, coincident with the disappearance of *Chaoborus*. Apparently, *Mesocyclops* was much less effective in regulating prey populations than *Chaoborus*, consistent with the arguments of Kerfoot (1987a,b).

The addition of golden shiners and removal of remaining piscivores in Peter Lake in 1990 resulted in variable but generally high planktivory by fishes (Chapter 6). The golden shiners, unlike the trout, did not feed effectively on migratory daphnid populations. *D. pulex* regained high densities. However, individuals were restricted to deep waters (7 m) for most of the day, rising to only 5 m during the night (Chapter 9). *D. pulex* were able to coexist with the planktivorous golden shiners for at least part of the summer by migrating to the deep-water refuge. They were not able to take advantage of this refuge when rainbow trout were present, since trout feed deeper in the water column (Chapter 4). *D. dubia* were also found in high numbers in early summer, but were found higher in the water column, between 2 and 7 m (Chapter 9), so the *Daphnia* species were almost completely segregated spatially. Apparently the small size and relatively rapid growth rates of *D. dubia* allowed it to coexist with planktivorous golden shiners. Birth rates of all *Daphnia*, however, remained lower than in Paul Lake, although they were slightly higher than in 1989. Although death rates were similar to those in Paul Lake, they were consistently high and exceeded those of 1989. The combination of reduced birth rates and consistently high death rates led to midsummer declines in *Daphnia*. Size at first reproduction of all *Daphnia* species, and of *D. pulex* alone, continued to be low, especially compared with Paul Lake.

Many taxa were found in the highest densities of the study in the first part of 1990, yet declined substantially by midsummer. In 1990, zooplankton populations were more variable than in any year of the study. Concentrations of edible algae were also quite low compared with other years, despite a few large peaks (Chapter 11). The extent to which the midsummer declines were due to food limitation is unclear. However, the combination of low birth rates and high death rates for *Daphnia* suggest that food may have limited the capacity of the zooplankton to withstand planktivory by golden shiners.

Zooplankton community change in Tuesday Lake

The manipulation of 1985 caused a nearly discontinuous step change in the zooplankton community of Tuesday Lake. Reversal of that manipulation, however, caused more gradual responses by the zooplankton. As planktivores were slowly reestablished in Tuesday Lake after 1987, the zooplankton community was experiencing different constraints than those found before manipulation (1984). From 1987 to 1990, food levels were greatly reduced, as were populations of the dominant invertebrate predator, *Chaoborus*. Interspecific competition may have played an increased role in zooplankton dynamics during this period. Nevertheless, the community changed little from 1986 to 1987.

1988 proved to be a transition year for the zooplankton in Tuesday Lake, as some formerly common species returned, including the dominant small-bodied cladocerans. The predaceous copepod *Mesocyclops*, which had not previously been found in the lake, appeared. The smaller-bodied *D. rosea* replaced *D. pulex* as the dominant daphnid. Food levels were also higher in 1988 (but still below those of 1984) and the total biomass of zooplankton approached premanipulation conditions.

Increases in smaller species of zooplankton may have been facilitated by the shift to the smaller daphnid. Competition between *Daphnia* and small zooplankton has been found to be sensitive to small changes in daphnid size. (MacIsaac & Gilbert, 1989). Whatever the mechanism, these dynamics suggest a 'species sorting' phenomenon: following changes in top predators, species appear in pulses before the system reaches some sort of balance with the predator species (Black & Hairston, 1988).

By 1989, the zooplankton community approximated that of 1984, but with a few key differences. First, food levels were still rather low, suggesting increased competition. Cladocerans made up a much larger percentage of the total biomass than in 1984. In addition, *Chaoborus* densities were exceptionally low compared with 1984 (Chapter 7) and the rotifer populations were higher. Suppression of *Chaoborus* by minnows may have allowed rotifer populations to increase in 1989–90. However, the reduced densities of the smaller *Chaoborus punctipennis* in Tuesday Lake appeared to have weaker effects on the zooplankton community than reduced densities of the larger chaoborids in Peter Lake. Unfortunately, this comparison is confounded by the changes in algal resources that occurred in Tuesday Lake.

By 1990, although the total biomass approached levels found in 1984 (Chapter 10), the community structure still differed. Rotifer biomass was elevated owing to the dominance of a previously rare predaceous species. Populations of juvenile and adult copepods and *Chaoborus* were reduced, while those of cladocerans were slightly elevated. These changes may be attributable to a combination of reduced algal resources and changes in invertebrate predation pressure.

Lessons from a large-scale approach

By definition, a stable community is one in which there are few appearances or disappearances of species (Black & Hairston, 1988). Zooplankton populations of Paul Lake appear stable and free from large oscillations in population sizes and appearances or disappearances of species. In contrast, the changing predation levels in both Peter and Tuesday Lakes caused relatively large changes in the zooplankton populations. In 1985, in Peter Lake, there were compensatory shifts in the cladoceran populations as the dominant species declined and a new species appeared as a result of increased planktivory by fishes. On the other hand, as *Chaoborus* declined and fish planktivory increased in

1988–9, the dominant cladoceran species was replaced by a new species, increases were observed in species previously found at low levels, and some species disappeared. These responses demonstrate the important role that planktivory by fishes, in concert with planktivory by *Chaoborus*, plays in maintaining the structure and diversity of the zooplankton community. In Tuesday Lake, a reduction of planktivory that occurred with the removal of minnows led to the dominance of large daphnids, and complete exclusion of previously dominant cladocerans as well as other species. After reintroduction of minnows these shifts were reversed, although not completely. Predation by minnows strongly influenced the species that were present in the community, especially the dominants. However, the slow return of the zooplankton community in Tuesday Lake suggests that after two seasons (1985 and 1986) the community may have reached a relatively stable state in which it was resistant to perturbation.

Can we predict the trophic cascade from fish to zooplankton? The answer is a qualified yes. Fish predation clearly regulates populations of *Chaoborus*, the invertebrate planktivore (Chapter 7), and *Daphnia*, the dominant herbivore, and thereby potentially affects the community structure of zooplankton and phytoplankton. The rates of these responses are affected by the types of invertebrate planktivores present (e.g. effects of contrasting *Chaoborus* species in Peter and Tuesday Lakes) and the fish planktivores manipulated (e.g. effects of rainbow trout vs. golden shiners in Peter Lake). Also, food limitation may affect the rate of community response to changes in top predators (e.g. Tuesday Lake 1986–90). All of these processes are potentially predictable. The most intractable source of surprise appears to be the appearance, and sometimes dominance, of previously unknown species (e.g. *Daphnia dubia* in Peter Lake).

Our ecosystem approach cannot answer certain mechanistic questions at the community level (Tilman, 1989). Such questions can be addressed by focused, small-scale experiments on selected interactions. For example, we have independently checked some inferences using consumption estimates (Chapter 6; Elser *et al.*; 1987*b*) or enclosure experiments (MacKay *et al.*, 1990; Dini & Carpenter, 1992). However, the number of small-scale experiments needed to disentangle all of the direct and indirect links in the zooplankton communities of these lakes is daunting. Even if the experiments were feasible, they would be extremely complex, perhaps difficult to interpret, and most likely could not include all relevant factors (Yodzis, 1988). Finally, and most importantly, some phenomena crucial to the lake-wide responses could not have become evident at smaller spatial and temporal scales. Some examples from this chapter and Chapter 9 are (1) delayed responses that were longer than one season length (e.g. Tuesday Lake from 1985 to 1986); (2) the colonization and dominance by new species (e.g. *D. dubia* in Peter Lake); (3) the long-term stabilizing effect of antipredator defences (e.g. migration and the persistence of daphnids in Tuesday Lake in 1987–8); and (4) the spatial and temporal segregation of competing species of zooplankton (e.g. *D. pulex* and other zooplankton populations in Peter Lake in 1990).

Summary

Pelagic predators influence the size structure and community composition of zooplankton (Brooks & Dodson, 1965; Hall *et al.*, 1976; Lynch, 1979; Zaret, 1980). Manipulations of the fish communities in Peter and Tuesday Lakes allowed us to address how planktivory by fishes regulates zooplankton community structure as defined by (1) the relative biomass of zooplankton functional groups, (2) the dominant species of the community and of each group, and (3) the size structure of the community and key taxa (Chapter 10). Planktivory by fishes directly affected vulnerable populations of zooplankton and the community as a whole. However, indirect effects of invertebrate predators, resource limitation and interspecific competition were evident as well. Most likely, all of these factors contributed to the often surprising responses of the community.

Manipulations of fish populations in Peter Lake resulted in a series of relatively short pulses of zooplanktivory. The large year class of large-mouth bass recruited in 1985 generated a wave of predation as the cohort developed from planktivorous YOY to piscivores in 1987. *Daphnia* declined later in 1985 and *Diaphanosoma* appeared for the first time in the lake. However, the community structure was not severely altered by these dynamics of the fish community. Zooplankton community structure may remain relatively stable in the face of highly variable recruitment of piscivores like largemouth bass.

Introduction of a different type of planktivore, rainbow trout, altered the zooplankton community in Peter Lake more profoundly in 1988 and 1989. Size-selective predation by trout on large *Daphnia* and *Chaoborus* had significant effects on the zooplankton community. Predation on small zooplankton was reduced as *Chaoborus* populations declined.

Large-bodied daphnids declined to the lowest levels of the study, and were eventually replaced by the smaller *D. dubia*. Increases in other small taxa occurred as well, and the relative contribution of cladocerans to the total zooplankton biomass declined.

The addition of golden shiners to Peter Lake in 1990 also altered the zooplankton community. Although high densities of *D. dubia* were sustained in the first part of the summer, so were high densities of *D. pulex* as vertical migration allowed them to coexist with planktivorous fish. However, by the latter half of the summer, declines were observed in all taxa of zooplankton. Both elevated planktivory by fishes and intensified competition for limiting food resources may account for these declines.

Manipulations of fish population in Tuesday Lake resulted in changes in planktivory that were sustained for more than one season. In 1985, zooplankton responded to the rapid decrease in predation within a few weeks, as large cladocerans replaced the smaller ones that had been present. Many of the changes in the community were sustained even after piscivores had been removed and planktivores reintroduced (1987). The zooplankton responded gradually to the increase in planktivory that occurred from 1987 to 1988. By 1989 and 1990 the zooplankton community approached, but had not completely reached, premanipulation conditions. Decreased invertebrate planktivory by *Chaoborus* and decreased food resources may account for the differences between the 1984 and 1990 zooplankton communities.

Our findings demonstrate the effects of size-selective predation by fishes and *Chaoborus*, as well as some surprising community responses influenced by changes in competition and algal resources. Although indirect interactions could not be disentangled in this study, some are clearly important at whole-lake, multiyear scales and appear to be fertile ground for further research.

Acknowledgements

We thank Bart DeStasio, Jim Kitchell, Bill Neill and Peter Leavitt for helpful reviews of the draft.

9 · *Effects of predators and food supply on diel vertical migration of* Daphnia

Michael L. Dini, Patricia A. Soranno, Mark Scheuerell and Stephen R. Carpenter

Introduction

The Cascade Project provided a perfect opportunity to study one of the great puzzles of limnology, a puzzle which has occupied numerous aquatic biologists over the past century-and-three-quarters. Hardy (1956) called it 'the planktonic problem No. 1' and more recent research has still not succeeded in providing an ultimate explanation that satisfies all cases. The adaptive significance of nocturnal diel vertical migration (DVM), a phenomenon whereby organisms throughout 15 aquatic phyla (Kerfoot, 1985) ascend through the water column around dusk and descend before dawn on a daily basis, is one of limnology's longest-standing enigmas.

Initial research on DVM was not published until almost sixty years after Baron Cuvier (1817) first documented its existence among freshwater crustaceans. The phenomenon received attention from many of the crowned heads of nineteenth century European biology. August Weismann (1874, 1877), one of Darwin's strongest supporters on the Continent, and Forel (1876) both speculated on the causes of the behavior, as did Thienemann in the next century (1919). Most explanations from the nineteenth century, as well as from the first half of the twentieth, focused on the proximal causes of the behavior. Many abiotic factors were proposed as cues for the initiation of migratory behavior, including diel changes in temperature, pH, light intensity and density (Kikuchi, 1930). It eventually became apparent that no single factor could explain the many behavioral variations exhibited by migrators, including differences in the same species' migratory behavior in lakes near each other, and even substantial differences in migration by the same species in the same body of water (Juday, 1904; Kikuchi, 1930). As a result, researchers in the early part of the twentieth century attempted syntheses of two or more of the abiotic factors to make sense of what was obviously a complex phenomenon (Juday, 1904; Dice, 1914). These attempts were still limited in their ability to explain the behavior's variability. Researchers eventually arrived at a sort of consensus regarding the cue to DVM: rate of change in relative light intensity at dusk and dawn (Ringleberg, 1964; Daan & Ringleberg, 1969; Ringleberg, 1991).

As the proximal cause of the behavior became clearer, researchers shifted their attention to the ultimate cause of the behavior. Knowing what induces zooplankton to migrate is quite different from knowing what makes the migration worth while or, in other words, knowing how migration contributes to the reproductive fitness of migrators.

Many hypotheses for DVM's adaptive significance have been tendered. Most are based on factors which vary with depth: light intensity, temperature, transmission of different wavelengths, and food abundance, quality and distribution. For example, part-time residence in cooler, deeper waters has been seen by some as a means of achieving metabolic or demographic advantage (McLaren, 1974; Enright, 1977). An earlier explanation (McLaren, 1963) of the demographic consequences of DVM incorporated an unnecessary and probably erroneous metabolic model in which cooler temperatures were said to bring about a shift in metabolic pathways and allow energy reserves to be shunted away from respiratory heat loss to egg production. A variant hypothesis explains migration into cold waters as a means to conserve energy during times of food limitation, rather than as a means of achieving an energy bonus (Geller, 1986). These hypotheses presume food quantity or quality to be greater in surface waters, thus providing a reason for animals to ascend.

This stratification of food resources has itself been the basis of several hypotheses. Fuhrmann (1900) proposed that increased algal abundance in surface waters caused significant shading, used as a cue by zooplankters to ascend and take advantage of this abundance. The same abundance was said to result in descent when herbivory increased the concentration of algal toxins in surface waters (Hardy & Gunther, 1935). More recently, McAllister (1969) proposed that the quality of phytoplankton is highest at the end of the photosynthetic day when the store of photosynthate has not yet been reduced by cellular respiration occurring in the absence of the light-dependent photosynthetic reactions. Zooplankters were said to take advantage of this high-quality food by migrating upward at dusk to feed and then downward before dawn to afford their algal food source the opportunity to replenish itself before the next feeding bout. Within the past two decades, many researchers have associated diel vertical migration with conditions of phytoplankton abundance (Huntley & Brooks, 1982; Stich & Lampert, 1981, 1984; Dagg, 1985; Gliwicz, 1985; Johnsen & Jakobsen, 1987) and also with phytoplankton paucity (Giguere & Dill, 1980; George, 1983; Geller, 1986; Hoenicke & Goldman, 1987).

The harmful effects of various wavelengths of the electromagnetic spectrum have been proposed as an explanation for downward migration at dawn, with blue light the culprit in some scenarios (Harvey, 1930; Hairston, 1980) and UV radiation in others (Thomas, 1977; Siebeck, 1978).

The most prominent hypothesis today, however, involves not the quality of light, but its intensity as it varies with depth. Visually orienting predators of zooplankton require light to feed. Downward migration before dawn reduces the likelihood of capture as zooplankters take advantage of the predation refuge offered by deeper, darker waters. Ascent at dusk provides access to food-rich surface waters. The origin of this predation-avoidance hypothesis is obscure (see Morgan, 1903; Murray & Hjort, 1912; Russell, 1927), but much recent work has made it the preeminent explanation (Zaret & Suffern, 1976; Stich & Lampert, 1981, 1984; Gliwicz, 1986; Dini & Carpenter, 1988; Lampert, 1989; Neill, 1990; Dini & Carpenter, 1991).

Though preeminent, it is by no means unchallenged, for it is not without anomalies. There are, for example, oceanic organisms which perform migrations at such depth that they never encounter light (Longhurst, 1976). Then, there is the existence of reverse migration whereby organisms spend the day at the surface and descend to deeper water at night. Other explanations than visual predation pressure are required for this behavior (Dumont, 1972; Ohman, Frost & Cohen, 1983; Bayly, 1986; Neill, 1990). Thus, many researchers have adopted the same outlook taken by early researchers faced with the tremendous variability of the behavior: they have eschewed single-factor explanations. Most of the current multifactor hypotheses involve predation avoidance to some extent, ranging from McLaren (1963) who uses it merely as a synchronizing agent, to Lampert (1989), Clark & Levy (1988), Gabriel & Thomas (1988) and others who view it as the primary forcing factor.

Until recently, most evidence for the predation-avoidance hypothesis relied on comparison of migratory intensity to measures of predator abundance or fish-gut analysis (Zaret & Suffern, 1976; Stich & Lampert, 1981; Gliwicz, 1985, 1986) without immediate or deliberate manipulation of the system. Our whole-lake manipulations provided a unique opportunity to study responses of zooplankton migration to dramatic changes in the intensity of planktivory (Chapter 6). Our interpretation of migratory behavior subsequent to the various manipulations supports the predation-avoidance hypothesis and also suggests the importance of more subtle factors than predation. A mesocosm experiment conducted in 1988 (Dini & Carpenter, 1992) confirmed the importance of a multifactorial approach to DVM, indicating the importance of phytoplankton abundance and distribution.

In the remaining sections of this chapter, migration of *Daphnia* in response to the whole-lake fish manipulations will be treated first, followed by a brief summary of the mesocosm experiment. This order matches not only the actual chronological sequence, but makes sense insofar as it reflects a transition on our part from single-factor to multifactor understanding of this complex behavior.

Diel vertical migration has important consequences for the trophic cascade (Lampert & Taylor, 1985; Dorazio, Bowers & Lehman, 1987; Shapiro, 1990). If the deepwater predation refuge is effective, DVM can prevent planktivores from eliminating key herbivores, which can continue to impact the phytoplankton assemblage through their night-time grazing. Nocturnal DVM can further affect the trophic cascade by increasing the rate of downward translocation of mineral nutrients, like phosphorus, when herbivores excrete a greater proportion of ingested nutrients in the hypolimnion than in the surface waters.

Daphnia migration: response to fish manipulation

Our goal was to study DVM in the three lakes at least once a month during summers throughout the manipulation period to detect possible changes in migratory behavior of *Daphnia*. Preliminary studies in May 1985 indicated the daytime depth below which sampling was unnecessary (because of scarcity of animals) and indicated that twilight migration (Hutchinson, 1967) was not occurring, making noon and midnight sampling adequate for determining maximal migratory amplitude. It also indicated substantial horizontal heterogeneity in daphnid distribution and permitted the design of an appropriate sampling protocol (Dini & Carpenter, 1988, 1991).

There were minor differences in sampling from year to year and from lake to lake, but in broad outline, lakes were sampled at 3–5 horizontal stations along a cross-lake transect line. Each station was sampled consistently at 5–6 depths with duplicate hauls of a 12 l Schindler–Patalas trap. Duplicate hauls were pooled and preserved in the field with sugared, buffered formalin to a final concentration of about 4%. Samples were counted at 7–10 × under a binocular dissecting microscope; samples with fewer than 80 animals were counted in their entirety, and larger samples were subsampled. Migratory amplitude of the daphnid assemblage was determined as the difference between mean midday and midnight depths, both calculated after the formula suggested by Worthington (1931) which calculates a density-weighted mean depth:

$$\text{mean depth} = \left(\sum X_i Z_i\right) / \left(\sum X_i\right),$$

where X_i is the number of Daphnia and Z_i is the depth of sample i.

Migration in Tuesday Lake

Before the manipulation, the zooplankton assemblage of Tuesday Lake was dominated by small-bodied animals (Chapter 8). There was intense predation by planktivorous fish (Chapter 6): northern redbelly dace, finescale dace and central mudminnow (Table 2.1), collectively referred to as minnows in the following discussion. This zooplanktivory was sufficient to prevent large-bodied zooplankton like Daphnia from successfully colonizing the lake.

Within three months of the removal of the minnows, the zooplankton community underwent a profound change; Daphnia, specifically D. pulex and D. rosea, dominated the zooplankton as was already true in both of the other lakes (Chapter 8). By mid-August of that same summer, Daphnia was sufficiently abundant in this lake for us to conduct an initial study of its migratory behavior. Four previous sampling episodes in Peter and Paul Lakes, all showing consistent night-time ascents averaging about 2 m in Peter Lake and 1 m in Paul Lake, did not prepare us for the results from Tuesday Lake. Data collected at a single horizontal station at the end of the field season indicated that the newly established daphnids had performed a weak reverse migration, i.e. they were higher in the water column at midday than at midnight (Fig. 9.1)! Hutchinson (1967) had vouched for the relative rarity of reverse migration and, having sampled at only one station, it seemed appropriate to dismiss these results as a fluke. Yet, the first two sampling episodes of the 1986 field season, each carried out at five horizontal stations, validated the previous data: reverse migrations were indeed occurring (Fig. 9.2). Three sampling episodes subsequent to these gave inconsistent results: on two occasions, daphnids performed weak nocturnal migrations (up at night, down at day) and on the third, the mean depth of the assemblage did not change from midday to midnight, nor did it change on the first sampling episode in 1987.

In May 1987, the reduced–planktivory phase of the fish manipulation in Tuesday Lake ended with the reintroduction of minnows (Chapter 2). Within six weeks of this reintroduction, daphnid migration exhibited a radical departure from the highly variable and unpredictable migrations which had occurred during the reduced–planktivory phase. From July 1987 to August 1989, each of eight sampling episodes found normal nocturnal migration by *Daphnia* (Fig. 9.1). Two of these were weak

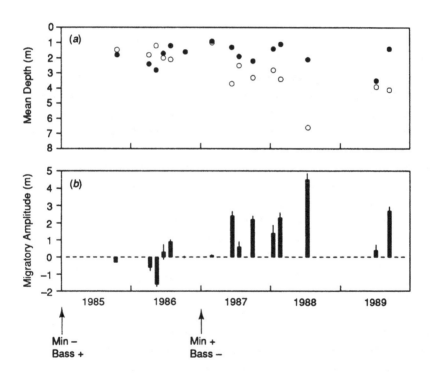

Fig. 9.1. (*a*) Mean midday depth (open circles) and mean midnight depth (filled circles) of the daphnid assemblage in Tuesday Lake on 15 dates in June–August of 1985–9. The minnow reintroduction occurred in late May 1987. *Daphnia* was too rare in 1990 to study its migration. (*b*) Migratory amplitude measured as change in mean depth from midday to midnight. Error bars represent the standard error of the differences.

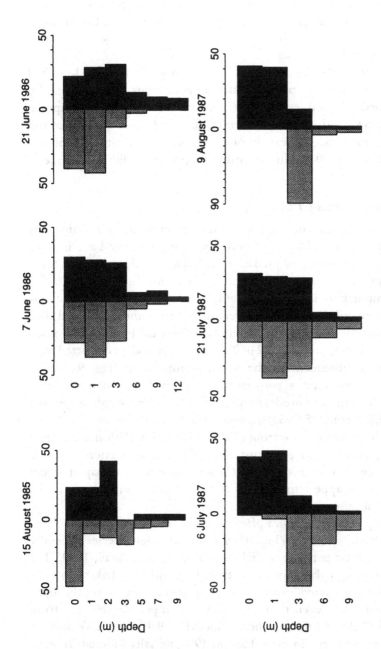

Fig. 9.2. Profiles of percentage *Daphnia* versus depth in Tuesday Lake. Midday distributions are solid. The upper three profiles are the first from this lake following the initial manipulation and portray instances of reverse migration: the assemblage was centered higher in the water column at midday than at midnight. The three lower profiles represent the first sampling episodes after the minnow reintroduction in late May 1987, all indicating instances of nocturnal migration. The August 1985 profile is from only one horizontal station; others are the pooled results of five horizontal stations. Note the change in depths sampled from one year to the next.

migrations, but six were strong, surpassing the amplitudes of those occurring in Paul Lake (Fig. 9.5) and, on occasion, rivaling the strongest migrations in Peter Lake (Fig. 9.3), where transparency was 2–3 times greater than in Tuesday Lake.

By summer 1990, the zooplankton assemblage of Tuesday Lake had essentially reverted to premanipulation conditions characterized by small-bodied zooplankters (Chapter 8). This shift occurred in response to the recovery of the minnow assemblage to its former abundance and to the accompanying increase in planktivory. The paucity of daphnids during summer 1990 precluded further migration studies in this lake.

Migration in Peter Lake

The removal of most of this lake's adult largemouth bass (Table 2.1) population and the addition of minnows from Tuesday Lake in May 1985 should have increased predation on *Daphnia* in this lake. Yet, there was no significant change in migratory behavior associated with this initial manipulation, i.e. migratory amplitudes before the original manipulation were not statistically different (*t*-test) from those recorded after the manipulation. Instead, the daphnids in Peter Lake exhibited consistent nocturnal migrations on 22 of 23 sampling episodes conducted over the six years subsequent to the initial manipulation (Fig. 9.3). The variability in migrations performed by Peter Lake's daphnids was associated with the amplitudes of migrations between sampling episodes which ranged from 0.5 to 5.0 m, and averaged about 3.8 m.

The addition of rainbow trout (Table 2.1) in May 1988 and 1989 had strong impacts on the size structure, size of first reproduction, biomass, and the species composition of the *Daphnia* assemblage (Chapters 8 and 10) but did not appear to coincide with any noticeable change in the amplitude of migration.

Daphnia pulex and *D. rosea* often co–occur in our experimental lakes and other nearby lakes. Where they co–occur, these species generally have very similar patterns of diel vertical migration (Dini, 1989). For example, in Long Lake (about 1 km from Paul and Peter Lakes) *D. pulex* and *D. rosea* maintained virtually identical migratory patterns throughout summer 1988, even though migratory amplitude changed from about 7 m to about 2 m over the summer (Fig. 9.4). The newcomer *D. dubia*, which appeared in Peter Lake in 1989, initially differed from *D. pulex* in both vertical distribution and migratory amplitude (Fig. 9.3). *D. dubia* occupied shallower depths than *D. pulex* during migration studies in late summer 1989 and early summer 1990. At these times, planktivory

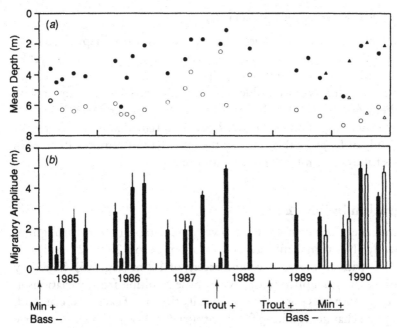

Fig. 9.3. (a) Mean midday depth (open circles) and mean midnight depth (filled circles) of the daphnid assemblage in Peter Lake on 23 dates throughout summers 1985–90. *Daphnia dubia* became established in 1989 and is represented by triangles. (b) Migratory amplitude with symbols for *D. pulex* as in Fig. 9.1. In 1989 and 1990, *D. dubia*'s amplitude is indicated by the open bars.

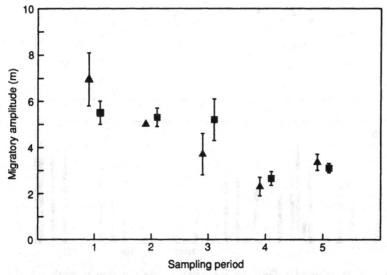

Fig. 9.4. Migratory amplitude of *Daphnia pulex* (triangles) and *D. rosea* (squares) in Long Lake on five dates between 10 June and 1 August 1988. Migratory amplitude is measured here as the change in the populations' modal depths from midday to midnight. Error bars represent 95% confidence intervals.

rates by both fishes and *Chaoborus* were relatively low (Chapters 6 and 7). Owing to its small size, *D. dubia* may have low apparency to fishes yet be highly vulnerable to *Chaoborus* (Chapter 8). In the presence of high densities of planktivorous golden shiners (Chapters 4 and 6), the migratory behavior of *D. dubia* changed during the summer of 1990. During the two later migration studies, *D. dubia* had attained the high migratory amplitude of *D. pulex* (Fig. 9.3). In the latter half of summer 1990, both species followed similar depth distributions.

Migration in Paul Lake

In Paul Lake, *Daphnia* migrated in a very predictable fashion throughout the study. Nocturnal migration was observed on 20 of 21 sampling episodes, with an average evening ascent of 1.2 m (Fig. 9.5). The only variations in visual planktivory were those resulting from variations in the largemouth bass population, especially the size of each year class and, perhaps, in changes resulting from ontogenetic diet shift over the course of each field season.

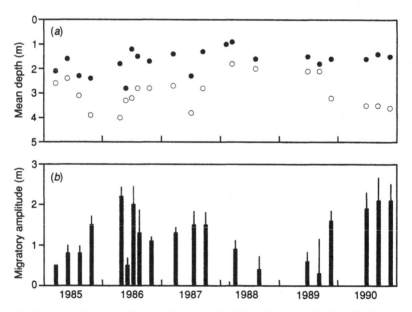

Fig. 9.5. (*a*) Mean midday depth (open circles) and mean midnight depth (shaded circle) of the daphnid assemblage in Paul Lake on 21 dates throughout summers 1985–90. (*b*) Migratory amplitude; all symbols as for Fig. 9.1.

Discussion of migration pattern in the lakes

If the predation-avoidance hypothesis is to derive support from the fish community manipulations in these lakes, we would expect that in Peter Lake, where predation intensity was increased in 1985 and 1988–90, vertical migration by *Daphnia* should have been intensified in some way. In Tuesday Lake, during the reduced-planktivory phase of 1985–6, migration should have been weak or nonexistent and should have intensified during the increased-planktivory phase of 1987–90. In Paul Lake, we would expect to see the familiar, consistent pattern of nocturnal migration continue essentially unchanged. While there were deviations from these expectations, we do not believe these present any convincing challenges to the predation-avoidance hypothesis.

Daphnids appearing in Tuesday Lake in 1985–6 were not exposed to high levels of predation (Chapter 6). Minnows that had been left behind were eaten or driven into the bog mat by the introduced bass. The bass themselves fed mainly on these minnows, benthos, and *Chaoborus* and essentially ignored the daphnids (Chapters 4 and 6). In the near-absence of visual predation, there was apparently no factor operating to cause consistent migration. Instead, migratory direction and intensity of *D. pulex* and *D. rosea* varied tremendously while the same species in the two neighboring lakes exhibited consistent nocturnal migrations.

Factors other than changes in fish abundance were considered as possible causes of the dramatic change in daphnid migratory behavior in Tuesday Lake. Such a change might also be accounted for by a shift in the species composition of the daphnid assemblage, by variations in temperature stratification, light penetration and/or oxygen distribution, by change in mean daphnid body length, by change in abundance or behavior of invertebrate predators or by change in distribution, concentration or quality of food resources, algal or otherwise. We were able to discount the importance of some of these, namely changes in species composition, daphnid body length, algal abundance in various parts of the water column, and the effect of vertically migrating chaoborids (Dini & Carpenter, 1988). Others remain unaccounted for, such as depth distribution of preferred food items or the distribution of oxygen relative to temperature or light profiles.

Theoretically, increased water transparency and the resulting increase in light penetration would be expected to be associated with increased migratory amplitude since migrating plankters would take deeper daytime stations (Dodson, 1990). This expectation was not fulfilled in

Tuesday Lake where Secchi depths ranged from 1.5 to 3.5 m, but which had migrations after May 1987 that were nearly as strong as in Peter Lake, where Secchi depths ranged from 3.5 to 7.0 m (S. Carpenter, unpublished data; see Chapter 13 for a comparison of transparency in the two lakes). The highest water transparencies in Tuesday Lake, as measured by Secchi depth and depth of 1% light transmission, occurred in 1987. This was also the summer of the dramatic shift to strong nocturnal DVM among this lake's *Daphnia*. However, transparency decreased in subsequent summers while migratory amplitudes intensified. Thus, there was no simple relationship between migratory amplitude and water transparency.

One possibly important change in *Daphnia* accompanied the change in DVM: egg ratios (the product of mean brood size and proportion of adult females carrying broods) significantly increased at the time migration in Tuesday Lake intensified. This may have been an incidental advantage accruing to migrators or may itself have been the ultimate reason for the behavioral change.

The results from Tuesday Lake provide the strongest support for the predation-avoidance hypothesis. Evidence from the other lakes is less conclusive. In Peter Lake, daphnids migrated consistently through years of very low and very high fish predation (Chapter 6). Perhaps these results can be explained with information about the genetics of daphnid migration. It has been demonstrated that clones of females within the same species have different migratory behaviors, with some clones migrating and others not (Weider, 1984; Dumont *et al.*, 1985; DeMeester & Dumont, 1988; Dodson, 1988). As long as the relative proportion of migrating to non-migrating clones does not change, we expect to see no change in behavior. Such lack of change would not be likely when visual predators are introduced to a habitat containing a genetically heterogeneous population, such as we believe existed in Tuesday Lake. Lack of change would *not* be surprising, though, if visual predators were numerically unimportant in a habitat containing primarily migrating clones and under a set of conditions where DVM presented no substantial costs. In Peter Lake, the selection pressure against migrating clones, in the near absence of fish predation, was perhaps nil. When fishes were introduced to the lake in numbers sufficient to significantly increase predation rate on *Daphnia* (Chapter 6), migrating clones already predominated, so no change in migration was observed.

In Paul Lake, the origin and maintenance of consistent DVM are more mysterious. Bass exerted consistent, but very low, predation on daphnids (Chapter 6). Minnows apparently exerted strong predation pressure on zooplankton in Paul Lake in the 1970s (Chapter 15). Perhaps low, consistent predation pressure by bass was sufficient to maintain DVM for many years after the episode of minnow predation.

While not eliminating other hypotheses, these results give strong support to the predation-avoidance hypothesis, especially the results from the reduced- and increased-planktivory phases in Tuesday Lake. We proposed the following hierarchy to account for the adaptive significance of the behavior (Dini & Carpenter, 1988), based on the finding that diel vertical migration in *Daphnia* has a strong genetic component. Given a daphnid population with variability at the gene locus or loci governing migratory behavior, the observed behavior should depend in large part on the intensity of predation pressure by visually orienting predators. If predation pressure is consistent and sufficiently intense, non-migrating clones will be preyed upon more often than migrating clones, and migratory behavior will come to predominate in such a lake as migrating clones make up a greater proportion of the organisms present. On the other hand, if predation pressure is weak and/or variable, some other, more subtle factor or suite of factors will determine whether migration occurs, in which direction it occurs and to what extent it occurs.

Under conditions of low planktivory, as might follow an especially drastic winterkill of fishes or introduction of an effective piscivore, the rate of change in migratory behavior will be affected by the genetic variability present within the zooplankton assemblage. Change may be expected to occur slowly in an extant assemblage which has long been subject to visual predation (Luecke, 1986). More rapid change could be expected where the composition of the assemblage changes, especially if this change involves introduction of non-migrating clones.

Does food availability affect migration?

The whole-lake results show that fish predation is a factor controlling DVM by *Daphnia*. However, as discussed above, migration does occur in circumstances were fish predation is quite low. Such migrations may be genetically fixed consequences of past predation episodes; this argument is difficult to test experimentally. Moreover, fish abundance cannot change fast enough to explain distinct changes in migratory amplitude occurring from one day to the next (M. L. Dini, unpublished data). Certain alternative hypotheses, several of which were discussed in this chapter's introduction, are readily tested experimentally. These include effects of food availability (algal concentration), which also affects light transmission, and of the water column's thermal stratification on DVM. To gauge the importance of these factors compared to predation avoidance, an enclosure experiment using a modified 2 × 2 × 2 factorial design was carried out in Peter Lake during summer 1988 (Dini & Carpenter, 1992). The experimental design included two levels of predation pressure (fish present/fish absent); two enclosure depths (10 m/ 4 m) and thus two temperature regimes; and two levels of algal abundance (high/ambient). This design permitted us to test elements of the predation-avoidance, metabolic and food-availability hypotheses. The 2 × 2 × 2 factorial design was modified by eliminating shallow, fish-containing enclosures. There was no point in running this treatment, since fish rapidly eliminated daphnids from shallow enclosures (M. L. Dini, unpublished data). Thus, there were six treatments, each in triplicate. The experiment ran for 25 d during which time daphnid midday and midnight depth distribution, daphnid Lipid–Ovary–Egg index (Hoenicke & Goldman, 1987) and chlorophyll *a* depth distribution were monitored weekly.

We found substantial differences in both adult daphnid daytime depth and migratory amplitude between treatments with fish and those without fish (Dini & Carpenter, 1992). Where fish were present, mean midday depths were substantially deeper and vertical migrations had greater amplitudes. These differences were not the result of mere cropping by fishes during daytime, but reflected behavioral changes by *Daphnia*.

We also observed a food effect, more subtle than the fish effect, on migratory amplitude. Migrations were stronger in treatments with lower chlorophyll *a* concentrations than in those with chlorophyll concentrations maintained above ambient by nutrient additions. There was also evidence that vertical distribution of chlorophyll *a* influenced the vertical distribution of *Daphnia* and, consequently, the amplitude of migration.

The enclosures themselves reduced the advection currents which help keep phytoplankton suspended at the surface. Rapidly sinking algae accumulated toward the bottoms of the enclosures and may have been partly responsible for damped migrations observed in fishless treatments.

We found no evidence of metabolic advantage to animals whose migrations exposed them to cold water for most of the day (those in the deep enclosures), since there was no depth effect on daphnid Lipid–Ovary–Egg indices of triacylglycerol content.

Synthesis

The results of the whole-lake manipulation, unique for its scope and for the duration and intensity of its monitoring period, support the predation-avoidance hypothesis as an explanation for the adaptive significance of DVM in the daphnid assemblages of these lakes. Thus, they are in agreement with much of the recent research in this field (Zaret & Suffern, 1976; Stich & Lampert, 1981; Gliwicz, 1985, 1986; Gliwicz & Pijanowska, 1988; Bollens & Frost, 1989; Neill, 1990). However, as suggested throughout this chapter, predation-avoidance does not operate to the exclusion of other factors which can also confer reproductive fitness upon migrating organisms. Great variability in migratory intensity within the space of just weeks, as in Peter Lake during June 1986 and again in May 1988 (Fig. 9.3), suggests that more than predation pressure is at work in determining migratory behavior.

Our enclosure experiment indicated that, whether predators were present or not, food availability played a significant role in shaping migratory behavior. In particular, it indicated the possible importance of deep algal maxima in reducing the intensity of migration. Such maxima are certainly common in Paul and Peter Lakes (Chapter 12) and are well documented in other aquatic environments as well (Ortner, Wiebe & Cox, 1980; Scrope-Howe & Jones, 1986; Pijanowska & Dawidowicz, 1987; Daro, 1988). The transitory nature of these deep maxima is likely to introduce considerable variability into the vertical migrations of herbivores, perhaps including the variability in Peter Lake.

Results of the enclosure experiment allow for some expansion of the regulatory hierarchy for diel vertical migration (Dini & Carpenter, 1988). Ultimate control over migration rests with the genome. If the genetic constitution permitting migration is not present, migration will never occur. If it is present, migratory behavior will depend on an array of environmental factors, the least subtle of which is visual predation. When predation is intense, DVM should occur consistently. When predation is not intense, DVM may or may not occur depending on food availability. If food is limited, DVM may continue, allowing organisms to locate food-rich strata or to increase their encounter rate with diffuse food particles, as first suggested by Weismann (1874). If food is abundant, there would be little use in migrating unless the food source were at depth and daphnids needed to ascend into warm water, possibly to take advantage of its ability to speed metabolism or growth rate. DVM may continue in such a case if selection pressure against it is low. Where food availability is variable, migratory variability should be expected.

Outside of this hierarchy, there are yet other factors which may influence DVM. Not only the quantity of food but its quality may be important. Vertical distribution of food quality with respect to food quantity is a potentially important but basically unfathomed factor. Tactile invertebrate predators, themselves performing vertical migrations, have been shown to be a possible influence (Ohman *et al.*, 1983; Neill, 1990; Leibold, 1990).

Stich & Lampert (1981) documented a case with parallels to that of *D. pulex* and *D. dubia* in Peter Lake. In Lake Constance, planktivorous fishes are abundant yet only one daphnid species, *D. hyalina*, migrates during summertime. Its slightly larger congener, *D. galeata*, never migrates but remains near the surface where it is vulnerable and experiences high mortality due to predation. This situation illustrates that fish predation alone is not an adequate predictor of migratory behavior. Other factors are often involved. In this case, *D. galeata* is able to keep pace with loss to predation by remaining near the surface where it has a faster development rate and larger brood size, permitting it higher birth rates. It is likely that, if food were not of substantially greater quality or quantity near the surface, *D. galeata* would not be able to persist. In Peter Lake, the smaller *D. dubia* remained near the surface while the larger *D. pulex* migrated to greater depths. In this case, the low apparency of *D. dubia* to fishes and its rapid population growth in the warmer surface waters probably allowed it to persist. As summer progressed, however, *D. dubia* assumed migratory patterns similar to those of *D. pulex*. There was no decline in edible algal biomass (Chapter 11) or primary production (Chapter 13) at this time. Sustained planktivory by golden shiners is the most plausible explanation for the intensification of migration by *D. dubia*.

These results, and those of other studies, portray DVM as a multifactorial phenomenon dependent upon historical contingencies. Whether DVM occurs may depend upon past fish communities and the genetic composition of the daphnid subpopulations extant in a given lake. These complexities may confound the prediction of DVM (Dodson, 1990). Continued efforts to predict and understand migration are likely to lead to progress on two fronts: the etiology of migration, and its ecosystem significance.

While migration is not likely to have a simple, global explanation, numerous testable possibilities have been raised and examinations of some of them have begun (Dodson, 1988); Bollens & Frost, 1989). Does visual predation act by changing the relative proportions of migrating and non-migrating individuals, or by inducing changes in the behaviours of all individuals? Are those individuals capable of performing nocturnal migrations the same as those that perform reverse migrations? What are the environmental conditions that make reverse migration likely or worthwhile? Which food resources make the largest food contribution to migrating organisms' fitness, how are these distributed in the water column and how do changes in this distribution affect DVM? The experimental tools are in place to answer these and other important questions about the etiology of DVM.

From an ecosystem perspective, the significance of DVM may be profound. When planktivory by fishes increases, DVM can allow large daphnids to persist, thereby delaying or mitigating ecosystem responses to fish manipulation (Chapters 8, 10 and 13; Shapiro, 1990). At a shorter time scale, DVM has potentially conflicting implications for algal growth and productivity. Migration reduces the amount of time spent grazing in warm epilimnetic waters, and may therefore decrease the impact of *Daphnia* on phytoplankton (Lampert & Taylor, 1985; Dorazio *et al.*, 1987). On the other hand, migration can increase the rate of phosphorus loss from epilimnetic waters (Wright & Shapiro, 1984; Dini *et al.*, 1987). Reduced nutrient concentrations in the epilimnion may offset any advantages to algae of lower grazing rates. The relative importance of these effects is not clear and may well be site-specific. There is no doubt, however, that DVM of *Daphnia* can be a key process in the response of lake ecosystems to fish manipulations. This behavioral process, usually studied by population biologists, has pivotal effects on food web dynamics in lake ecosystems.

Summary

Large-bodied herbivores like *Daphnia* occupy a fulcral position in the trophic cascade. Any behavior that affects their exercise of this role can potentially affect the response of phytoplankton. Diel vertical migration (DVM) is such a behavior; it affects food availability for planktivores, grazing pressure on phytoplankton and the rate of nutrient regeneration by herbivores.

The fish manipulations, especially the minnow reintroduction in

Tuesday Lake, confirmed the primary importance of visual predation in determining migratory behavior. Once planktivorous fish were removed from Tuesday Lake in 1985, formerly rare daphnids dominated the zooplankton and exhibited highly variable DVM. Low planktivory was associated with inconsistent DVM. Soon after removal of all fish and subsequent reintroduction of minnows in 1987, *Daphnia* performed strong nocturnal migrations. These intensified in amplitude over two subsequent summers until 1990, when rebounding minnow populations made *Daphnia* in this lake too rare to study.

Manipulations in Peter Lake in 1985 had little effect on daphnid DVM because minnows were quickly driven from the lake. Additions of rainbow trout (1988 and 1989) and golden shiners (1990), while affecting daphnid biomass and body length, did not alter DVM of *Daphnia pulex*. These results would make sense if migrating clones had been the sole occupants of this lake during the seven-year study and if migration did not impose substantial metabolic costs.

In Paul and Peter lakes, *Daphnia* migrated consistently under consistently low predation pressure. Food availability can also affect DVM, and there are also persistent claims for the importance of temperature stratification. In an attempt to resolve the matter, we conducted a modified $2 \times 2 \times 2$ factorial enclosure experiment featuring two levels of predation pressure (fish present/absent); two temperature regimes (4 m deep enclosure/10 m deep enclosure); and two concentrations of algal food (ambient/high). This experiment confirmed the importance of fish predation. Migration was stronger in enclosures with fish than in those without. Migrations were also stronger under food-limiting conditions. There was no evidence that exposure to cold water, permitted in deep enclosures, gave daphnids a metabolic advantage over those restricted to warm water.

Observations of DVM over this seven-year span and the results of the enclosure experiment support the predation-avoidance hypothesis. The presence of planktivorous fish induced and maintained DVM by *Daphnia* in these lakes. They also support a multifactorial etiology for DVM with food availability and distribution playing important roles.

Acknowledgements

Very helpful reviews of this chapter were written by Stanley Dodson and Bill Neill. Thanks go to the members of the UNDERC classes of 1986–9 for their assistance in the field, especially to Renato DaMatta who conducted the study of DVM in Long

Lake during summer 1988. We thank Marty Berg, Peter Leavitt and Ann St. Amand for their help. Indispensable were the many Cascade technicians. We thank the following U.W.-Madison personnel: Mike Vanni provided invaluable advice and assistance in the field; Tom Frost lent the pump and other equipment; Xi He *et al.* provided fish on demand. The Indiana Academy of Science provided partial financial support for the enclosure experiment.

10 · Zooplankton biomass and body size

Patricia A. Soranno, Stephen R. Carpenter and Xi He

Introduction

Zooplankton affect lake ecosystem processes by grazing on phytoplankton, recycling nutrients through excretion, and serving as prey for both vertebrate and invertebrate planktivores (Brooks & Dodson, 1965; Peters, 1975; Neill, 1981). Consequently, the zooplankton can be analyzed from two points of view: as a dependent variable with respect to planktivores, and as an independent variable with respect to algal community dynamics and nutrient recycling. In this chapter, we focus primarily on the responses of zooplankton biomass and body size to manipulations of fish populations. Responses of the phytoplankton to changes in the zooplankton community are addressed in Chapters 11 and 13.

Planktivorous fish feed selectively on larger and more conspicuous zooplankton (Zaret, 1980; Neill, 1984; Vanni, 1986). Zooplankton community changes associated with fish manipulations were discussed in Chapter 8. Here, we examine the implications for the total biomass and the size structure of the zooplankton, which are associated with rates of key ecosystem processes. Rates of grazing or nutrient excretion per unit biomass are proportional to the body mass of individual zooplankters (Peters, 1983). The range of particle sizes consumed by filter-feeding cladocerans is also proportional to body size (Burns, 1968).

Total rates of grazing or nutrient excretion scale directly with zooplankton biomass. Therefore, zooplankton body size and biomass indicate the potential rates of ecosystem processes, such as grazing and nutrient excretion, that are performed by the herbivorous zooplankton.

Methods

Here, we use the data base presented in Chapter 8 to analyze the mean lengths of: (1) crustaceans (cladocerans plus copepods), (2) cladocerans, the dominant grazers in these lakes (Chapter 8), and (3) *Daphnia*, which often dominate the zooplankton community, are especially vulnerable to vertebrate planktivory, and are keystone grazers. We also present the total biomass of the zooplankton community as well as the relative biomass of key groups of zooplankton, both of which have implications for ecosystem process rates.

To assess the relationships between fish populations and zooplankton, we applied transfer functions and intervention analysis (Chapter 3) to time series of fish consumption of zooplankton and zooplankton variates. Consumption by fishes was estimated from the diet, population size, and growth rates of the fish populations and temperature of the lake using the bioenergetics model (Chapter 6).

Results

Paul Lake

Seasonal variations of zooplankton length in Paul Lake were subtle. Crustacean length tended to decline throughout the summers (Figure 10.1*a*). Cladoceran length, on the other hand, showed no consistent pattern (Figure 10.1*b*). Some years exhibited midsummer declines (1986, 1987 and 1990) whereas peaks in midsummer occurred in other years (1984, 1988 and 1989). Daphnid mean lengths varied between 1 and 1.5 mm, and exhibited trends similar to cladoceran mean lengths (Figure 10.1*c*).

Total biomass of zooplankton in Paul Lake was dominated by daphnids only in 1985 and 1986, at the same time that the biomass of rotifers and copepods was lowest (Figure 10.2*a*). Few seasonal trends from the summer months are apparent for the major groups of zooplankton: cladocerans, copepods and rotifers (Figure 10.2*b–d*).

Transfer functions were used to assess possible relationships between fish consumption and several zooplankton variates (Chapter 2). For Paul Lake, no models were significant except that between fish consumption and cladoceran length (Table 10.1). However, the ten-week lag in this model is difficult to understand biologically. There were no significant correlations at any time lags between fish consumption and biomass estimates of any aggregation of zooplankton.

Peter Lake

Lengths of crustaceans, cladocerans and *Daphnia* responded to the fish manipulations (Figure 10.3*a–c*). As in Paul Lake, mean lengths of crustaceans, cladocerans, and *Daphnia* showed a general tendency to decline through the course of the summer in some of the years between 1984 and 1987, and in 1990 (Figure 10.3*a–c*). However, upon introduction of rainbow trout in 1988 and 1989 (Chapter 4), mean lengths departed from previous trends. Cladoceran and *Daphnia* mean lengths showed sustained declines in midsummer of 1988 owing to the virtual elimination of the previously dominant *Daphnia pulex*. In 1989, after additional rainbow trout were added, mean crustacean length was unusually low in early summer, but steadily increased throughout the summer. Low crustacean length in early summer was attributable to the dominance of juvenile copepods, increases in the abundance of other small taxa, and the replacement of *D. pulex* as the dominant cladoceran by *D. dubia*, a medium-sized daphnid (Chapter 8). In fact, mean lengths of both cladocerans and daphnids were consistently low during 1989 and varied little seasonally compared with the variability in lengths present in other years. Although planktivory was elevated in 1990, mean lengths

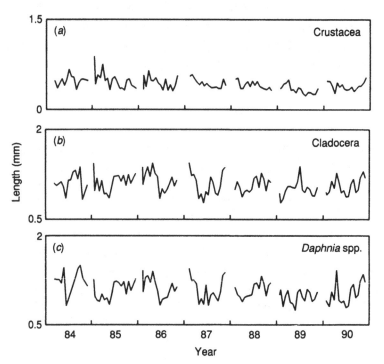

Fig. 10.1. Mean lengths of selected groups of zooplankton versus time during the summer stratified season in Paul Lake: crustaceans (all species excluding rotifers) (*a*), cladocerans (*b*), and *Daphnia* (all species) (*c*).

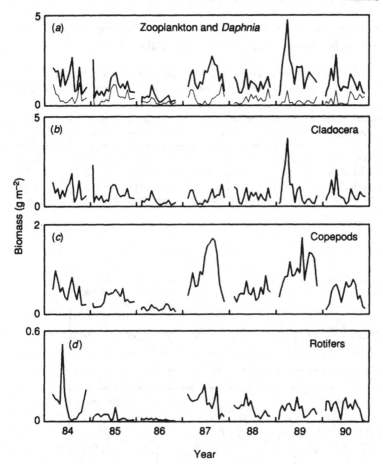

Fig. 10.2.   Paul Lake: zooplankton biomass versus time during the summer stratified season for total zooplankton biomass (thick line) and *Daphnia* biomass (thin line) (*a*), and selected functional groups: cladocerans (*b*), copepods including all adults and juveniles (*c*), and rotifers (*d*).

of zooplankton were high in the first half of the summer because of a major increase in *Daphnia pulex* density (Chapter 8). Furthermore, mean lengths of all species of *Daphnia* were high in early 1990, perhaps a result of low planktivory in spring and early summer before planktivory became intense (Chapter 6). The decline in mean length in all groups in the latter half of the summer coincides with reductions in the *D. pulex* population (Chapter 8).

Total zooplankton biomass showed little deviation from that of Paul Lake during the years that bass were present and dominated the fish

Table 10.1. *Coefficients of transfer functions with fish consumption of zooplankton as the independent variable for Paul Lake*

The delay, in weeks, is given for each numerator and denominator term in the model, in addition to the estimate, the standard error and the T ratio (significance at the 5% level is 1.96).

Dependent variable	Delay	Term	Estimates	s.e.	T
cladoceran length	10	ω_0	21.7	8.30	2.61
log *Daphnia* biomass	0	ω_0	0.037	0.022	1.73
Daphnia length	0	ω_0	−11.94	8.44	1.40

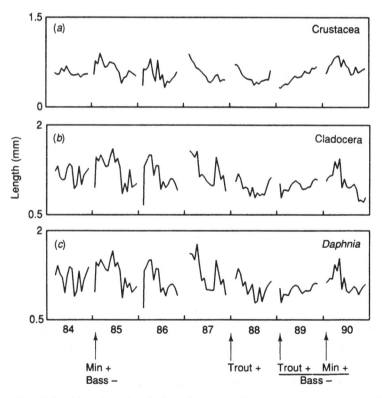

Fig. 10.3. Mean lengths of selected groups of zooplankton versus time during the summer stratified season in Peter Lake: crustaceans (all species excluding rotifers) (*a*), cladocerans (*b*), and *Daphnia* (all species) (*c*).

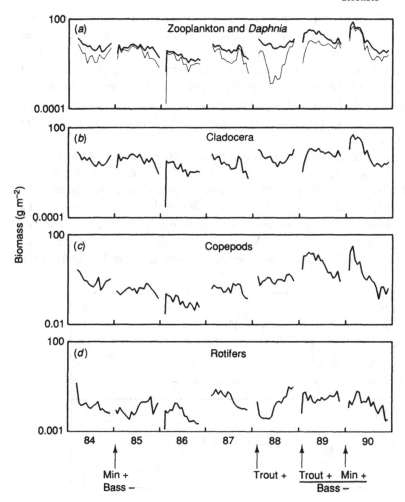

Fig. 10.4. Peter Lake: zooplankton biomass versus time during the summer stratified season for total zooplankton biomass (thick line) and *Daphnia* biomass (thin line) (*a*), and selected functional groups: cladocerans (*b*), copepods including all adults and juveniles (*c*), and rotifers (*d*). Note logarithmic scale of *Y* axis.

assemblage (1984–7), even when the 1985 year class was most plankti-vorous (late 1985 and 1986) (Figure 10.4*a*). Daphnids generally made up a large portion of the total biomass in 1985 and 1986, but only in certain times of the year in 1984 and 1987. In 1988, size-selective predation by the introduced rainbow trout reduced *Daphnia* populations to the lowest levels of the study. However, small zooplankton increased, resulting in an unusual increase in total biomass (Chapter 8). In 1989, total biomass increased again, despite moderate levels of vertebrate planktivory. This change in biomass is due to the remarkable success of *Daphnia dubia* and substantial increases in other small bodied taxa (Figure 10.4*c*,*d*) (Chapter 8). Increases in zooplankton biomass of even greater magnitude occurred in early 1990 despite elevated levels of planktivory by fishes. Most taxa declined in the latter half of the summer, however, perhaps because of a combination of food limitation and sustained planktivory by fishes (Chapter 8).

Transfer functions between fish consumption and zooplankton variates revealed several significant relations (Table 10.2). Significant models were found between fish consumption rates and biomass of crustaceans, cladocerans, and *Daphnia*. All correlations were positive at a lag of 0 weeks, which means that consumption was high when zooplankton biomass was high. Apparently, the fishes responded opportunistically to high biomass of zooplankton (Hodgson & Kitchell, 1987).

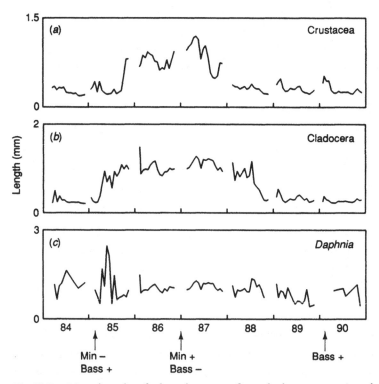

Fig. 10.5. Mean lengths of selected groups of zooplankton versus time during the summer stratified season in Tuesday Lake: crustaceans (all species excluding rotifers) (*a*), cladocerans (*b*) and *Daphnia* (all species combined) (*c*).

Table 10.2. *Coefficients of transfer functions with fish consumption of zooplankton as the independent variable for Peter Lake*

The delay, in weeks, is given for each numerator and denominator term in the model, in addition to the estimate, the standard error and the T ratio (significance at the 5% level is 1.96).

Dependent variable	Delay	Term	Estimates	s.e.	T
log crustacean biomass	0	ω_0	5.94	1.60	3.72
log cladoceran biomass	0	ω_0	7.08	1.67	4.24
log *Daphnia* biomass	0	ω_0	5.93	2.33	2.55
crustacean length	0	ω_0	1.42	0.56	2.55
		ω_2	-1.07	0.56	1.92
cladoceran length	6	ω_0	-2.74	0.97	2.82
		ω_0	-0.57	0.23	2.42
Daphnia length	11	ω_0	-2.64	0.98	2.68

Cladoceran length responded negatively to increases in fish consumption at a lag of 6 weeks, suggesting size-selective predation by fishes. *Daphnia* length responded negatively to fish consumption at a lag of 11 weeks. This long lag is difficult to understand biologically.

Tuesday Lake

The zooplankton of Tuesday Lake responded within a few weeks to the reduction of planktivores in early summer 1985, and more slowly to their gradual recovery from 1987 to 1990 (Chapters 6 and 8). Pronounced responses occurred in the mean lengths of crustaceans and cladocerans (Figure 10.5a,b). Within approximately six weeks of the removal of the minnow population in early summer of 1985, cladoceran mean length increased and stayed high for the remainder of the season and the following three years. This increase was a result of the replacement of smaller cladoceran species by large ones (Chapter 8). Crustacean length, however, did not increase until the final two weeks of the 1985 sampling season when smaller copepods were declining in density. Changes in daphnid mean lengths were difficult to detect since population densities in June and July were so low that estimates of mean length were quite variable (Figure 10.5c). Mean sizes of the *Daphnia* present in 1985 were higher and more variable than in the following years.

The return of planktivores to the system in May 1987 caused only

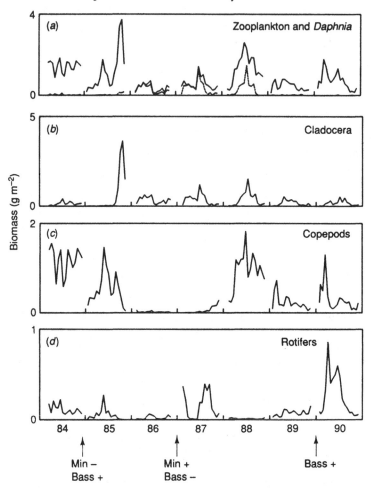

Fig. 10.6. Tuesday Lake: zooplankton biomass versus time during the summer stratified season for total zooplankton biomass (thick line) and *Daphnia* biomass (broken line) (*a*), and selected functional groups: cladocerans (*b*), copepods including all adults and juveniles (*c*) and rotifers (*d*).

slight decreases in crustacean length by the end of that summer. Sustained decreases occurred the following year, when cladoceran biomass declined relative to other taxa. Cladoceran length, however, did not decline until the end of 1988. At that time, planktivory by fishes had returned to premanipulation levels (Chapter 6) and the daphnid populations were replaced by the small cladocerans that were present in 1984, *Bosmina* and *Diaphanosoma* (Chapter 8).

Table 10.3. *Coefficients of transfer functions with fish consumption on zooplankton as the independent variable for Tuesday Lake*

The delay, in weeks, is given for each numerator term in the model, in addition to the estimate, the standard error and the T ratio (significance at the 5% level is 1.96).

Dependent variable	Delay	Term	Estimates	s.e.	T
log cladoceran biomass	0	ω_0	0.54	0.19	2.84
		ω_7	0.69	0.20	3.51
crustacean length	0	ω_0	−0.15	0.06	2.63
cladoceran length	6	ω_0	−0.15	0.06	2.41

Total zooplankton biomass responded unexpectedly to decreased planktivory levels (Figure 10.6a). Even though planktivory by fishes was reduced to nearly zero in 1985 and 1986, total biomass of zooplankton scarcely responded. Except for the initial increase in biomass at the end of 1985, total biomass was lowest during the years that planktivory was lowest, 1986 and 1987. During this time, large daphnids dominated the community, to the exclusion of copepods and rotifers (Figure 10.6a–d). Total biomass did not increase again until 1988, when planktivory approached premanipulation levels. However, this increase was partly due to major increases in copepods that were not sustained the following years. By 1990, three years after the return of the resident minnow populations, total biomass was only slightly lower than in 1984, although the community composition was different (Chapter 8).

Transfer functions between planktivory rates and zooplankton variates revealed significant relations similar to those for Peter Lake (Table 10.3). Cladoceran biomass was positively related to planktivory rate, suggesting that consumption was high when cladocerans were abundant. Also, anomalous increases in fish consumption led to declines in cladoceran length. The negative relation between fish consumption and crustacean length at a lag of 0 may indicate rapid compensatory shifts in the crustacean community.

Intervention analysis (Chapter 3) was also used to test for shifts in the time series of total zooplankton biomass or cladoceran length following both manipulations. Following the removal of minnows, a significant increase in cladoceran length occurred (T-ratio = 5.01). A significant decrease in cladoceran length followed the return of the minnow populations (T-ratio = 4.54). No significant effects of the two manipulations were detected on zooplankton biomass (T-ratios were 0.83 and 0.43 for the 1985 and 1987 manipulations, respectively).

Discussion

Changes in planktivory by fishes altered the body sizes, and sometimes the biomass, of the zooplankton. However, the direction of biomass responses was sometimes contrary to expectations. On the basis of models, we expected that total biomass and mean lengths of herbivores would be inversely related to planktivory by fishes (Carpenter & Kitchell, 1987; Chapter 16). More often than not, total biomass responded oppositely to predictions.

Mean lengths of herbivores, especially cladocerans, were a much more reliable indicator of planktivory by fishes. Crustacean length was not a good indicator of fish predation when juvenile copepods were abundant, because large numbers of these small crustaceans masked effects of planktivory on the less numerous large crustaceans. Daphnid length was not a good indicator of fish predation, because it could be large even when daphnids were rare. Cladoceran length was a more satisfactory indicator of fish planktivory because large daphnids tended to be abundant when planktivory was low, and be replaced by smaller cladocerans as planktivory increased.

Since cladocerans are the dominant herbivores in many lakes, changes in their dynamics have important implications for primary producers in the pelagic zone. Their capacity to graze algae and excrete nutrients is strongly size-dependent (Peters, 1975; Peters & Downing, 1984; Knoechel & Holtby, 1986). Although cladoceran mean lengths have been shown to vary seasonally owing to a combination of factors that include water temperature, food quality and quantity, and size-selective predation (Threlkeld, 1979; Zaret, 1980), we found no consistent seasonal patterns in cladoceran length, even in unmanipulated Paul Lake. Whereas in some years cladoceran lengths exhibited midsummer declines, in most they did not. The shifts in cladoceran mean length that occurred in Peter and Tuesday lakes following fish manipulations demonstrate the important role of size-selective planktivory in determining cladoceran lengths. When planktivory is low, as in Paul Lake, cladoceran length is less variable and shows no consistent seasonal trends. Relations between cladoceran length and algal dynamics are explored in Chapters 11 and 13.

In both Peter and Tuesday Lakes, transfer functions fitted to time series of fish consumption and cladoceran length revealed significant negative effects on mean length after unusually high episodes of vertebrate planktivory. The time lag of six weeks is approximately two to three generations of cladocerans (Lei & Armitage, 1980). The following scenario suggests how such lags may occur. Densities of herbivores increase in the presence of elevated algal resources. The transfer functions show that fish feed opportunistically on the largest animals when biomass of herbivores is high. Initially, elevated birth rates can offset mortality from predation. However, as food levels are depleted and the largest grazers are selectively preyed upon, birth rates cannot compensate for the increased mortality due to predation. The birth rates of the remaining small-bodied individuals may actually increase as competition with the larger animals is alleviated. For example, in Peter Lake, birth rates of daphnids increased five weeks after pulses of fish consumption (Chapter 8). As a result, the total biomass of the population will not necessarily be negatively affected, but the remaining individuals will be those that reproduce at a smaller size, have a smaller overall body size, and possibly have elevated birth rates. Further evidence for this scenario lies in the significant relations between cladoceran lengths and algal variates such as cell size and biomass of edible algae (Chapter 11) which suggests a tight coupling of grazers and algae.

Although most fluctuations in cladoceran length appear consistent with the above scenario, several unusual responses in mean length occurred. For example, in Peter Lake in 1990, the mean lengths of crustaceans, cladocerans and *Daphnia* were high even though planktivory by fishes was high (Chapter 6). Diel vertical migration by the dominant grazer, *D. pulex*, allowed it to withstand high levels of planktivory by golden shiners and to increase in mean length as well as biomass, at least for a few weeks.

Total biomass of zooplankton proved to be a complex and unpredictable response variable. In Peter Lake, the total biomass of zooplankton deviated little from long-term means during the years that largemouth bass dominated the fish assemblage. Total zooplankton biomass may not respond to variations in planktivory by unexploited piscivore populations, even when an exceptionally large year class is recruited (Chapter 8). However, upon introduction of rainbow trout in 1988 and 1989 and golden shiners in 1990, zooplankton biomass actually increased in spite of elevated planktivory. These increases are attributable to changes in the zooplankton and invertebrate predator communities (Fig. 10.7), especially: (1) the elimination of the dominant herbivore, *D. pulex*, and its replacement by the smaller *D. dubia*; (2) the virtual elimination of the invertebrate predator, *Chaoborus*; (3) compensatory increases in smaller taxa (probably caused by decreased competition with large daphnids and release from invertebrate predation); and (4) in 1990, the reappearance of *D. pulex*.

Total zooplankton biomass in Tuesday Lake responded in an equally complex manner. Upon removal of the minnow populations in 1984, total biomass increased at the end of the first season as large cladocerans replaced the previously dominant smaller cladocerans. However, this increase in biomass was not sustained the following year. Biomass

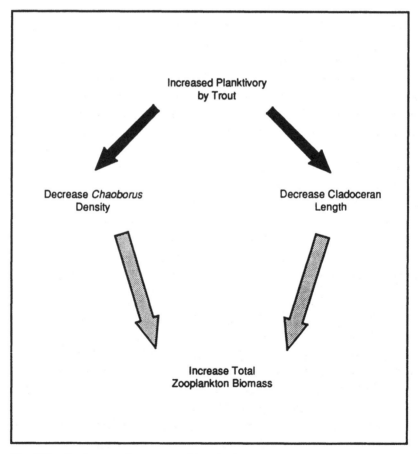

Fig. 10.7. Summary of responses of zooplankton and *Chaoborus* in Peter Lake to the addition of rainbow trout in 1988–9.

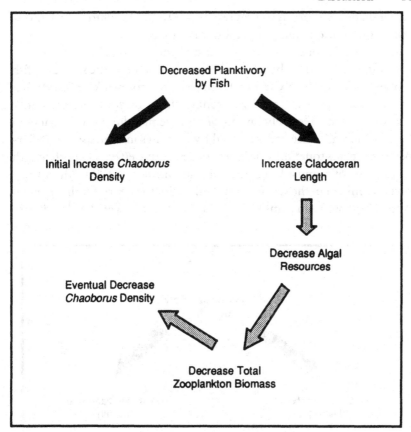

Fig. 10.8. Summary of responses of zooplankton in Tuesday Lake to the addition of largemouth bass in 1985–6.

actually declined, owing to substantial declines in edible algae (Figure 10.8) (Chapters 8 and 11). Declines in biomass of smaller species of zooplankton may have contributed to the declines in *Chaoborus* populations that occurred at the same time (Chapter 7).

Total zooplankton biomass in Tuesday Lake was slow to recover following reintroduction of minnows in 1987 and did not approach premanipulation levels until well into 1988, when a smaller species of *Daphnia* dominated the community (Chapter 8) and cladoceran length declined (Figure 10.9). The continued declines in *Chaoborus* densities may also have contributed to increases in small zooplankton taxa (Chapter 7). By 1990, although total zooplankton biomass was similar to that in 1984, the zooplankton community structure was different (Chapter 8). The response of total zooplankton biomass to changes in vertebrate planktivory appeared responsive to food resources and interspecific competition, and not planktivory alone.

Statistical relations between fish consumption and cladoceran biomass and length are obscured by the fact that they depend on the species of fish involved (Chapter 8). YOY bass briefly affected zooplankton length but not biomass. Rainbow trout had stronger effects on zooplankton length because they fed effectively on *D. pulex* in the metalimnion. Declines in *D. pulex* biomass were compensated by increases in biomass of smaller-bodied zooplankton. Golden shiners had negative effects on both length and biomass of zooplankton through the summer of 1990. All of these effects are mixed in the statistical analysis, with the net result that general relationships between planktivory and zooplankton were hard to detect.

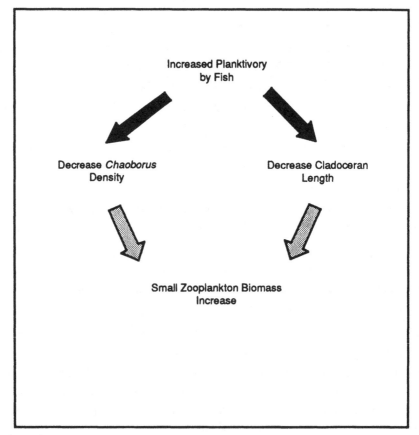

Fig. 10.9. Summary of responses of zooplankton in Tuesday Lake to the reintroduction of minnows in 1988–90.

Nevertheless, intervention analysis revealed significant shifts in clado-ceran length, but not total biomass, following the manipulations in Tuesday Lake.

In subsequent chapters, we will use cladoceran length as an index of size-selective predation. The analyses detailed above substantiate that choice in Paul, Peter and Tuesday Lakes. However, other indicators may be preferable in other lakes. For example, a trophic cascade in Lake Mendota, Wisconsin, depends on whether *Daphnia pulicaria* or *Daphnia galeata* dominates the zooplankton (Lathrop & Carpenter, 1992; Vanni, Carpenter & Luecke, 1992*a*). *D. galeata* is about the same total length as *D. pulicaria*. However, *D. galeata* is a much slimmer animal and a translucent helmet makes up part of its body length. Consequently *D. galeata* is less apparent to fishes and a less effective grazer than *D. pulicaria*. In Lake Mendota, size-selective predation is best indicated by mean zooplankter mass, modified length (e.g. length from the eyespot to the base of the tail spine), or the ratio of *D. pulicaria* to *D. galeata*.

This chapter shows many potential links between population interac-tions, community structure, and ecosystem processes. Shifts in fish predation triggered complex series of indirect effects including changes in intensity of planktivory by invertebrates, food limitation, and compe-tition among herbivores (Chapter 8). Time lags that are inherent in many of these interactions have the potential to lead to oscillations and cycles that may be destabilizing (Murdoch & McCauley, 1985) and lead to even more complex dynamics at the ecosystem level.

Summary

This chapter describes the responses of zooplankton biomass and body size to the fish manipulations. These variates are correlated with key ecosystem process rates, including grazing rate and nutrient recycling rate.

Biomass and body size of zooplankton fluctuated from week to week in Paul Lake, but showed no sustained trends. In the manipulated lakes, mean lengths of cladocerans were the most sensitive indicators of vertebrate planktivory.

In Peter Lake, declines in zooplankton length occurred in response to the trout additions in 1988 and 1989. When golden shiners were added to the lake, cladoceran length actually increased in the first part of the summer of 1990. In this case, vertical migration by *Daphnia pulex* allowed it to coexist with high levels of planktivory for at least part of the summer. In Tuesday Lake, sustained increases in cladoceran mean length occurred when minnows were absent from the lake and largemouth bass were present. This size shift was reversed over two years after planktivores were reintroduced into the lake.

Total zooplankton biomass dynamics often contradicted the expectations of the cascade hypothesis. In Peter Lake, total biomass increased during the years that planktivory was highest (1988 and 1990). The increase was associated with community shifts due in part to decreased predation by *Chaoborus* and decreased density of large *Daphnia* (Chapters 7 and 8). In Tuesday Lake, total zooplankton biomass declined when planktivory was low (1986 and 1987), apparently as a result of severe food limitation and competitive exclusion of smaller zooplankton species by *D. pulex*.

Crustacean or cladoceran body size, but not biomass, were useful indicators of trophic interactions. Biomass dynamics are confounded by fluctuations in algal resources. Increased planktivory by fishes, on the other hand, is generally followed by reductions in body size of cladocerans and other crustaceans. These size shifts have important implications for ecosystem processes, because smaller cladocerans have higher rates of grazing and nutrient excretion per unit mass, but feed on a smaller range of particle sizes.

Acknowledgements

We thank David Armstrong, Bart DeStasio, Jim Kitchell and Peter Leavitt for helpful reviews of earlier drafts.

11 · Phytoplankton community dynamics

Stephen R. Carpenter, John A. Morrice, James J. Elser,
Ann St. Amand and Neil A. MacKay

Introduction

Limnologists have long appreciated the potential importance of grazing
to phytoplankton community composition (Reynolds, 1984a). At
certain times during seasonal succession, grazers have clear-cut effects on
algal assemblages (Lampert et al., 1986; Sommer et al., 1986; Vanni &
Temte, 1990). Grazers affect phytoplankton communities through
several mechanisms, including direct suppression of edible algae, enhancement of inedible algae via nutrients excreted by grazers, and shifts in
the outcome of competition caused by grazer effects on nutrient supply
ratios (Sterner, 1989). The complexity of the mechanisms may explain
the multifarious and individualistic outcomes of field experiments on
zooplankton–phytoplankton interactions (Lehman & Sandgren, 1985;
Bergquist & Carpenter, 1986; Elser et al., 1986a; Vanni & Temte, 1990).

Organism size may provide important organizing principles for
understanding zooplankton–phytoplankton interactions (Peters &
Downing, 1984). With the exception of calanoid copepods capable of
feeding selectively on certain algae, the size of the herbivore largely
determines the range of algal sizes upon which it can feed (Burns, 1968;
Reynolds, 1984a). Feeding rates also depend on herbivore size (Peters &
Downing, 1984). Size-selective predation by fishes strongly influences
size structure of the herbivore community (Brooks & Dodson, 1965;
Chapter 8). Therefore, size-structured interactions of zooplankton and
phytoplankton have important implications for the trophic cascade
(Bergquist et al., 1985).

We developed a simulation model to determine the potential effects of
herbivore size on phytoplankton size structure, biomass, and primary
production (Carpenter & Kitchell, 1984; Chapter 17). The model
predicted that increased body size or biomass of zooplankton would lead

to dominance by larger phytoplankton. Because larger phytoplankton have lower metabolic rates per unit biomass (Reynolds, 1984a), these structural changes in the phytoplankton community have significant implications for primary production (Carpenter & Kitchell, 1984; Chapter 17).

In this first of three chapters on responses of phytoplankton to our whole-lake experiments, we focus on changes in community composition and size structure in the epilimnion. Metalimnetic communities are contrasted with those of the epilimnion in Chapter 12. In Chapter 13, we turn to the responses of ecosystem processes mediated by phytoplankton, including primary production and nutrient limitation.

This chapter has two goals. First, we examine the effects of the food web manipulations on total phytoplankton biomass, biomass of selected groups of algae, and phytoplankton size structure. Second, we test the predictions of the simulation model that increased body size or biomass of zooplankton shifts the algal community to one dominated by larger phytoplankton. These results allow us to evaluate the extent to which size can serve as a useful organizing principle for interpreting these food web manipulations.

Methods

Weekly samples of the epilimnetic algae of Paul, Peter and Tuesday Lakes were collected during summer stratification from 1984 to 1990. Collection procedures have been detailed elsewhere (Soranno, 1990) and will be summarized briefly here. Water samples from depths corresponding to 100%, 50% and 25% of surface light irradiance were collected and pooled. From 1984 to 1987, the pooled samples were preserved in Lugol's iodine solution, settled in towers, and counted at 200 × magnification with an Olympus inverted microscope. In 1988 we adopted a procedure in which samples were preserved with glutaraldehyde, filtered, and mounted in a water-soluble methacrylic resin (Crumpton, 1987). This method allowed us to store a permanent record of mounted phytoplankton samples from the lakes. We have also archived color photographs of most of the algal taxa to document our identifications. Samples were counted on an Olympus BH-2 microscope at 200 × and 400 × magnifications. Counts at 400 × were added to enumerate nanoplankton in more detail.

Phytoplankton biomass was expressed as biovolume, in microliters of cell volume per liter of lake water (μl l^{-1}). Assuming a cell density of

10^3 kg m^{-3} (a reasonable approximation for many algae) (Reynolds, 1984a), biovolume of 1 μl l^{-1} is equivalent to a wet biomass of 1 mg l^{-1}. Biovolumes were calculated for algal protoplasm exclusive of loricae and sheaths (Elser et al., 1986a). Calculations were based on individual measurements of a subsample of cells or colonies of each algal taxon (Elser et al., 1986a).

Phytoplankton were identified to species whenever possible. While this level of detail provided valuable information, certain broad trends were more evident at coarser levels of aggregation. Also, more aggregated data are less dependent than species-level data on the expertise of the microscopist. Four microscopists counted the samples presented in this book: J. Elser (1984–6), N. MacKay (1987), A. St. Amand (1988–9) and J. Morrice (1990). While we made every effort to sustain consistent procedures throughout the seven-year period, it is likely that there were some inconsistent identifications at the species level, especially in the nanoplankton. However, at coarser levels of aggregation, data from different microscopists were highly consistent. Therefore, we devised a functional group classification scheme based on morphological properties of the algae such as motility, investment, colony type and size (Soranno, 1990). These properties are correlated with important ecological processes such as floating and sinking, diffusion limitation of physiological rates, and susceptibility to grazing (Reynolds, 1984a).

This chapter will emphasize changes in functional groups and algal size distributions. Dynamics of selected species, known to be identified consistently throughout the study, will be discussed. As an index of algal size, we calculated weighted algal length as

$$\sum(BV_i\ GALD_i)/\sum(BV_i),$$

where BV_i is the biovolume of species i, $GALD_i$ is the greatest axial linear dimension of species i, and the summations are taken over all species. The weighted length is the length of the average unit of biovolume, and represents the centroid of the biovolume length distribution (Elser et al., 1988). The mean length of all algae is always quite small because nanoplankton are dominant numerically. Consequently, mean length shows little week-to-week variation. Weighted length is sensitive to shifts in the biovolume length distribution, and sometimes varied considerably among samples.

Time series analyses (Chapter 3) were used to test for effects of herbivores on the phytoplankton communities. Transfer functions were fitted to test for relations between cladoceran length and phytoplankton biovolume in all three lakes. Cladoceran length was used as an index of grazer community structure that is influenced by fish predation and not seriously confounded by changes in resource levels (Chapter 10). To test the hypothesis that increases in grazer size and biomass increase phytoplankter size, we calculated transfer functions for effects of cladoceran length and biomass on weighted length of phytoplankton. In Tuesday Lake, where manipulations caused relatively abrupt changes in the herbivore community (Chapter 8), we applied intervention analysis to test for nonrandom changes in biovolume of algal functional groups and weighted length of phytoplankton.

Results

Most of the algal biovolume in all three lakes was contributed by four functional groups: dinoflagellates, colonial chrysophytes, gelatinous colonies, and edible algae less than 20μm in length. The most interesting algal dynamics involved a relatively small number of taxa from these functional groups (Fig. 11.1). With respect to the community classifications of Reynolds (1984b), the phytoplankton most closely resembles the mesotrophic assemblage. At certain times, taxa characteristic of eutrophic lakes in summer (*Anabaena, Aphanocapsa, Ceratium, Microcystis*) dominated the phytoplankton. The remainder of this chapter will examine changes in the four dominant functional groups, total biovolume, and algal size.

Paul Lake

Total algal biovolume in the reference lake exhibited no unusual trends during the study (Fig. 11.2a). The phytoplankton community displayed no dramatic species or size shifts (Fig. 11.2b–e). The biovolume maximum occurred in mid or late summer, except in 1985, and was often preceded by a period of low biovolume. In 1987, the total biovolume of Paul Lake was only slightly higher than usual but the successional pattern was unusual, with blooms of several groups of algae reaching their highest biomass of the seven year study (Fig. 11.2a). Both Peter (Fig. 11.4) and Tuesday (Fig. 11.6) Lakes had relatively large algal blooms in 1987 but these were not as pronounced as those in Paul Lake.

Dinoflagellates were rarely a dominant group in the Paul Lake phytoplankton community. *Ceratium* sp. was the most abundant dinoflagellate in Paul Lake. On the date of its peak abundance in each year

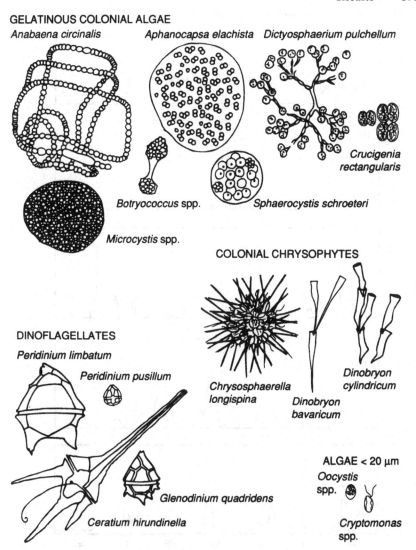

GELATINOUS COLONIAL ALGAE

Anabaena circinalis

Aphanocapsa elachista

Dictyosphaerium pulchellum

Crucigenia rectangularis

Botryococcus spp.

Sphaerocystis schroeteri

Microcystis spp.

COLONIAL CHRYSOPHYTES

DINOFLAGELLATES

Peridinium limbatum

Peridinium pusillum

Chrysosphaerella longispina

Dinobryon bavaricum

Dinobryon cylindricum

ALGAE < 20 μm

Oocystis spp.

Glenodinium quadridens

Ceratium hirundinella

Cryptomonas spp.

Fig. 11.1. The predominant phytoplankton taxa referred to in the text, arranged by functional group.

except 1985, *Ceratium* sp. accounted for over 25% of total algal biovolume and reached a maximum of 80% in early June 1987.

The most common colonial chrysophytes in Paul Lake were *Dinobryon* sp. and *Chrysosphaerella longispina*. These species were present in low numbers throughout the study. Large blooms of colonial chrysophytes were infrequent and of short duration (Fig. 11.2c). The only years

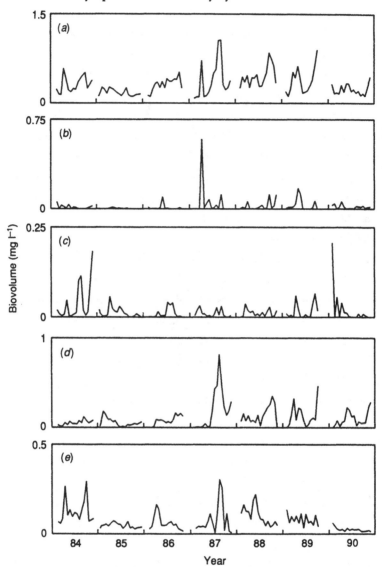

Fig. 11.2. Paul Lake: phytoplankton biovolume versus time for total biovolume (*a*) and selected functional groups: dinoflagellates (*b*), colonial chrysophytes (*c*), gelatinous colonies (*d*) and edible algae (*e*), defined as cells < 20 μm in length. Data are shown for the summer stratified season in each year.

Table 11.1. *Transfer functions for algal groups in Paul Lake*

Except where indicated, the input variable was mean cladoceran length. For each dependent variate, the delay, model terms, estimates, standard errors, and T ratios are presented. Values of T greater than 1.96 indicate significant anomalies in the dependent variate after anomalies in cladoceran length. All noise models were AR(1), except for dinoflagellates where no noise model was necessary.

Response variate	Delay	Term	Estimate	s.e.	T ratio
total biovolume	3	ω_0	-0.292	0.132	2.21
dinoflagellates	3	ω_0	-0.023	0.049	0.48
colonial chrysophytes	1	ω_0	0.024	0.015	1.60
gelatinous colonies	4	ω_0	-0.154	0.073	2.11
edible algae (<20 μm)	0	ω_0	0.049	0.042	1.17
weighted length	1	ω_0	38.31	17.92	2.14
weighted length[a]	2	ω_0	11.85	4.73	2.50

Note:
[a]Input variable is cladoceran biomass.

in which colonial chrysophytes made up more than 25% of total biovolume on any sampling date were 1984, 1987 and 1990. In May 1990, a bloom of *Dinobryon* sp. and *Tabellaria* sp. (a diatom) accounted for 75% of total biovolume.

The most abundant group of algae in Paul Lake was the gelatinous colonies. While in some years these algae were dominant throughout the sampling season, they usually reached their peak in mid to late summer (Fig. 11.2*d*). During this time of year gelatinous colonies dominated the algal community. The species composing this group are the green algae *Sphaerocystis schroeteri* and *Gloeocystis* sp. and the blue green algae *Anabaena* sp. and *Nostoc* sp. The large peak in biovolume in 1987 was a bloom of *Sphaerocystis* sp. (Fig. 11.2*d*).

Paul Lake has a relatively constant population of edible nanoplankton. Flagellates such as *Cryptomonas* sp. are the most important component of the algae <20 μm in Paul Lake (Fig. 11.2*e*). *Oocystis* sp. were also important nanoplankters. The large peak in 1987 (Fig. 11.2*e*) was due to *Chrysochromulina* sp., a species that was not prevalent in other years, and *Crucigenia* sp. The remaining peaks can be attributed to *Oocystis* sp. and *Cryptomonas* sp.

Size distributions of the phytoplankton community were computed for combined samples for each year (Fig. 11.3). While there was some year-to-year variation in the algal size distribution, there are no clear trends.

Time series analyses detected some significant relations between

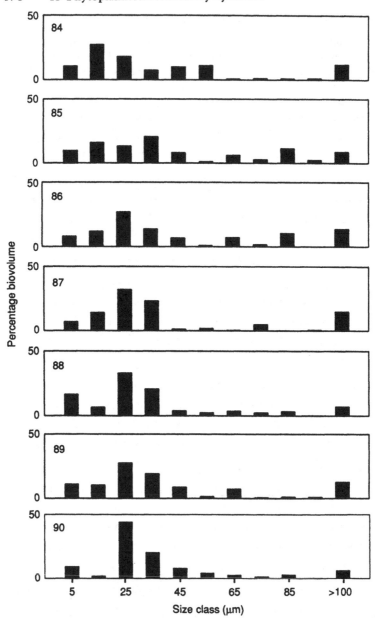

Fig. 11.3. Size distributions of algal biovolume in Paul Lake for the summers of 1984–90. Vertical axis is percentage of biovolume summed over all sampling dates in June–August; horizontal axis is midpoint of size class.

cladoceran length and biovolume of functional groups in Paul Lake (Table 11.1). Anomalous increases in cladoceran length were followed by anomalous decreases in total algal biovolume and biovolume of gelatinous colonies. These relations indicate significant negative effects of grazers.

Grazers had significant effects on weighted length of phytoplankton (Table 11.1). Anomalous increases in cladoceran length and biomass were followed by anomalous increases in weighted length of phytoplankters. These relationships indicate that increases in grazer length or biomass shifted the algal size distribution toward larger phytoplankters.

Peter Lake

Grazers in Peter Lake were affected by fish manipulations in 1985 (minnows were added and there was a strong bass year class), 1988 and 1989 (trout were added), and 1990 (piscivores were removed and minnows were stocked) (Chapter 10). This section will focus on the shifts in the phytoplankton that followed these manipulations.

Planktivory by a strong cohort of young-of-the-year largemouth bass caused a rapid decline of *Daphnia* size in late summer 1985 (Carpenter *et al.*, 1987) (Chapter 10). This event was followed by a bloom of the gelatinous colonial green alga *Sphaerocystis schroeteri*, resulting in the highest observed algal biovolume of the seven-year study (Fig. 11.4*a*). The increase in biovolume of algae < 20 μm (Fig. 11.4*e*) was primarily small colonies of *S. schroeteri*. Dinoflagellates and colonial chrysophytes also reached their annual maxima at this time (Fig. 11.4*b,c*).

In 1986, 1987 and early 1988 the zooplankton community was dominated by large *Daphnia* (Chapter 8). This period saw a relatively stable phytoplankton community with low total biovolume (Fig. 11.4*a*). Populations of all functional groups of phytoplankton were low in 1986 (Fig. 11.4*b–e*). A moderate bloom of *Dinobryon* sp. occurred in May 1987 and a large bloom of the gelatinous colonial blue green alga *Aphanocapsa* sp. came at the end of the summer (Fig. 11.4*c,d*).

In 1988, 2000 rainbow trout fingerlings were introduced into Peter Lake to increase planktivory (Chapter 2). In midsummer the mean cladoceran length and total *Daphnia* biomass declined (Chapter 8). This event was accompanied by a bloom of *Aphanocapsa* sp. The total algal biovolume in 1988 was higher than in the previous two years (Fig. 11.4*a*). There was also a sharp increase in algae < 20 μm in length (Fig. 11.4*e*).

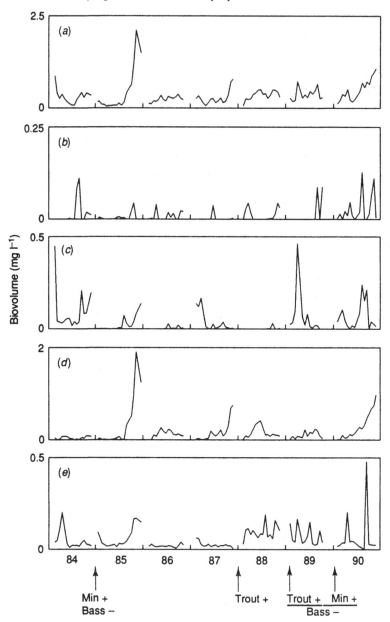

Fig. 11.4. Peter Lake: phytoplankton biovolume versus time for total biovolume (*a*) and selected functional groups: dinoflagellates (*b*), colonial chrysophytes (*c*), gelatinous colonies (*d*) and edible algae (*e*), defined as cells <20 μm in length. Data are shown for the summer stratified season in each year.

Rainbow trout of a larger size were stocked in 1989 (Chapter 2). The beginning of the summer saw a high *Daphnia* biomass but low mean cladoceran length resulting from a species shift from large *Daphnia pulex* to smaller *Daphnia dubia* (Chapter 8). *Daphnia* biomass decreased throughout the summer (Chapter 8). Total algal biovolumes in 1989 were similar to those in 1988 (Fig. 11.4*a*). There was a large bloom of *Dinobryon* sp. early in the summer (Fig. 11.4*c*). The dinoflagellate *Ceratium* sp. appeared in small numbers late in the summer (Fig. 11.4*b*). There was a gradual increase in the biovolume of gelatinous colonial algae throughout the summer (Fig. 11.4*d*).

Twenty thousand golden shiners were introduced to Peter Lake in 1990 (Chapter 2). At the time of stocking *Daphnia* biomass was at an unprecedented high and did not decline until midsummer (Chapter 8). Total algal biovolume increased through the summer as *Daphnia* biomass declined (Fig. 11.4*a*). The early summer algal community was composed of the colonial chrysophyte *Dinobryon* sp. and the small *Oocystis* sp. After the midsummer *Daphnia* decline there was a second, larger bloom of *Dinobryon* sp. (Fig. 11.4*c*). Late summer *Ceratium* sp. populations in Peter Lake were the highest observed during the study (Fig. 11.4*b*) but accounted for less than 25% of total algal biovolume. A large population of the gelatinous colonial alga *Botryococcus* sp. developed late in the summer (Fig. 11.4*d*). The increase in algae < 20 μm (Fig. 11.4*e*) was mainly due to small *Botryococcus* sp. colonies.

In periods of high grazing pressure, the algal size distribution was centered around the 30–40 μm size class (1985, 1986, 1987 and 1990) (Fig. 11.5). The prominence of this size class reflects the dominance of gelatinous colonial algae in Peter Lake during periods of increased grazing pressure. The different gelatinous colonial algae were usually about 30–40 μm in length. In 1990, however, the population of the colonial alga *Botryococcus* sp. grew through a range of 10–50 μm over the course of the summer (Fig. 11.5). The integrated biovolume of *Botryococcus* sp. accounted for a majority of the Peter Lake biovolume in 1990. In 1984, 1988 and 1989 other groups of algae attained dominance at certain times. Consequently the relative importance of the gelatinous colonial algae was diminished, and the distribution of size classes broadened. The colonial chrysophytes *Dinobryon* sp. and *Chrysosphaerella* sp. and the dinoflagellate *Ceratium* sp. are in the > 100 μm size class (Fig. 11.6).

Time series analyses revealed many significant relations between cladoceran length and phytoplankton biovolume (Table 11.2). Anomalous increases in cladoceran length were followed by anomalous

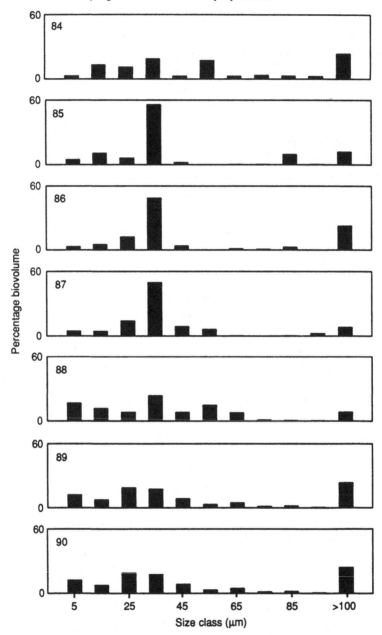

Fig. 11.5. Size distributions of algal biovolume in Peter Lake for the summers of 1984–90. Vertical axis is percentage of biovolume summed over all sampling dates in June–August; horizontal axis is midpoint of size class.

Table 11.2. *Transfer functions for algal groups in Peter Lake*

Except where indicated, the input variable was mean cladoceran length. For each dependent variate, the delay, model terms, estimates, standard errors, and T ratios are presented. Values of T greater than 1.96 indicate significant anomalies in the dependent variate after anomalies in cladoceran length. All noise models were AR(1).

Response variate	Delay	Term	Estimate	s.e.	T ratio
total biovolume	1	ω_0	-0.312	0.102	3.05
		ω_1	0.260	0.103	2.52
dinoflagellates	3	ω_0	-0.008	0.017	0.48
colonial chrysophytes	0	ω_0	0.009	0.041	0.21
gelatinous colonies	2	ω_0	-0.191	0.085	2.24
		ω_2	0.202	0.084	2.40
edible algae (<20 μm)	5	ω_0	-0.082	0.033	2.51
		ω_1	-0.064	0.032	2.00
weighted length	3	ω_0	-48.9	20.01	2.44
weighted length[a]	2	ω_0	0.782	1.37	0.57

Note:
[a] Input variable is cladoceran biomass.

decreases in total biovolume and biovolume of gelatinous colonies. The impulse weights were -0.312 at lag 1 week and -0.260 at lag 2 weeks for total biovolume, and -0.191 at lag 2 weeks and -0.202 at lag 4 weeks for biovolume of gelatinous colonies. The response of edible biovolume was more complex, with impulse weights of -0.082 at lag 5 weeks and 0.064 at lag 6 weeks. Significant effects at such long lags are surprising for such fast-growing organisms, and may represent indirect effects that were not included in the statistical analyses.

The relations between cladoceran size and phytoplankton weighted length was negative (Table 11.2). Anomalous decreases in cladoceran size were followed by increases in phytoplankton mean length. Examples occurred in 1985, 1987, and 1990 when cladoceran size was unusually small and biomass of large gelatinous colonies was unusually great. This pattern is consistent with the negative relationship between cladoceran size and biovolume of gelatinous colonies. However, the effect of cladoceran size on phytoplankton size in Peter Lake is the opposite of that found in Paul Lake, and contrary to our prediction from theory (Carpenter & Kitchell, 1984).

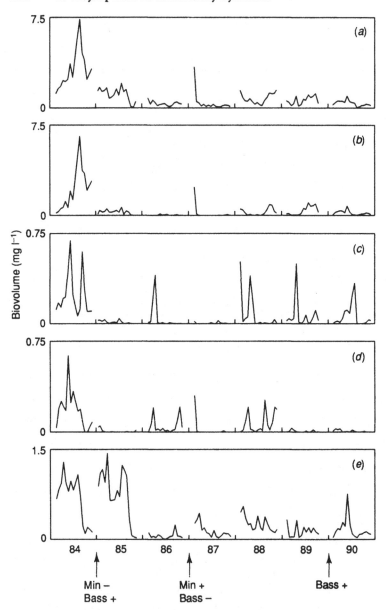

Fig. 11.6. Tuesday Lake: phytoplankton biovolume versus time for total biovolume (*a*) and selected functional groups: dinoflagellates (*b*), colonial chrysophytes (*c*), gelatinous colonies (*d*) and edible algae (*e*), defined as cells < 20 μm in length. Data are shown for the summer stratified season in each year.

Tuesday Lake

In 1984 Tuesday Lake had no piscivores (Chapter 4), a large population of planktivores (Chapter 4), a zooplankton community dominated by copepods and small cladocerans (Chapter 8), and high biovolumes of phytoplankton (Fig. 11.6a). The phytoplankton biovolume was dominated by dinoflagellates (Fig. 11.6a,b). Small *Peridinium pusillum* were abundant early in the summer and large *Peridinium limbatum* became dominant later in the season. The algae <20 μm in length were primarily composed of microflagellates (Fig. 11.6e). This group provided a large and relatively constant contribution to the total algal biovolume through early August when its population declined sharply. The gelatinous colonial blue green algae *Microcystis* sp. and the colonial chrysophyte *Chrysosphaerella longispina* were both present in moderate numbers (Fig. 11.6c,d).

Tuesday Lake's phytoplankton changed dramatically after the introduction of largemouth bass in May 1985 (Fig. 11.6a–e). Significant dinoflagellate populations did not become established and *Peridinium limbatum* was absent (Fig. 11.6b). Neither *Chrysosphaerella* sp. nor *Microcystis* sp. was observed, and biovolumes of colonial chrysophytes and gelatinous colonies were negligible (Fig. 11.6c,d). There was no immediate effect on the biovolume of algae <20 μm in length (Fig. 11.6e) but there were species shifts within this group. Populations of microflagellates were greatly reduced and the prymnesiophyte *Chrysochromulina* sp., absent in 1984, was dominant. Algae <20 μm in length account for the moderate total biovolume of 1985.

A considerable *Daphnia* population was established by August 1985 and persisted through August 1988 (Chapter 8). During this period the phytoplankton community was characterized by low total biovolume punctuated by brief blooms (Fig. 11.6a). *Peridinium* species were absent from Tuesday Lake until 1988. There was a large bloom of the dinoflagellate *Ceratium* sp. in May 1987 (Fig. 11.6b). This was the only notable bloom of this species in Tuesday Lake during the study. There were blooms of colonial *Dinobryon* sp. in 1986 and 1987 and both *Chrysosphaerella* sp. and *Tabellaria* sp. in 1988 (Fig. 11.6c). There were small blooms of the gelatinous colonial bluegreen alga *Microcystis* sp. in 1986 and 1987 (Fig. 11.6d). In 1988 there were three small blooms of gelatinous colonial green algae (Fig. 11.6d). Populations of algae <20 μm in length were greatly reduced during this time (Fig. 11.6e), 1986 being the low point. Microflagellates returned as the dominant member of this group in 1986. The *Daphnia* developed a diel vertical migration pattern in August 1987, which probably reduced their grazing effect in the epilimnion (Lampert & Taylor, 1985) (Chapter 9). Total biovolume and the frequency of algal blooms increased in 1988 after the *Daphnia* began migrating (Fig. 11.6*b–e*).

Between the 1986 and 1987 field seasons, bass were removed from Tuesday Lake and the minnow population was reintroduced (Chapter 2). The *Daphnia* population did not collapse until August 1988 (Chapter 8). There was an increase in total biovolume at this time due to growing populations of small *Peridinium* sp. and the gelatinous colonial alga *Dictyosphaerium pulchellum* (Fig. 11.6*a,b,d*).

In 1989 and 1990 dinoflagellates returned as the dominant group in the Tuesday Lake phytoplankton community (Fig. 11.6*a,b*). As in 1984, dinoflagellates made up over 50% of total algal biovolume for a majority of the sampling period. However, dinoflagellate biovolume did not approach its 1984 level (Fig. 11.6*b*). The absence of *Peridinium limbatum* accounts for much of this difference. There were blooms of *Chrysosphaerella* sp. in 1989 and 1990 and a bloom of *Dinobryon* sp. in 1990 (Fig. 11.6*c*). Populations of gelatinous colonial algae were low in 1989 and 1990 (Fig. 11.6*d*). During this time the algae < 20 μm in length were dominated by microflagellates with contributions from *Peridinium pusillum*.

The aggregated phytoplankton community shifts in Tuesday Lake are illustrated by the changes in annual phytoplankton size distributions (Fig. 11.7). In 1984, the community was widely divided among several size classes (Fig. 11.7) but was dominated by large algae (especially *Peridinium limbatum*). In 1985 all species but the algae < 20 μm in length were immediately affected by the manipulation and a majority of total annual biovolume was concentrated in the 10–20 μm size class (Fig. 11.7). Total biovolume was low in 1986 and the wide distribution of size classes can be attributed to brief blooms of opportunistic species (Fig. 11.7). This pattern was repeated in 1988 although total biovolume was higher (Fig. 11.7). In the absence of the succession of dominant algal groups that existed prior to the 1985 manipulation the size structure of the yearly phytoplankton community is determined by the sequence of short blooms. In 1989 and 1990, the reestablishment of dinoflagellates as the dominant group of algae is reflected in the size distributions (Fig. 11.7). In 1984 the highest concentration of biovolume was in the 80–90 μm size class while in 1989 and 1990 biovolume was concentrated in size classes under 40 μm. This difference results from the dinoflagellate

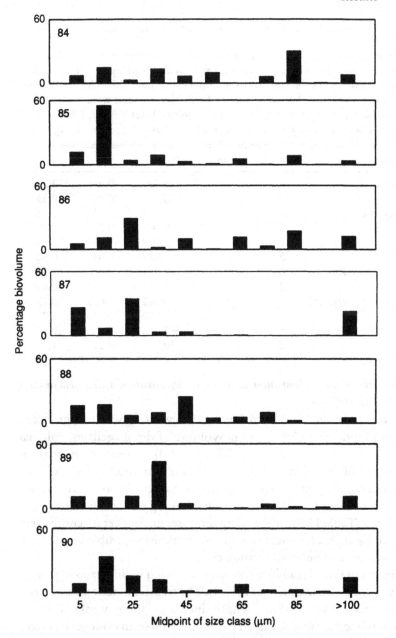

Fig. 11.7. Size distribution of algal biovolume in Tuesday Lake for the summers of 1984–90. Vertical axis is percentage of biovolume summed over all sampling dates in June–August; horizontal axis is midpoint of size class.

Table 11.3. *Results of intervention analyses in Tuesday Lake*

For each dependent variate and manipulation, the step change effect, standard error, and *T* ratio are reported. *T* ratios greater than 1.96 are significant at the 5% level. Manipulation 1 is the onset of large *Daphnia* in August 1985. Manipulation 2 is the collapse of large *Daphnia* in August 1988. Noise models were AR(1) for total biovolume, colonial chrysophytes, and weighted length; AR(1,3) for gelatinous colonies; AR(1,5) for dinoflagellates; and AR(1,10) for edible algae

Dependent variate	Manipulation	Effect	s.e.	*T* ratio
total biovolume	1	−1.40	0.46	3.24
	2	0.15	0.44	0.33
dinoflagellates	1	−1.19	0.33	3.63
	2	0.30	0.30	1.00
colonial chrysophytes	1	−0.109	0.051	2.11
	2	0.025	0.049	0.50
gelatinous colonies	1	−0.064	0.046	1.40
	2	−0.041	0.043	0.95
edible algae (< 20 μm)	1	−0.142	0.046	3.12
	2	0.049	0.038	1.27
weighted length	1	−5.46	8.66	0.63
	2	−10.71	8.27	1.29

species shift from *P. limbatum* in 1984 to *P. pusillum* and *Glenodinium quadridens* in 1989–90.

The onset of large *Daphnia* in August 1985 was followed by significant declines in total biovolume and biovolume of dinoflagellates, colonial chrysophytes, and edible algae (Table 11.3). Intervention analysis detected no other significant effects. Because our intervention analyses are conservative tests for step changes (Chapter 3), we used transfer functions to search for alternative models linking cladoceran size to algal biovolume (Table 11.4). Anomalous increases in cladoceran length were followed by anomalous decreases in total biovolume, edible biovolume, and biovolume of gelatinous colonies.

Increases in cladoceran biomass were followed by increases in phyto-plankton weighted length (Table 11.4). The transfer function parameters indicate that phytoplankton length increased three weeks after an anomalous increase in cladoceran biomass. Increases in cladoceran length were also followed by increases in phytoplankton weighted length, but the relationship was more complicated. Impulse weights were 34.1 at lag 8 weeks, −62.2 at lag 9 weeks, and 30.8 at lag 12 weeks. This complex pattern and long delay suggest the involvement of indirect effects that were not included in the statistical models.

Table 11.4. *Transfer functions for selected algal groups in Tuesday Lake*

Except where indicated, the input variable was mean cladoceran length. Transfer functions could not be fitted for dinoflagellates and colonial chrysophytes because of the large number of zeroes after 1985. For each dependent variate, the delay, model terms, estimates, standard errors, and T ratios are presented. Values of T greater than 1.96 indicate significant anomalies in the dependent variate after anomalies in cladoceran length. Noise models were AR(1) for total biovolume and weighted length, AR (1,2) for gelatinous colonies and AR(1,12) for edible algae.

Response variate	Delay	Term	Estimate	s.e.	T ratio
total biovolume	8	ω_0	-0.972	0.396	2.45
gelatinous colonies	4	ω_0	-0.113	0.059	1.92
edible algae ($<20\ \mu m$)	1	ω_0	-0.378	0.151	2.49
weighted length	8	ω_0	34.14	15.12	2.26
		ω_1	62.15	15.66	3.97
		ω_4	-30.84	14.85	2.08
weighted length[a]	3	ω_0	17.89	5.03	3.55
		δ_1	-0.476	0.232	2.05

Note:
[a] Input variable is cladoceran biomass.

Discussion

Shifts in the zooplankton community had substantial effects on phytoplankton community structure. Even unmanipulated Paul Lake showed negative effects of increased grazer size on biovolume of gelatinous colonies and of all algae combined. Effects of grazers were stronger and more widespread among algal groups in manipulated Peter and Tuesday Lakes. Statistically significant effects of grazer size on algal groups were consistent across lakes. Increased grazer size was followed by decreased biovolume of every algal group in one or more lakes.

We are surprised by Tuesday Lake's failure to recover from the first manipulation. The rise of large herbivores in 1985 had negative effects on every functional group of algae, and eliminated several prominent taxa from the lake. Our earlier papers documented peaks in abundance of certain grazing-resistant taxa during the period of *Daphnia* dominance (Carpenter *et al.*, 1987; Elser & Carpenter, 1988). These pulses proved to be short-term phenomena. More than two years after disappearance of *Daphnia* from the lake, total biovolume remained at relatively low levels. Dinoflagellates were the functional group most strongly affected. Bag experiments performed in Tuesday Lake in 1982 and 1983 indicated that large dinoflagellates, including *Peridinium limbatum*, were readily eliminated by large zooplankters (Bergquist, 1985; Bergquist *et al.*, 1985). If *P. limbatum* were present at premanipulation levels, algal biovolume in 1989 and 1990 would approach that of 1984. We can only speculate about why *P. limbatum* has not returned. Perhaps sedimentary reservoirs of cysts were depleted over the years as excysting cells were killed by grazing before they could reproduce. Whatever the cause, the slow dynamics of the *P. limbatum* population have limited the community recovery and ecosystem resilience of Tuesday Lake.

The simulation model (Carpenter & Kitchell, 1984; Chapter 18) correctly predicted plankton size shifts in Paul and Tuesday Lakes: increases in zooplankter size were followed by increases in phytoplankter size. However, results in Peter Lake contradicted the model. Decreases in zooplankter size were followed by increases in phytoplankter size on several occasions. In all cases, the increases in phytoplankter size were caused by blooms of gelatinous, colonial green or bluegreen algae.

Plankter size distributions may fail to represent community dynamics for several reasons. Size is an imperfect, though significant, predictor of key rate processes. Important physiological, morphological and life history properties of algae are sometimes independent of size (Sandgren, 1988). Biological details may dictate responses to perturbation in important ways. For example, gelatinous colonial algae were able to respond rapidly to decreases in grazer size in Peter Lake. In Tuesday Lake, large dinoflagellates did not respond to decreases in grazer size in more than two years. These groups of algae are about the same size, yet have very different morphologies and life histories that may explain their contrasting responses to similar manipulations.

Our results show that size-based models alone cannot adequately predict the response of phytoplankton communities to food web perturbations. However, it would be unwise to categorically reject size-based models of plankton community dynamics. A considerable body of rigorous experimentation suggests that grazers' effects on phytoplankton community composition are complex and idiosyncratic (Porter, 1973; Lynch & Shapiro, 1981; Lehman & Sandgren, 1985; Bergquist *et al.*, 1985; Bergquist & Carpenter, 1986; Elser *et al.*, 1986a; Sommer, 1988; Vanni, 1987; Vanni & Temte, 1990). A significant fraction of this variability is plausibly explained by size (Bergquist *et al.*, 1985). Abandonment of size-based models leaves us with few unifying principles. At the very least, size-based models explain many examples and provide a conceptual framework for organizing the exceptions.

Summary

Phytoplankton in our reference lake, Paul Lake, were dominated by gelatinous colonial algae throughout the seven-year study. Small, edible algae (< 20 μm) had relatively stable concentrations, whereas dinoflagellates and colonial chrysophytes had variable concentrations. There were no notable trends in the phytoplankton of Paul Lake during the study. Some statistical relations were detected between cladoceran length and phytoplankton. Increases in cladoceran length were followed by decreases in total algal biovolume and biovolume of gelatinous colonies, and by increases in phytoplankter size.

The phytoplankton of Peter Lake were far more variable than those of Paul Lake. Brief periods of high planktivory that caused decreases in grazer size were followed by increases in biovolume of colonial algae in 1985 (large bass year class followed by blooms of *Sphaerocystis*), 1988 (stocking of rainbow trout followed by blooms of *Aphanocapsa*) and 1990 (stocking of golden shiners followed by blooms of *Botryococcus*). Many statistical relations were detected between cladoceran length and phytoplankton. Decreases in cladoceran length were followed by increases in phytoplankter size and biovolume of total algae, gelatinous colonies, and edible algae (< 20 μm).

The transformation of Tuesday Lake from planktivore domination to piscivore domination in 1985 caused dramatic changes in the phytoplankton. Reductions occurred in total biovolume and biovolume of dinoflagellates, colonial chrysophytes, and edible algae (< 20 μm). However, there were no significant changes in the algal community after the removal of piscivores and reestablishment of planktivores in 1987. The failure of large dinoflagellates, especially *Peridinium limbatum*, to recover their former abundance was notable. Slow dynamics of this phytoplankter limited the community recovery and ecosystem resilience of Tuesday Lake.

Acknowledgements

We thank Craig Sandgren for helpful comments on a draft.

12 · *Metalimnetic phytoplankton dynamics*

Ann St. Amand and Stephen R. Carpenter

Introduction

Temperature, light and oxygen change markedly with depth, thereby altering the habitats of phytoplankton. Often, the upper waters are the most favorable habitat for algae, so maximum algal biomass occurs in the epilimnion. Where there is adequate light penetration to near or below the thermocline, however, the maximum algal biomass may occur deep in the water column. Algal maxima in the metalimnion and hypolimnion are known from many lakes (Moll & Stoermer, 1982). This chapter is especially concerned with metalimnetic communities of algae found near the thermocline. Although metalimnetic algae can attain high biomass and can contribute substantially to whole-lake primary productivity (Moll & Stoermer, 1982), little is known about the mechanisms which control their dynamics.

Epilimnetic phytoplankton communities benefit from relatively favorable light and temperature, but can suffer great losses through sedimentation. In addition, cells are constantly mixed throughout the epilimnion and must cope with changing light conditions (Reynolds, 1984a). Nutrients are depleted as the summer progresses with little opportunity for regeneration from below before fall turnover. Also, migrating zooplankton and zooplankton feces can transport phosphorus into the hypolimnion, further depleting the nutrient pool (Dini *et al.*, 1987). On the other hand, grazing losses may be low for many hours each day if zooplankton migrate in and out of the epilimnion (Lampert & Taylor, 1985; Dorazio *et al.*, 1987; Chapter 9).

In contrast, metalimnetic phytoplankton communities must contend with lower light levels and temperatures, which can reduce growth rates (Healey, 1983; Raven & Geider, 1988; Reynolds, 1984a). However, sedimentation rates can be lower in the colder metalimnetic water and growth may actually be stimulated by nutrients diffusing upward from below the thermocline (Pedrós-Alió *et al.*, 1987).

In Tuesday Lake, humic staining impedes light transmission and a distinctive metalimnetic community does not consistently develop. Paul and Peter Lakes, however, are relatively clear and support dense populations of metalimnetic algae during summer stratification.

This two-year study was designed to determine the major forces structuring metalimnetic algal communities in Paul and Peter Lakes. Epilimnetic and metalimnetic communities were compared to test several hypotheses concerning the establishment and maintenance of deep algal maxima and to examine the implications for lake metabolism during the stratified period of the year. Weekly monitoring, combined with enclosure experiments both at depth and in the epilimnion, assessed metalimnetic and epilimnetic changes in chlorophyll *a* attributable to *in situ* growth, sedimentation, and grazing. Comparisons between the epilimnion and metalimnion were made using physiological indicators and nutrient enrichment bioassays.

Methods

Comparison of epilimnion and metalimnion

Comparative sampling of the epilimnion and metalimnion was conducted during the summer stratification period in 1987 and 1988. Most of the analytical methods are the same as those used in Chapter 13, and are described in detail in Soranno (1990) and St. Amand (1990).

Epilimnetic and metalimnetic measurements included chlorophyll *a*, pheopigments, primary production (^{14}C fixation), phytoplankton enumeration, AER (ammonium enhancement response), APA (alkaline phosphatase activity) and sedimentation rate. At biweekly intervals, nutrient enrichment experiments were conducted in the laboratory to determine nutrient status in the absence of grazers.

Sedimentation rate was measured using cylindrical traps. Sediment traps from the upper limit of the thermocline (shallow) and the lower limit of the thermocline (deep) were collected in duplicate and pooled. Each tube was poisoned with Lugol's iodine (plus salt to form a density gradient) which prevented cell degradation and allowed taxa-specific enumeration of sedimenting phytoplankton (Bloesch & Burns, 1980; Gardner, 1980).

Epilimnetic carbon fixation was integrated from measurements taken at depths corresponding to 100%, 50% and 25% of surface irradiance (Carpenter, Elser & Elser, 1986). Methodological details are described in Chapter 13. In 1987, linear interpolation was used to calculate metalimnetic primary production from the sampling depths between 5% and

1% surface irradiance. In 1988, additional primary production incubations were conducted at the depth of the biomass maximum.

Grazing experiments

Enclosure experiments were conducted in both lakes during 1987 and 1988 to assess grazing and nutrient effects on epilimnetic and metalimnetic phytoplankton. In this chapter, we present results of an experiment conducted in Peter Lake, 8–12 August 1987. Conclusions from other experiments were consistent with those presented here, although their objectives and designs differed (St. Amand, 1990).

The experiment reported here was designed to evaluate the responses of epilimnetic and metalimnetic phytoplankton to grazing and nutrients when exposed to migrating versus nonmigrating zooplankton assemblages. The zooplankton gradient ranged from 0.25 × ambient to 8 × ambient zooplankton concentration (seven enclosures in all). Native phytoplankton assemblages from the epilimnion and metalimnion were exposed to zooplankton characteristic of both diurnal (0600–2200) and nocturnal (2200–0600) grazing assemblages. In addition, two nutrient-enriched enclosures contained a 10:1 nitrogen (160 μm $NH_4 - N$ as NH_4Cl) to phosphorus (16 μm $PO_4 - P$ as KH_2PO_4) ratio and ambient densities of zooplankton.

Epilimnetic (1 m) and metalimnetic (biomass maximum) zooplankton were collected at 1200 and 2400 hours the day of the experiment with a 28 l Schindler–Patalas trap (80 μm net). The morning of the experiment, zooplankton were rinsed with filtered water from the appropriate lake (125 μm Nitex net) to prevent excretion products high in nutrients from being added to the carboys (Elser et al., 1988).

At the beginning and end of the experiment, chlorophyll a determinations, phytoplankton counts and zooplankton counts were made. Chlorophyll a was determined by fluorometry and corrected for pheopigments (Marker, Crowther & Gunn, 1980; Chapter 13).

Phytoplankton were prepared for counting as described in Chapter 11. Cells were enumerated on an Olympus BHTU compound microscope at 200–400 × . GALD (greatest axial linear dimension) and biovolume were measured on a subsample of each taxon. Phytoplankton biovolume was estimated by approximating the shape of the alga to composites of simple geometric figures (spheres, cylinders, prolate ellipsoids, etc.) (Bergquist, 1985; Elser et al., 1986a, 1987a; Pip & Robinson, 1982; Sournia, 1978).

Zooplankton were preserved in 4% sugared buffered formalin and refrigerated (Haney & Hall, 1975; Chapter 8). Enumeration and measurement followed the procedures given in Chapter 8.

Data analysis

To analyze the grazing experiments, net phytoplankton growth rates (r, d^{-1}) were calculated for chlorophyll a, total community biovolume, and biovolume of individual taxa. Growth rate was calculated assuming exponential growth:

$$r = \ln(B_i/B_f)/t.$$

B_i is the initial biomass (chl a as $\mu g\, l^{-1}$ or biovolume as $\mu m^3\, l^{-1}$), B_f is the final biomass (same units as initial biomass), and t is the duration of the experiment in days (Bergquist, 1985; Elser et al., 1987a; Lehman & Sandgren, 1985). Initial values were corrected for algal biovolume added with zooplankton. Calculations were made for all taxa present in the initial sample and at least five of the seven final samples.

Phytoplankton growth rates were regressed against zooplankton biomass, and 95% confidence intervals were calculated around each regression line (SAS Institute, 1985). Linear or quadratic regression models were calculated, depending on goodness of fit criteria. Grazing responses were classified as positive (linear fit with positive slope $p < 0.05$), negative (linear fit with negative slope, $p < 0.05$), unimodal (significant concave downward quadratic fit, $p < 0.05$), or nonsignificant ($p > 0.05$). Nutrient response was considered significant if both nutrient-enriched replicates were outside of the 95% confidence interval (Lehman & Sandgren, 1985).

Results and discussion

Phytoplankton standing stock and seasonal change

In Paul Lake, epilimnetic phytoplankton were dominated by chlorophytes (Crucigenia rectangularis and colonial chlorophytes) with sub-dominant cryptophyte populations (Cryptomonas phaseolus and C. marssonii) (Fig. 12.1a,b). Both chrysophytes (Dinobryon divergens, D. bavaricum, and Synura sp.) and cyanophytes (Nostoc sp. and Oscillatoria limnetica) were consistently present, though in variable concentrations.

Metalimnetic communities in Paul Lake were more diverse than epilimnetic communities, and contained several codominant divisions

(Fig. 12.1c,d). Dinoflagellates (*Gymnodinium* sp., *Ceratium hirundinella*), cryptophytes (*Cryptomonas* spp.), chlorophytes (*Crucigenia rectangularis*, *Shroedaria*), chrysophytes (*Dinobryon divergens*, *D. bavaricum*, *Mallomonas caudata* and *Synura* sp.) and (in spring 1988) diatoms, were well represented. In 1988, cryptophytes and chlorophytes became more abundant than in 1987.

In Peter Lake, epilimnetic communities were more diverse than those of Paul Lake (Fig. 12.1e,f). In both years, blooms of cyanophytes (*Chroococcus prescottii* and *Aphanocapsa elachista* in 1987 and *A. elachista* and *Gomphosphaeria lacustris* in 1988) dominated the assemblage in mid to late summer. A bloom of *A. elachista* developed at the end of 1987 in Peter Lake, and dominated the mid–summer period in 1988. *Aphanocapsa* is heavily grazed and cannot attain dominance unless populations of large *Daphnia* are low (St. Amand, 1990). Several species of heterocystous cyanophytes were also present, although these taxa tended to be more abundant later in the summer. In 1987, chlorophytes (dominated by *Sphaerocystis schroeteri*) and cryptophytes (*Cryptomonas phaseolus* and *C. marssonii*) were less abundant than in 1988, while chrysophytes

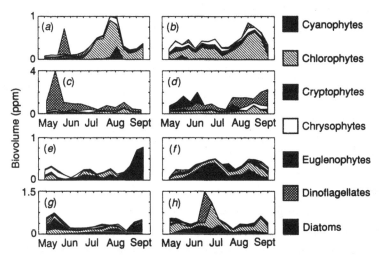

Fig. 12.1. Phytoplankton community composition in the epilimnion and metalimnion of Paul and Peter Lakes. (Biovolume in parts per million (ppm) is equivalent to mg fresh mass l^{-1}.) (*a*) Paul Lake epilimnion, 1987; (*b*) Paul Lake epilimnion, 1988; (*c*) Paul Lake metalimnion, 1987; (*d*) Paul Lake metalimnion, 1988; (*e*) Peter Lake epilimnion, 1987; (*f*) Peter Lake epilimnion, 1988; (*g*) Peter Lake metalimnion, 1987; (*h*) Peter Lake metalimnion, 1988. The chrysophyte category excludes diatoms, and includes mainly flagellated species.

(*Dinobryon divergens, D. bavaricum, Chrysosphaerella longispina* and *Mallo-monas acaroides*) were more abundant than in 1988.

Metalimnetic phytoplankton assemblages in Peter Lake were more diverse than epilimnetic ones (Fig. 12.1*g*,*h*). Cyanophytes (*Aphanocapsa elachista*) were less abundant in the metalimnion, but chlorophytes (colonial chlorophytes, *Shroedaria setigera*) were more abundant. Differences between 1987 and 1988 were distinct. Dinoflagellates (*Gymnodinium* sp.) peaked in mid-June in 1988, when cryptophytes (*Cryptomonas phaseolus* and *C. marssonii*) and chrysophytes were less abundant.

Metalimnetic assemblages differed between lakes in both years. Peter Lake's metalimnetic assemblages were dominated by chlorophytes and cyanophytes for most of the summer season. Metalimnetic assemblages in Paul Lake were dominated by dinoflagellates and chlorophytes while cyanophytes were rarely abundant. Cryptophytes were substantially more abundant in the metalimnion of both lakes in 1988 compared to 1987. Apparently, cryptophyte populations moved upward within the water column in 1988 because oxygen did not penetrate as deeply (St. Amand, 1990).

In Peter and Paul lakes, metalimnetic communities exhibited unique dynamics only loosely coupled to the overlying epilimnetic communities. This pattern was consistent over both years of the study. Several studies that measured *in situ* production and seasonal dynamics have concluded that deep biomass maxima are driven by autotrophic production (Priscu & Goldman, 1984; Moll, Brahce & Peterson, 1984; Gálvez, Niell & Lucena, 1988; Venrick, McGowan & Mantyla, 1973; Baker & Brook, 1971; Fahnenstiel & Scavia, 1987; Coon *et al.*, 1987). However, sinking could be an important source of phytoplankton biomass for the metalimnion. We examined sedimentation rates into and out of the metalimnion to assess the relative importance of sinking from the epilimnion as an input to the metalimnion.

Sedimentation

Sedimentation is a loss process for epilimnetic phytoplankton, but both an input and a loss process for metalimnetic phytoplankton. Ratios of deep to shallow trap sedimentation were usually greater than one (range: 0.30 to > 30), indicating that outputs by sinking exceeded inputs. At most times, sedimentation rates out of the metalimnion were greater than sedimentation rates into the metalimnion (St. Amand, 1990). This excess of output over input should be accounted for by new production within the metalimnetic layer, plus sediment focusing (Carpenter *et al.*, 1988).

Species composition in the shallow sediment traps was similar to that of the epilimnetic assemblages in both lakes. However, in the deep sediment traps several taxa often appeared in higher proportions than their representation in the metalimnetic assemblage. Overrepresentation was especially evident for cryptophytes in Paul Lake in 1987 and both lakes in 1988. There was a high concentration of euglenoids in the deep sediment traps, though these were rare in metalimnetic assemblages. In contrast, nonmotile taxa were not overrepresented in sediment traps.

Alloxanthin, a pigment of cryptophytes, is often found at high concentrations in sediment traps that are not poisoned (Leavitt & Carpenter, 1990*b*; Chapter 15). Zooplankton feces are the major source of alloxanthin in unpoisoned traps (Leavitt & Carpenter, 1990*b*; Chapter 15). Rates of grazing on cryptophytes are high within the metalimnion (see below). However, grazing probably cannot explain the high concentrations of cryptophytes found in the poisoned traps used in our study. Cryptophytes are relatively digestible, and are often not recognizable in zooplankton feces. Direct measurements of diel vertical migration showed that motile cryptophytes (as well as some dinoflagellates and chrysophytes) move downward in the water column by night (St. Amand, 1990). It is therefore likely that our poisoned traps overestimated sedimentation of these motile algae from both layers.

The composition of phytoplankton sinking from the epilimnion to the metalimnion differs substantially from the composition of the metalimnetic phytoplankton in both lakes (Table 12.1). In Paul Lake, inputs to the metalimnion contained higher proportions of chlorophytes and chrysophytes, and lower proportions of dinoflagellates, than were found in the metalimnetic phytoplankton. In Peter Lake, inputs to the metalimnion were relatively enriched in cyanophytes and depauperate in dinoflagellates, compared with the metalimnetic phytoplankton. These differences indicate that active growth and loss processes within the metalimnion are very important to metalimnetic community structure. While sedimentation is a significant input to the metalimnetic biomass, inputs from the epilimnion appear to have limited effects on the structure of the metalimnetic community.

Nutrient limitation

Nutrient limitation was much more apparent in epilimnetic than in metalimnetic communities. Specific alkaline phosphatase activity (i.e.

Table 12.1. *Composition (as percentages) of phytoplankton inputs to the metalimnion by sinking from the epilimnion compared with composition of metalimnetic phytoplankton*

Annual means are presented for 1987 and 1988 in both Paul and Peter Lakes. The chrysophyte category excludes diatoms, and comprises mainly flagellated species.

| | Paul Lake | | | | Peter Lake | | | |
| | 1987 | | 1988 | | 1987 | | 1988 | |
Taxonomic group	Input from epi.	Meta.	Input from epi.	Meta.	Input from epi.	Meta.	Input from epi.	Meta.
chlorophytes	35.8	13.6	35.3	19.5	25.5	23.3	42.6	27.1
chrysophytes	33.6	9.4	18.0	9.5	11.7	10.5	3.1	1.6
cryptophytes	6.8	3.5	9.6	10.1	2.3	19.4	9.0	3.6
cyanophytes	6.4	3.4	6.8	19.5	52.1	23.3	33.9	27.1
diatoms	0.9	0.2	3.2	17.6	1.8	3.9	7.1	2.8
dinoflagellates	16.1	69.8	25.9	42.1	6.6	12.5	3.1	47.5
euglenophytes	0.4	0.1	1.2	0	0	0.2	1.2	0.1

alkaline phosphatase activity per unit chlorophyll) was almost always higher in the epilimnion than in the metalimnion, indicating greater phosphorus limitation in the epilimnion (Fig. 12.2).

Nitrogen limitation, measured by ammonium enhancement response, also tended to be greater for epilimnetic algae (Fig. 12.3). However, differences in nitrogen deficiency between depths were more variable than differences in phosphorus deficiency.

Laboratory growth bioassays and *in situ* growth bioassays also indicated greater nutrient limitation in the epilimnion (St. Amand, 1990). Nitrogen and phosphorus often appeared to be colimiting in growth bioassays (St. Amand, 1990). Conclusions from growth bioassays were generally consistent with the physiological bioassays reported here (St. Amand *et al.*, 1989).

In sum, epilimnetic phytoplankton are more nutrient-limited than metalimnetic phytoplankton, especially by phosphorus. Nutrients are regenerated in the hypolimnion, and concentrations of inorganic N and P are therefore higher there (St Amand, 1990). As nutrients diffuse upward, metalimnetic phytoplankton encounter higher nutrient availability than their epilimnetic counterparts.

Grazing and its interactions with nutrients

In the grazing experiment, cladocerans dominated both the daytime and nighttime metalimnetic assemblages, and the nighttime epilimnetic assemblage (St. Amand, 1990). These cladoceran communities were dominated by *Daphnia pulex* individuals greater than 1 mm in length. Large *D. pulex* were absent from the daytime epilimnetic assemblage due to their diel vertical migration (Chapter 9). The daytime epilimnetic grazers were dominated by copepods, especially immature instars, and rotifers (St. Amand, 1990).

In the epilimnion, chlorophyll growth rate was stimulated by grazing (Fig. 12.4*a,b*). Chlorophyll growth rate was also strongly limited by nutrients. Apparently, the positive effects of nutrient recycling by grazers offset the negative effects of grazing in the nutrient-deficient epilimnion. Responses to both daytime and nighttime grazer assemblages were similar, except that the *Daphnia*-dominated nighttime

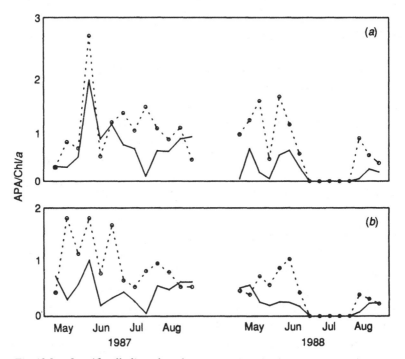

Fig. 12.2. Specific alkaline phosphatase activity (APA) (see text) versus sample date for the epilimnion (dashed line) and metalimnion (solid line) of (*a*) Peter Lake and (*b*) Paul Lake.

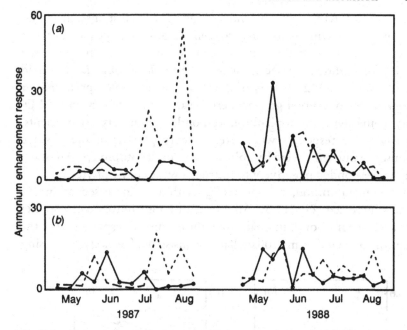

Fig. 12.3. Ammonium enhancement response versus sample date for the epilimnion (dashed line) and metalimnion (solid line) of (*a*) Peter Lake and (*b*) Paul Lake.

assemblage yielded much more variable growth rates than the daytime assemblage.

In the metalimnion, chlorophyll growth rate decreased as grazer biomass increased (Fig. 12.4*c,d*). Chlorophyll growth rate was not limited by nutrients. Responses to both daytime and nighttime grazer assemblages were similar.

In other experiments not presented here, grazers had weak negative effects on growth rates of epilimnetic chlorophyll (St. Amand, 1990). However, the response of metalimnetic algae to grazers was always more negative than the response of epilimnetic algae to grazers.

In the epilimnion, effects of grazers on species-specific growth rates included increasing, decreasing, and unimodal responses (Fig. 12.5). In the unimodal case, growth rate increases at low levels of grazers because the benefits of nutrient recycling offset the debits of grazing (Chapter 16, Fig. 16.2). At higher levels of grazers, grazing losses exceed the benefits of nutrient recycling. Several taxa that responded positively or unimo- dally to daytime grazers either decreased or displayed no significant response to nighttime grazers. Small taxa (*Cryptomonas phaseolus* and

Oocystis sp. 2, GALD < 15 μm) showed the opposite response, declining in growth rate with increasing concentrations of daytime, but not nighttime, grazers. Of the taxa that did decline in growth rate in the presence of nighttime grazers, all but one (*Aphanocapsa elachista*) had GALD less than 30 μm. Many of the taxa that responded positively or unimodally to nocturnal grazers were greater than 52.5 μm in GALD. Many epilimnetic phytoplankton responded positively to nutrients, while negative responses were rare. Seventy percent of the positive responses to nutrients were by chlorophytes or cryptophytes. All of the taxa that responded negatively to nutrients were chrysophytes.

In the metalimnion, no species' growth rate increased as grazer biomass increased (Fig. 12.5). All effects of zooplankton at the species level were negative or unimodal. All of the unimodal responses, with the exception of that for microflagellates, consisted of weakly increasing

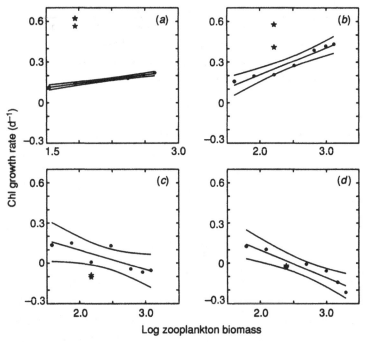

Fig. 12.4. Chlorophyll growth rate versus zooplankton biomass in the enclosure experiment. For unenriched enclosures (solid dots), linear regression fits with 95% confidence intervals are shown. Stars denote enriched enclosures incubated at the zooplankton concentrations found in the lake. (*a*) Epilimnetic daytime assemblages; (*b*) epilimnetic nighttime assemblages; (*c*) metalimnetic daytime assemblages; (*d*) metalimnetic nighttime assemblages.

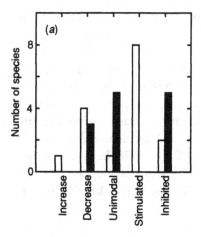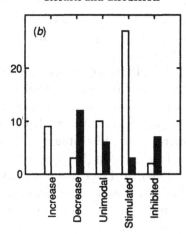

Fig. 12.5. Statistically significant ($p < 0.05$) responses of phytoplankton species to grazers (increase, decrease, or unimodal) and nutrient enrichment (stimulated or inhibited by nutrient addition) in experiments performed in the epilimnion (open bars) and metalimnion (dark bars) of (*a*) Paul Lake and (*b*) Peter Lake. This figure is based on all grazing experiments reported by St. Amand (1990).

growth rates at low zooplankton biomass and strongly decreasing growth rates at medium to high zooplankton biomass (St. Amand, 1990). Several taxa that responded positively or unimodally in the epilimnion decreased or did not respond to grazing in the metalimnion (e.g. *Chrysosphaerella longispina, Dinobryon bavaricum, D. divergens* and *Oocystis* sp. 1). Lack of response to grazers was not related to cell or colony size. In contrast to the epilimnetic algae, most species of metalimnetic phytoplankton did not respond to nutrients. Of those that did respond to nutrients, the majority were inhibited rather than stimulated.

In Paul and Peter Lakes, grazing has much more severe effects on metalimnetic algae than on epilimnetic algae. Grazing loss rates are more rapid in the metalimnion, and both daytime and nighttime zooplankton assemblages have negative effects on a wide range of species. In the epilimnion, net effects of grazers are weaker, and range from mildly positive to mildly negative. The strongest effects of grazing in the epilimnion occur at night when large migratory *Daphnia* are present. The consequence of these differences for biomass dynamics are profound. Grazing loss rates for metalimnetic phytoplankton range from 30% to 60% per day, while grazing loss rates for epilimnetic phytoplankton are less than 10% per day (Fig. 12.6). These figures are annual averages for 1987 and 1988. Calculations of grazing losses were based on all experiments reported by St. Amand (1990).

Considerable interest has focused on the importance of grazing in the maintenance of deeply stratified phytoplankton communities (Brahce, 1980; Longhurst, 1976; Ortner *et al.*, 1980; Fahnenstiel & Scavia, 1987; Coon *et al.*, 1987; Jamart *et al.*, 1977; Venrick *et al.*, 1973). In one of the few studies to directly measure metalimnetic grazing rates, Fahnenstiel & Scavia (1987) concluded that grazing was a significant source of loss for deep algal communities in Lake Michigan, at least during the late summer season. Other studies have suggested that the metalimnion is a refuge from grazing (Richerson, Lopez & Coon, 1978; Fee, 1976; Coon *et al.*, 1987). In Peter and Paul Lakes, this was not the case. Similar patterns may exist in other lakes where non-migratory *Daphnia* populations subsist near the biomass maximum.

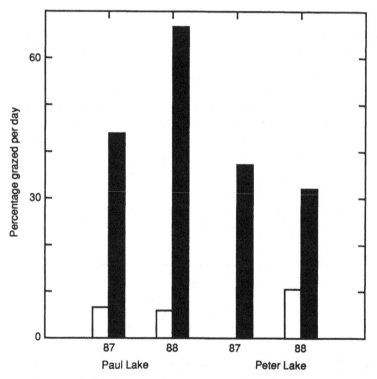

Fig. 12.6. Calculated grazing losses (as percentage of standing crop per day) during summer stratification from the epilimnion (open bars) and metalimnion (dark bars) of Paul and Peter Lakes in 1987 and 1988.

Conclusions

In Paul and Peter Lakes, the differences between epilimnetic and metalimnetic algal communities can be explained by differences in nutrient limitation of growth rates and loss rates to grazers. In the epilimnion, nutrient limitation is relatively severe, and the negative effects of grazing mortality are nearly compensated by the beneficial effects of nutrients recycled by grazers. In the metalimnion, nutrients are not limiting and grazers have a marked negative effect on phytoplankton. The greater effect of grazers in the metalimnion is compounded by the fact that large daphnids graze in the metalimnion throughout the diel cycle, but graze in the epilimnion only at night.

Elser & Goldman (1991) propose that net effects of grazers are weak in oligotrophic lakes because consumption and nutrient excretion are compensatory, but strong in mesotrophic lakes because the benefits of nutrient recycling are moot. Our results support this idea. Instead of occurring between lakes, our nutrient gradient occurs vertically in a single lake.

The relatively strong effects of grazing in the metalimnion suggest that effects of the trophic cascade in Peter Lake will not be uniform with depth. We would expect decreases in grazing to have little effect on epilimnetic algae but positive effects on metalimnetic algae. Metalimnetic dynamics may be pivotal in the transmission of cascading effects from zooplankton to phytoplankton. This possibility is among those addressed in Chapter 13.

Summary

In Paul and Peter Lakes, the greatest phytoplankton biomass occurred in the metalimnion, not the epilimnion. Tuesday Lake's water was stained, so light penetration was not sufficient to support a metalimnetic algal maximum.

Composition of the metalimnetic algal communities differed from that of the overlying epilimnetic communities. Metalimnetic communities were more diverse than epilimnetic ones.

Sedimentation of phytoplankton out of the metalimnion exceeded sedimentation into the metalimnion. Species composition of algae sinking into the metalimnion was different from that of the metalimnetic standing stock. While algae sinking from the epilimnion influenced metalimnetic biomass and community composition, that was not the dominant factor regulating the metalimnetic phytoplankton.

Epilimnetic algae were more limited by nutrients, especially phosphorus, than metalimnetic algae. Nutrient availability for metalimnetic algae was relatively high in Paul and Peter Lakes.

Grazing losses in the metalimnions were relatively high, and occurred throughout the diel cycle because large daphnids were nearly always present. In the epilimnion, net effects of grazers were near zero because losses were balanced by the stimulatory effect of nutrients excreted by grazers.

Despite relatively low levels of light and temperature and high grazing losses in the metalimnion, the algae maintained relatively fast growth rates because nutrient availability was high. Grazing and sedimentation losses were the most important factors limiting the metalimnetic algae in Paul and Peter Lakes.

Acknowledgements

We thank J. J. Elser for a helpful and thorough review.

13 · Primary production and its interactions with nutrients and light transmission

Stephen R. Carpenter, John Morrice, Patricia A. Soranno, James J. Elser, Neil A. MacKay and Ann St. Amand

Introduction

The trophic cascade from fish to ecosystem processes is the central theme of this book. Previous chapters have documented the changes in piscivore, planktivore and herbivore trophic levels in our experimental lakes. Effects of changes in herbivory on phytoplankton communities have been discussed (Chapter 11) and we have compared the distinctive habitats for algae provided by the epilimnion and metalimnion (Chapter 12). This chapter turns to the central ecosystem variates of our study: biomass and metabolism of primary producers.

In lakes, dynamics of herbivores, primary producers nutrients, and light have strong feedbacks. Excretion by herbivores is a major source of nutrients for phytoplankton. Limnetic grazers affect their prey negatively, through grazing, and positively, through nutrient recycling. The phytoplankton also respond to inputs of nutrients from outside the lake, and to mixing of nutrients upward from the hypolimnion. Light drives photosynthesis, and phytoplankton attenuate light as it passes downward through the water column. In lakes where substantial aggregations of algae exist in deep, low-light habitats, feedbacks between algal biomass and light extinction may strongly influence primary production (Chapter 12).

These strong feedbacks imply that primary production cannot be addressed without considering physical–chemical factors such as mixing depth, irradiance, and nutrient limitation. This chapter focuses on dynamics of algal biomass (measured as chlorophyll) and production, in the context of selected physical and chemical characteristics of the water column. These include indicators of nutrient deficiency, light penetration, and mixing depth.

Methods

Primary production is calculated from vertical profiles of carbon fixation, chlorophyll, dissolved inorganic carbon, irradiance and temperature. Measurements of all of these variates will be discussed in this chapter. We also present and analyze indicators of phytoplankton nutrient limitation. Because all of our methods have been presented elsewhere (Carpenter *et al.*, 1986; Elser *et al.*, 1988), only a brief summary will be given here. Step-by-step analytical procedures are given by Soranno (1990).

Sampling and routine limnological methods

Sampling occurred weekly in each lake during the summer stratified season (mid-May to mid-September). Profiles (8–12 depths) were measured of irradiance (LiCor submersible spherical quantum sensor), dissolved oxygen and temperature (YSI meter). Samples were taken at six depths (100%, 50%, 25%, 10%, 5% and 1% of surface irradiance) for dissolved inorganic carbon (DIC), chlorophyll *a* and alkaline phosphatase activity (APA) analyses. Samples from the epilimnion (usually corresponding to the three upper depths) were pooled for determination of ammonium enhancement response (AER) and total phosphorus (TP). Surface irradiance was monitored continuously with a Belfort pyrheliometer. Digitized pyrheliometer data were converted to quantum units using an empirically determined equation (Carpenter *et al.*, 1986).

DIC was determined from pH and total inflection point alkalinity (Gran, 1952) from 1984 to 1987, and occasionally from 1988 to 1990. From 1987 to 1990, DIC was determined by gas chromatography (Stainton, Capal & Armstrong, 1977). Samples analyzed by both methods were used to calculate empirical conversion factors to place data from the two methods on an equal footing (Soranno, 1990).

Chlorophyll was determined fluorometrically. Samples collected on Whatman GF/F filters were frozen, extracted in methanol, and analyzed before and after acidification to correct for pheopigments (Marker *et al.*, 1980). The fluorometer was calibrated using commercial chlorophyll standards. Concentrations of the standards were verified by spectrophotometry.

Every one or two weeks, profiles of carbon fixation were measured *in situ* at the same depths used for DIC and chlorophyll measurements. Incubations ran from approximately 0930 to 1530 hours on each sampling date. At each depth in each lake, we incubated two light bottles and a DCMU (dichlorophenol dimethylurea) control bottle. The DCMU corrects for nonmetabolic absorption of labeled carbon by the algae (Legendre *et al.*, 1983). In 1984 only, we ran an additional dark bottle control at 1% of surface irradiance to correct for possible effects of bacterial carbon fixation (Parkin & Brock, 1980). Bacterial carbon fixation was not a source of error in the photic zones of these lakes, and dark bottles were discontinued after 1984. Each 125 ml bottle was injected with 185 kBq of $NaH^{14}CO_3$. Incubations were ended by collecting the algae on Whatman GF/F filters which were then rinsed with 1 N HCl and dried overnight before liquid scintillation counting.

Calculation of primary production

Variables that govern primary production (chlorophyll, DIC, irradiance, temperature) were measured more frequently than carbon fixation. While chlorophyll, DIC and temperature show only modest changes from day to day, irradiance fluctuates considerably. Daily irradiance exhibited a coefficient of variation (standard deviation/mean) of 59% over the seven-summer study. Fluctuations in light intensity are a significant source of variance that would affect comparisons of carbon fixation profiles measured on different dates.

Other researchers have shown that the error of primary production estimates can be reduced substantially by using a model that accounts for fluctuations in surface irradiance and vertical inhomogeneities in the water column (Bower *et al.*, 1987). We adopted a similar approach to interpolate primary production between sampling dates and across sampling depths. Empirical equations were fitted to data from each lake to predict carbon fixation rate from irradiance, temperature, depth, DIC concentration and chlorophyll. Daily depth profiles of light extinction coefficient, temperature, DIC and chlorophyll were interpolated from the weekly measurements. The empirical equations were then used to calculate primary production from interpolated daily profiles and daily surface irradiance measurements. Since the actual profiles of irradiance, temperature, DIC and chlorophyll were measured weekly, daily primary production estimates were summed to calculate weekly time series for the analyses reported in this book.

Table 13.1. *Sources of variance in the calculation of primary production from observations of chlorophyll, light intensity, depth and temperature, 1984–90*

For each lake, the number of data points analyzed (n), the multiple coefficient of determination (R^2) and the standard deviation (s.d.) of residuals are presented for the empirical regression equations used to calculate primary production. Also shown are standard deviation expected from variation in irradiance alone, and the standard deviation expected from variation in chlorophyll alone.

Lake	n	R^2	s.d. of residuals	s.d. from variation in surface irradiance	s.d. from variation in chlorophyll
Paul	318	0.766	0.544	0.250	0.238
Peter	328	0.758	0.625	0.175	0.771
Tuesday	342	0.632	0.755	0.272	0.430

Similar equations were obtained for each lake (Table 13.1). Predictors in all equations included day of the year, chlorophyll concentration, log of irradiance, and squared log of irradiance. Additional predictors were log of temperature (Paul and Peter Lakes), depth (Peter and Tuesday Lakes), and depth squared (Tuesday Lake). Temperature and depth covaried strongly, and so only one of the two was necessary for prediction in Paul and Tuesday Lakes. DIC varied little and was not a useful predictor for any lake. All equations were fitted by backwards elimination (Draper & Smith, 1981). Iterations were controlled by the analyst (S. R. Carpenter), who included marginally significant predictors that might have been discarded by an automated procedure. Plots of residuals were examined using the criteria of Draper & Smith (1981), and all fits appeared satisfactory.

Chlorophyll and irradiance were by far the most important sources of variance in the primary production regressions. To examine the implications of variance in chlorophyll and surface irradiance, we used the regressions to calculate the standard deviation of integrated water column production derived from variance in surface irradiance only, and from variance in chlorophyll only. To compute the standard deviation of production due to variance in surface irradiance, 1000 surface irradiance values were generated randomly from the observed distribution of surface irradiance for 1984–90, while all other predictors were held constant at their means for 1984–90. Integrated primary production was calculated for each surface irradiance value, and the standard deviation of primary production resulting from variance in irradiance was calculated from this sample. A similar procedure was followed to estimate the standard deviation in primary production resulting from variance in chlorophyll. In this case, all predictors except chlorophyll were held constant at their means for 1984–90, while chlorophyll at each depth was drawn randomly from the observed distribution at that depth over 1984–90.

In all three lakes, fluctuations in surface irradiance result in substantial variation in primary production (Table 13.1). This variation is a major reason for using a model that accounts for irradiance to interpolate over intervals between observations. In Paul Lake, surface irradiance and chlorophyll variations have about equal effects on primary production. In Tuesday Lake, and especially in Peter Lake, variation in chlorophyll has even greater effects than variation in irradiance. This high variance reflects the high variance of the observed distribution of chlorophyll from which the random values were drawn. The high variance of chlorophyll in the manipulated Peter and Tuesday Lakes, relative to the reference Paul Lake, is a consequence of the manipulations (see below). Table 13.1 suggests that any changes in chlorophyll resulting from manipulation of the lakes are likely to correspond with changes in primary production.

Indicators of nutrient deficiency

Of the many indicators of nutrient deficiency in algae (Healey & Hendzel, 1980; Vincent, 1981), we focus here on alkaline phosphatase activity and ammonium enhancement response. These physiological indicators can be rapidly assayed, so it was possible to measure them frequently throughout our experiments. Elsewhere we have compared the physiological indicators with more traditional growth bioassays (Elser *et al.*, 1988; St. Amand *et al.*, 1989). While the relationships show considerable scatter, the physiological indicators are significantly correlated with growth bioassays and lead to the same ecological conclusions (Elser *et al.*, 1988; St. Amand *et al.*, 1989). In growth bioassays, algae are incubated for several days under enriched conditions, so responses reflect community changes as well as nutrient deficiency. In contrast, the physiological indicators are measured immediately on the extant algal assemblage. Thus a perfect correspondence between the methods is not expected, and the immediacy of the physiological indicators may be advantageous (St. Amand *et al.*, 1989).

This chapter reports nutrient deficiency indicators and nutrient concentrations for the epilimnion only. Since metalimnetic algae are not nutrient-limited (presumably because adequate nutrient supplies diffuse from the hypolimnion), food web effects on nutrient limitation were not anticipated there (Chapter 12). In the epilimnion, where nutrient limitation was evident, shifts in food webs may cause changes in nutrient limitation (Elser *et al.*, 1988; Sterner, 1990).

Alkaline phosphatase is an enzyme induced by phosphorus shortage, so increases in alkaline phosphatase activity (APA) can be used to indicate phosphorus deficiency (Petterson, 1980). We determined APA fluorometrically (Elser *et al.*, 1986*a*). The fluorometer was calibrated for the fluorescent end product, methylumbelliferone, using commercial standards. To correct for variations in alkaline phosphatase activity due simply to variations in algal biomass, we report chlorophyll-specific alkaline phosphatase activity, APA/chlorophyll (Elser *et al.*, 1986*a*).

Under nitrogen limitation, phytoplankton will fix carbon in the dark if ammonium is available (Yentsch, Yentsch & Strube, 1977). The carbon is used to form amines, thereby detoxifying the ammonium. The rate of dark carbon fixation in the presence of ammonium, or the ammonium enhancement response (AER), can be used to indicate nitrogen limitation (Yentsch *et al.*, 1977; Elser *et al.*, 1988). For each water sample, radiocarbon fixation in the dark was measured in three aliquots enriched with ammonium, and three control aliquots. AER was calculated as a *t*-statistic, $(A_E - A_C)/$s.e., where A_E is the mean radiocarbon activity in the enriched samples, A_C is the mean activity of the control samples, and s.e. is the pooled standard error (Box, Hunter & Hunter, 1978).

Measurements of total phosphorus (TP) were performed within six months on samples preserved by freezing. TP was determined by the phosphomolybdate method after persulfate digestion (Wetzel & Likens, 1979).

Results

Paul Lake

The reference ecosystem displayed no trends or unusual events to suggest regional shifts in variates related to primary production. The variability exhibited by Paul Lake is similar to that known from many lakes (Carpenter & Kitchell, 1987).

Total phosphorus concentrations in the epilimnion and physical conditions in Paul Lake showed no trends throughout the experiment (Fig. 13.1). Thermocline depth declined through the course of each

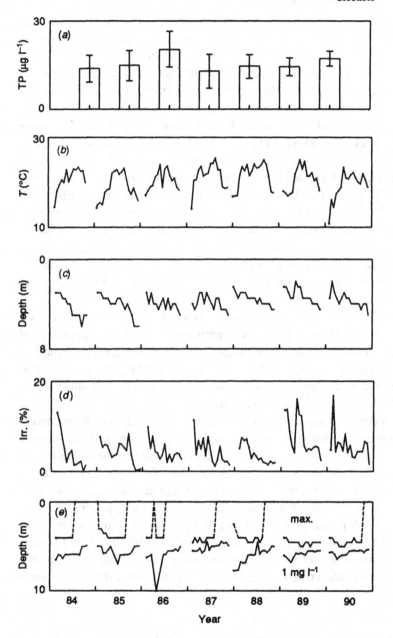

Fig. 13.1. Characteristics of the water column of Paul Lake during the summer stratified seasons of 1984–90. (*a*) Mean total phosphorus in the epilimnion (TP; error bar shows standard deviation); (*b*) epilimnetic temperature (*T*); (*c*) depth to thermocline; (*d*) percentage of surface irradiance (Irr) at the thermocline; (*e*) depth of maximum dissolved oxygen concentration (max.) and depth to 1 mg l^{-1} dissolved oxygen.

summer as the epilimnion warmed. Irradiance at the thermocline fluctuated between 1% and 10% of surface irradiance, largely as a result of variations in epilimnetic chlorophyll concentration (Elser, 1987). Depth of the photic zone (to 1% of surface irradiance) was almost constant at 5.3 m (s.e. = 0.05, $n = 110$) because metalimnetic peaks in algal concentration sharply attenuated irradiance. The hypolimnion of Paul Lake was anoxic throughout our study. Photosynthesis by metalimnetic algae often produced maximal oxygen concentrations near the thermocline, especially in the first halves of the summers.

Epilimnetic chlorophyll fluctuated consistently about a mean of 4.3 μg l^{-1} (Fig. 13.2a). When integrated over the entire photic zone, chlorophyll was more variable because concentrations in the metalimnion were sometimes quite high (Fig. 13.2b,c). Both alkaline phosphatase activity (Fig. 13.2d) and ammonium enhancement response (Fig. 13.2e) were variable around roughly constant means. There is no indication of sustained changes in nutrient limitation in Paul Lake during the experiment.

Primary production rates varied from week to week but showed no systematic trends throughout the experiment (Fig. 13.3). This pattern of fluctuation about a stationary mean is similar to the patterns found in chlorophyll concentration, chemical variates, and physical variates. The epilimnion and metalimnion made roughly equal contributions to integrated water column primary production.

Cross correlations of selected time series from Paul Lake do not detect relationships among nutrient, phytoplankton or zooplankton variates (Fig. 13.4). The time series were filtered to remove any autoregressive or seasonal effects before cross correlations were calculated (see Chapter 3). The only cross correlation function that suggests causal relationships is that between alkaline phosphatase activity and primary production (Fig. 13.4a). Alkaline phosphatase activity and primary production are negatively correlated at lag 0, suggesting phosphorus limitation of primary production. The positive correlations at negative lags suggest that periods of high primary production are followed by phosphorus limitation.

Peter Lake

In Peter Lake, herbivory was altered in several episodes during 1985, 1988, 1989 and 1990 (Chapter 10). Therefore, we tested for short-term relations between herbivory and algal variates using transfer functions

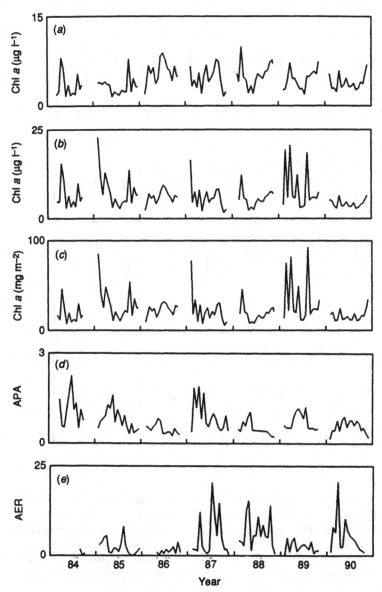

Fig. 13.2. Chlorophyll and indicators of nutrient limitation in Paul Lake during the summer stratified seasons of 1984–90. (a) Chlorophyll (Chl a) concentration in the epilimnion; (b) chlorophyll concentration in the photic zone; (c) areal chlorophyll density integrated over the photic zone; (d) specific alkaline phosphatase activity (APA) in the epilimnion (nmol PO_4 μg Chl^{-1} min^{-1}); (e) ammonium enhancement response (AER) in the epilimnion (T statistic, dimensionless).

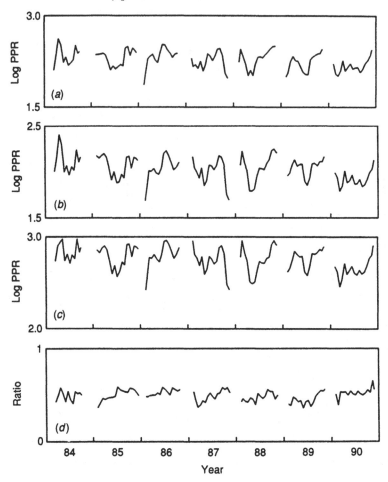

Fig. 13.3. Primary production (PPR) of Paul Lake during the summer stratified seasons of 1984–90. (Note logarithmic scales of panels *a–c*.) (*a*) Volumetric primary production of the epilimnion (mg C m^{-3} d^{-1}); (*b*) volumetric primary production of the photic zone (mg C m^{-3} d^{-1}); (*c*) areal primary production integrated over the photic zone (mg C m^{-2} d^{-1}); (*d*) primary production of the epilimnion as a fraction of photic zone primary production (dimensionless ratio).

(Chapter 3). Such relations may not be readily apparent from plots of the time series.

Concentrations of total phosphorus in the epilimnion of Peter Lake showed little variability from year to year during the experiment (Fig. 13.5*a*). Within each summer, week-to-week patterns of epilimnion temperature and thermocline depth in Peter Lake appeared similar to those of Paul Lake (Fig. 13.5*b,c*). The percentage of surface light reaching

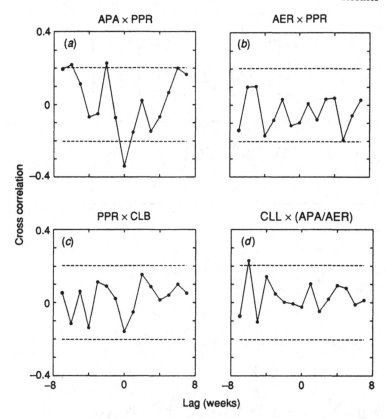

Fig. 13.4. Cross correlation functions for filtered time series for selected variates measured in Paul Lake during the summer stratified seasons of 1984–90. (*a*) Specific alkaline phosphatase activity (APA) vs. volumetric primary production of the epilimnion (PPR); (*b*) ammonium enhancement response (AER) vs. volumetric primary production of the epilimnion (PPR); (*c*) volumetric primary production of the photic zone (PPR) vs. cladoceran biomass (CLB); (*d*) cladoceran length (CLL) vs. the ratio of APA to AER.

the thermocline varied from 1% to more than 30% (Fig. 13.5*d*). This range is greater than observed in Paul Lake because Peter Lake is less stained by organic acids (Elser *et al.*, 1986*b*; Elser, 1987). As in Paul Lake, the depth of the photic zone (to 1% of surface irradiance) is nearly constant (mean = 7.2 m, s.e. = 0.06, *n* = 109) because of high light attenuation by dense populations of deep–dwelling algae. Metabolism of these algae produces a deep oxygen maximum at almost all times during the stratified period in Peter Lake (Fig. 13.5*e*). The hypolimnion of Peter Lake was anoxic throughout our experiment (Fig. 13.5*e*).

Few patterns are apparent in time series of chlorophyll and nutrient

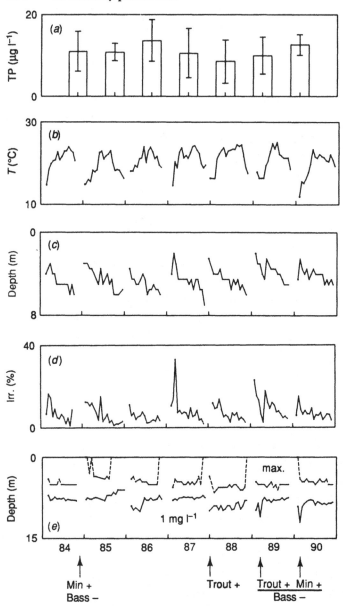

Fig. 13.5. Characteristics of the water column of Peter Lake during the summer stratified seasons of 1984–90. (*a*) Mean total phosphorus in the epilimnion (TP; error bar shows standard deviation); (*b*) epilimnetic temperature (*T*); (*c*) depth to thermocline; (*d*) percentage of surface irradiance (Irr.) at the thermocline; (*e*) depth of maximum dissolved oxygen concentration and depth to 1 mg l^{-1} dissolved oxygen.

deficiency indicators (Fig. 13.6). As in Paul Lake, photic zone chlorophyll is more concentrated and more variable than epilimnetic chlorophyll because of dense layers of algae in the metalimnion and upper hypolimnion. Alkaline phosphatase activity, though variable, appears higher in 1984, early 1985 and 1987 (Fig.13.6*d*). These are periods when large *Daphnia pulex* dominated the zooplankton (Chapter 8). Ammonium enhancement response reached unusually high values in late 1987 (Fig. 13.6*e*), coincident with blooms of the bluegreen alga *Aphanocapsa elachista* in the lake (Chapter 11).

Primary production rates, like chlorophyll concentrations, have few apparent patterns (Fig. 13.7). The sharp rise in production associated with the collapse of large *Daphnia* in 1985 (discussed by Carpenter *et al.*, 1987) is apparent (Fig. 13.7*a–c*). As in Paul Lake, the epilimnion and metalimnion make roughly equal contributions to integrated water column production (Fig. 13.7*d*).

Transfer functions revealed several significant relationships between anomalies in cladoceran length and anomalies in algal variates (Table 13.2). We used herbivore length, rather than biomass, as the independent variate for transfer functions. Cladoceran length is expected to be highly responsive to size-selective predation (Brooks & Dodson, 1965; Chapter 10) and is closely correlated with grazing rates (Peters & Downing, 1984). Herbivore biomass, on the other hand, responds to both predation and food availability (Chapter 10) and the transfer functions would therefore be harder to interpret. In the transfer functions of Table 13.2, numerator (ω) terms cause a shift in the dependent variate, while denominator (δ) terms, if present, affect the rate of the shift (Chapter 3).

Anomalous increases in cladoceran length had negative effects on chlorophyll and primary production (Table 13.2). The algal responses are significant biologically as well as statistically. For example, the effect on photic zone chlorophyll is -1.66 μg l^{-1} per 1 mm increase in cladoceran length. This change represents a 41% decline relative to the long-term mean concentration of 4.1 μg l^{-1}. The pulse effects on primary production of about 0.2 log units represent a 38% decline relative to the long-term mean. (Chapter 3 explains the interpretation of transfer function coefficients for log-transformed variates.) However, the proportion of primary production in the epilimnion was not affected by anomalies in cladoceran length.

Increases in cladoceran length also affected physiological indicators of nutrient limitation. A 1 mm increase in mean cladoceran length caused alkaline phosphatase activity to increase by a factor of 2.2, but had no

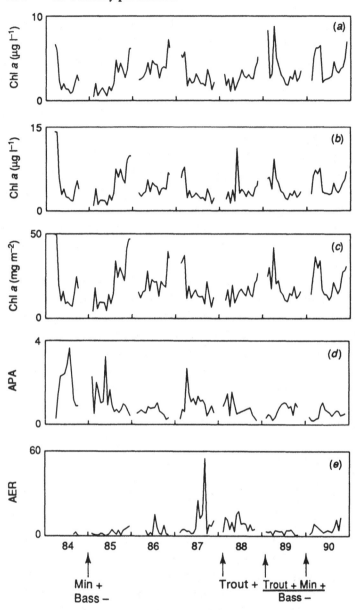

Fig. 13.6. Chlorophyll and indicators of nutrient limitation in Peter Lake during the summer stratified seasons of 1984–90. (*a*) Chlorophyll concentration in the epilimnion; (*b*) chlorophyll concentration in the photic zone; (*c*) areal chlorophyll density integrated over the photic zone; (*d*) specific alkaline phosphatase activity (APA) in the epilimnion (nmol PO_4 μg Chl^{-1} min^{-1}); (*e*) Ammonium enhancement response (AER) in the epilimnion (*t*-statistic, dimensionless).

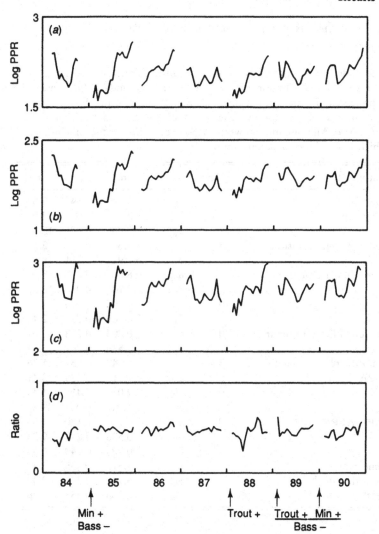

Fig. 13.7. Primary production of Peter Lake during the summer stratified seasons of 1984–90. (Note logarithmic scales of *a–c*.) (*a*) Volumetric primary production (PPR) of the epilimnion (mg C m^{-3} d^{-1}); (*b*) volumetric primary production of the photic zone; (*c*) areal primary production integrated over the photic zone (mg C m^{-2} d^{-1}); (*d*) Primary production of the epilimnion as a fraction of photic zone primary production.

Table 13.2. *Coefficients of transfer functions relating cladoceran length to algal variates in Peter Lake*

For each dependent variate, the delay in weeks, transfer function coefficients (numerator terms ω and denominator terms δ), estimates, standard errors and T ratios are presented. T ratios greater than 1.96 indicate significant effects (at the 5% level) of anomalies in cladoceran length on subsequent anomalies in the dependent variate. All noise models were AR(1) except that for thermocline depth, which also included a seasonal term.

Dependent variate	Delay	Term	Estimate	s.e.	T ratio
epilimnion Chl	0	ω_0	-1.74	0.84	2.07
metalimnion Chl	6	ω_0	-3.20	1.06	3.01
		δ_1	-0.886	0.071	12.4
photic zone Chl concentration	6	ω_0	-1.66	0.65	2.55
		δ_1	-0.917	0.051	18.1
photic zone Chl per unit area	6	ω_0	-8.86	3.37	2.63
		δ_1	-0.904	0.056	16.2
log photic zone PPR per unit volume	0	ω_0	-0.204	0.076	2.67
		δ_1	0.566	0.231	2.45
log photic zone PPR per unit area	0	ω_0	-0.203	0.078	2.61
		δ_1	0.539	0.241	2.24
epilimnion PPR/photic zone PPR	3	ω_0	-0.042	0.031	1.36
specific APA	0	ω_0	1.11	0.325	3.41
AER	0	ω_0	2.04	1.41	1.44
		δ_1	-1.04	0.035	29.8
specific APA/AER	0	ω_0	0.729	0.180	4.04
		ω_1	-0.536	0.240	2.23
		δ_1	-0.929	0.140	6.63
thermocline depth	7	ω_0	1.51	0.435	3.48
irradiance at thermocline	1	ω_0	5.74	2.39	2.40
photic depth	2	ω_0	0.305	0.355	0.86

significant effect on ammonium enhancement response. The ratio of alkaline phosphatase activity to ammonium enhancement response was also increased significantly by increases in cladoceran length.

The reductions in chlorophyll that followed increases in cladoceran length caused irradiance at the thermocline to increase and, with a lag of seven weeks, deepened the thermocline. Photic zone depth was unaffected because it is controlled by the position of the deep algal maximum. These physical changes are limnologically significant. The increase in percentage of surface light reaching the thermocline (5.74%) after a 1 mm increase in cladoceran length represents an 82% increase in irradiance. The subsequent increment in thermocline depth of 1.5 m is

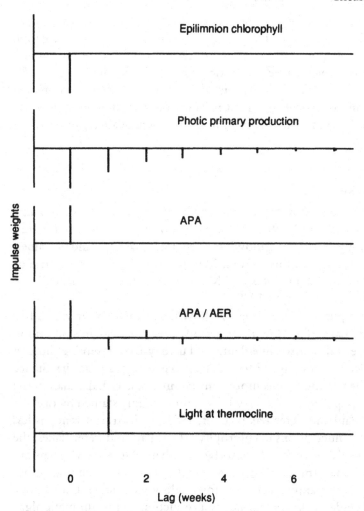

Fig. 13.8. Impulse response weights for selected transfer functions for time series from Peter Lake, 1984–90 (Table 13.2). In all cases, cladoceran mean length is the independent variate. Dependent variates are chlorophyll concentration in the epilimnion, volumetric primary production in the photic zone, specific alkaline phosphatase activity (APA), the ratio of specific alkaline phosphatase activity to ammonium enhancement response (APA/AER) and irradiance at the thermocline.

substantial relative to the long-term mean thermocline depth of 4.4 m.

The coefficients of the transfer functions can be converted to impulse weights, which illustrate the effects of pulse perturbations of cladoceran length on the response variates, as explained in Chapter 3 (Fig. 13.8). Transfer functions with a single numerator coefficient indicate a simple pulse effect with the sign and magnitude of the coefficient (epilimnion chlorophyll, Fig. 13.8*a*; alkaline phosphatase activity, Fig. 13.8*c*; irradiance at the thermocline, Fig. 13.8*e*). Transfer functions with negative numerator and positive denominator terms denote a decline which develops over several weeks (photic primary production, Fig. 13.8*b*). The more complicated 3-term model for the alkaline phosphatase activity/ammonium enhancement response ratio yields a pulse response followed by weights that decay in magnitude while alternating in sign (Fig. 13.8*d*).

Tuesday Lake

In Tuesday Lake, herbivory was altered by two major, sustained manipulations (Chapter 10). We used intervention analysis to test for step changes in algal variates following these rather abrupt and sustained shifts in herbivory. In this section, Manipulation 1 refers to the rise of large *Daphnia* in August 1985, and Manipulation 2 refers to the collapse of large *Daphnia* in August 1988.

Total phosphorus concentrations show little year-to-year variability in Tuesday Lake (Fig. 13.9*a*). Week-to-week changes in epilimnion temperature and thermocline depth in Tuesday Lake resemble those in Paul and Peter Lakes (Fig. 13.9*b,c*). The percentage of surface irradiance reaching the thermocline is more constant in Tuesday Lake than in the other lakes (Fig. 13.9*d*). Tuesday Lake is more highly stained by organic acids than Paul and Peter Lakes (Elser, 1987), so its transparency is less responsive to fluctuations in chlorophyll. Like Paul and Peter Lakes, the depth of the photic zone (1% of surface irradiance) is relatively constant in Tuesday Lake (mean = 3.3 m, s.e. = 0.07, $n = 109$). Staining, rather than deep algal maxima, accounts for the low variance in photic zone depth in Tuesday Lake. Nevertheless, production by metalimnetic algae is sometimes sufficient to produce a metalimnetic oxygen maximum in Tuesday Lake (Fig. 13.9*e*). Tuesday Lake never had oxygen below 5 m depth during our experiment (Fig. 13.9*e*).

Chlorophyll concentrations apparently declined after Manipulation 1 (Fig. 13.10*a–c*). Manipulation 1 was also followed by increased alkaline phosphatase activity (Fig. 13.10*d*) and decreased ammonium enhancement response. These changes indicate a shift toward greater phosphorus limitation and less nitrogen limitation (Elser *et al.*, 1988). Changes after Manipulation 2 are less apparent.

Primary production declined after Manipulation 1 and increased after Manipulation 2 (Fig. 13.11*a–c*). As in Paul and Peter Lakes, the epilim-

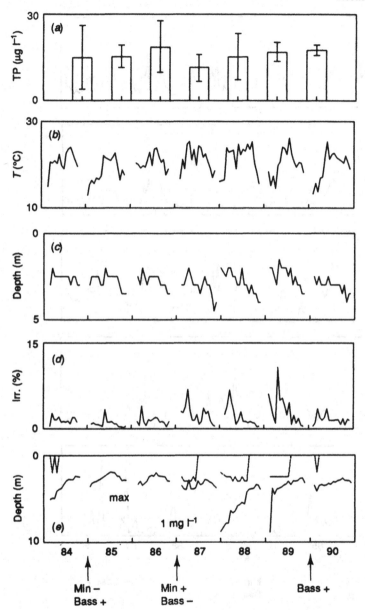

Fig. 13.9. Characteristics of the water column of Tuesday Lake during the summer stratified seasons of 1984–90. (a) Mean total phosphorus (TP) in the epilimnion (error bar shows standard deviation); (b) epilimnetic temperature (T); (c) depth to thermocline; (d) percentage of surface irradiance (Irr.) at the thermocline; (e) depth of maximum dissolved oxygen concentration and depth to 1 mg l^{-1} dissolved oxygen.

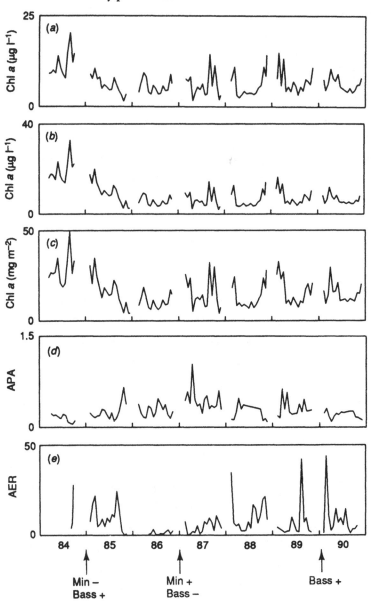

Fig. 13.10. Chlorophyll and indicators of nutrient limitation in Tuesday Lake during the summer stratified seasons of 1984–90. (*a*) chlorophyll concentration in the epilimnion; (*b*) chlorophyll concentration in the photic zone; (*c*) areal chlorophyll density integrated over the photic zone; (*d*) specific alkaline phosphatase activity (APA) in the epilimnion (nmol PO_4 μg Chl^{-1} min^{-1}); (*e*) ammonium enhancement response (AER) in the epilimnion (*t*-statistic, dimensionless).

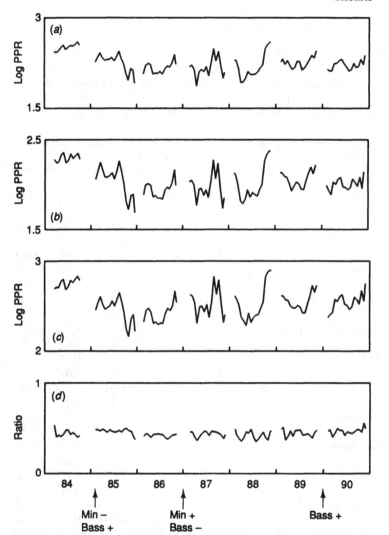

Fig. 13.11. Primary production of Tuesday Lake during the summer stratified seasons of 1984–90. (Note logarithmic scales of panels *a–c*.) (*a*) Volumetric primary production (PPR) of the epilimnion (mg C m^{-3} d^{-1}); (*b*) volumetric primary production of the photic zone; (*c*) areal primary production integrated over the photic zone (mg C m^{-2} d^{-1}); (*d*) Primary production of the epilimnion as a fraction of photic zone primary production (dimensionless).

Table 13.3. *Results of intervention analyses in Tuesday Lake*

For each dependent variate and manipulation, the estimated step change effect, standard error and T ratio are reported. T ratios greater than 1.96 are significant at the 5% level. Manipulation 1 is the onset of large *Daphnia* in August 1985. Manipulation 2 is the collapse of large *Daphnia* in August 1988. All noise models were AR(1) except that for thermocline depth, which also included a seasonal term.

Dependent variate	Manipulation	Effect	s.e.	T
epilimnion Chl	1	−5.76	1.10	5.15
	2	2.27	1.00	2.27
photic zone Chl concentration	1	−9.25	1.57	5.87
	2	1.42	1.36	1.04
photic zone Chl per unit area	1	−11.9	5.17	2.31
	2	2.34	4.49	0.52
log photic zone PPR per unit volume	1	−0.172	0.051	3.36
	2	0.117	0.046	2.53
log photic zone PPR per unit area	1	−0.214	0.104	2.05
	2	0.120	0.057	2.11
epilimnion PPR/photic zone PPR	1	−0.017	0.014	1.14
	2	0.006	0.014	0.41
specific APA	1	0.162	0.047	3.41
	2	−0.071	0.043	1.65
AER	1	−6.99	0.357	1.96
	2	3.35	2.43	1.38
specific APA/AER	1	0.0993	0.0442	2.24
	2	−0.0707	0.0304	2.33
thermocline depth	1	0.261	0.167	1.56
	2	−0.024	0.161	0.15
irradiance at thermocline	1	0.639	0.576	1.11
	2	−0.300	0.545	0.55
photic zone depth	1	−0.279	0.212	1.32
	2	−0.265	0.191	1.39

nion consistently accounted for about half of the primary production throughout the experiment.

Intervention analyses showed that responses to Manipulation 1 were significant statistically as well as biologically (Table 13.3). Photic zone chlorophyll concentration decreased 5.8 μg l⁻¹, comparable to the seven-year mean of 6.7 μg l⁻¹. Primary production declined on both a volumetric (33%) and areal (38%) basis. (Chapter 3 explains the interpretation of intervention analyses for log-transformed variates). Shifts in nutrient deficiency indicators also appeared significant. The increase in alkaline phosphatase activity was equivalent to 62% of the long-term mean, while the decline in ammonium enhancement response equalled 93% of the long-term mean. Physical variates (thermocline depth, irradiance at the thermocline, and photic depth) did not respond because they are controlled by the staining of the water rather than chlorophyll.

Responses to Manipulation 2 were weaker than responses to Manipulation 1 (Table 13.3). Epilimnion chlorophyll increased by 2.3 μg l^{-1}, a substantial change relative to the long-term mean of 6.7 μg l^{-1}. Metalimnetic chlorophyll was less responsive, and consequently changes in photic zone chlorophyll could not be detected. Primary production increased by 31%, on both areal and volumetric bases. The ratio of alkaline phosphatase activity to ammonium enhancement response decreased, suggesting less limitation by phosphorus and greater limitation by nitrogen. However, shifts could not be detected in either indicator alone.

Because several observations were missing from the time series for TP, the data were not suitable for time series analyses. Randomized intervention analysis (Chapter 3) was used to test for changes in total P following both manipulations. No significant changes were detected.

Discussion

Changes in herbivory altered algal biomass and primary production in directions consistent with the trophic cascade hypothesis in Peter and Tuesday Lakes. In Paul Lake, we did not detect the herbivore–alga relations evident in Peter Lake. Also, Paul Lake does not exhibit the shifts observed in Tuesday Lake in 1985 and 1988. The patterns observed in the manipulated lakes do not appear in the reference lake, suggesting that the responses were caused by the manipulations.

Manipulation of Peter and Tuesday Lakes produced quite different dynamics. The most striking difference is the duration of the ecosystem responses.

Fish manipulations in Tuesday Lake achieved sustained shifts in herbivory (Chapter 10). After the rise in *Daphnia* in August 1985, chlorophyll and primary production declined rapidly and fluctuated around a lower mean until the collapse of *Daphnia* in August 1988. During 1989 and 1990, chlorophyll and primary production of Tuesday Lake recovered only partly from the reductions that occurred in 1985. The lack of recovery in chlorophyll is largely explainable by reduced populations of the larger species of *Peridinium* (Chapter 11). Primary production recovered to a greater extent. The greater response of primary production is explained by regrowth of populations of small, relatively edible algae (Chapter 11). The smaller size classes of rapidly growing phytoplankton account for a disproportionately large share of the primary production in these lakes (Elser *et al.*, 1986*a*).

Peter Lake underwent short-term disturbances of herbivory lasting 2–10 weeks in the summers of 1985 and 1988–90 (Chapter 10). These pulses contrast with the sustained shifts that occurred in Tuesday Lake. Responses of chlorophyll and primary production were similar in duration to the pulse disturbances in herbivory.

Although the time scales of response were much different for Peter and Tuesday Lakes, the shifts in integrated chlorophyll and primary production are similar in magnitude. On a volumetric basis, Peter Lake has lower chlorophyll and primary production than Tuesday Lake. On an areal basis, however, chlorophyll and primary production are similar in the two lakes, because Peter Lake's photic zone is thicker. In Peter Lake, a 1 mm increase in cladoceran length decreases areal chlorophyll by 9 mg m^{-2} and areal primary production by 37%. The first manipulation in Tuesday Lake increased cladoceran length about 0.8 mm, decreasing areal chlorophyll by 12 mg m^{-2} and areal primary production by 39%. These responses are remarkably similar.

Zooplankton can change the light penetration and thermal structures of lakes by altering phytoplankton biomass and size distributions (Mazumder *et al.*, 1990). Evidence for this phenomenon has come from multilake comparisons and *in situ* enclosure experiments (Mazumder *et al.*, 1990). Results from Peter Lake provide further support for the existence of this feedback. Increases in cladoceran length in Peter Lake cause increases in irradiance at the thermocline, photic zone depth and thermocline depth. Because the staining of the water by humic substances controls light attenuation in Tuesday Lake, effects of herbivory on photic depth and mixing are not evident there. Food web structure can affect transparency and heating in lakes where light absorption is controlled mainly by the phytoplankton. In organically stained lakes, or lakes and reservoirs with high turbidity from suspended clays, these physical effects of food web structure may be less apparent.

In pelagic communities of stratified lakes, nutrient excretion by grazers is a major source of nutrients for the phytoplankton (Lehman, 1980). Nutrient excretion rates can therefore be viewed as nutrient supply rates in the context of phytoplankton competition theory (Tilman, 1982). The outcome of competition among phytoplankton depends on the supply ratio of limiting nutrients (Tilman, 1982; Sommer, 1989). Zooplankton destabilize the N : P supply ratio to their algal prey, and thereby accentuate the rate of community change due to competition for nutrients (Sterner, 1990). These effects of grazers on algal competition are independent of any compositional effects of selective grazing. They imply that changes in herbivory can cause changes in nutrient limitation of the phytoplankton (Elser *et al.*, 1988; St. Amand *et al.*, 1989). In Peter and Tuesday Lakes, increases in herbivore size accentuate phosphorus limitation, while decreases in herbivore size accentuate nitrogen limitation. Previously published enclosure experiments in these lakes show that shifts in density of large cladocerans are responsible for the shifts in nutrient limitation (Elser *et al.*, 1988). While mechanisms of herbivore-induced shifts in nutrient limitation await further analysis (see below), the significance of the phenomenon at the ecosystem scale is evident from our results. Biogeochemical processes and food web structure interact to determine nutrient supply rates to phytoplankton and the relative intensity of nitrogen and phosphorus limitation in these ecosystems.

Recent empirical and theoretical studies shed some light on the mechanisms by which herbivores alter nutrient limitation of algae. Sterner (1990) presents a model of the effects of zooplankton on the N : P ratio of regenerated nutrients as a function of the N : P ratio of the algae and the N : P ratio of the grazer. On the basis of mass balance, the N : P ratio of regenerated nutrients can diverge substantially from the N : P ratio of the food, if zooplankton physiologically regulate their body N : P ratio. Zooplankton feeding on seston with a N : P ratio greater than their body ratio, for example, are predicted to recycle N and P at a ratio even greater than that in the food. This high N : P ratio would accentuate the severity of P limitation in the algae by further shifting the supply ratio in favor of N. In turn, for a given food N : P, major differences in the N : P ratio of regenerated nutrients are predicted for grazers that differ in their body N : P ratios.

The validity of these predictions, and their importance in relation to the dynamics of N and P limitation in our lakes, rest on two assumptions. First, Sterner's (1990) stoichiometric theory depends on zooplankton grazers being homeostatic with respect to their body N : P ratio, regulating it within narrow bounds. Recent observations and experiments suggest that this assumption is met: despite extreme variability in food N : P ratios, the zooplankton taxa that have been studied have elemental compositions that are narrowly constrained (Andersen & Hessen, 1991;

Hessen, 1990; Sterner, Elser & Hessen, 1992). Apparently the N and P storage capacities of zooplankton, unlike those of phytoplankton and bacteria, are modest and the element in excess of immediate requirements is disposed of. Second, on the basis of Sterner's theory, changes in nutrient limitation such as those we have observed in Peter and Tuesday Lakes would be predicted only if the food-web manipulations caused changes in the body N : P of the zooplankton. Such changes may well have occurred. Hessen (1988) reports large taxon-specific differences in body N : P of freshwater zooplankton. In general, copepod taxa have N : P ratios higher than those in cladocerans. *Daphnia* has an especially low N : P (*ca.* 12 : 1 by mass, in contrast to two calanoid copepod species with ratios of 39–52). So, when food web manipulations induce shifts between dominance of small-bodied copepod taxa and large-bodied *Daphnia*, as occurred most dramatically in Manipulation 1 in Tuesday Lake, the N : P supply ratio due to zooplankton excretion also shifts (Sterner *et al.*, 1992). As size-selective predation is especially effective in changing zooplankton size distributions (Brooks & Dodson, 1965), food web manipulations may often shift the N : P supply ratio available to phytoplankton.

Summary

Paul Lake, the reference ecosystem, showed no trends that suggest regional changes in determinants of primary production. Relations among nutrient, phytoplankton and zooplankton variates in Paul Lake were weak and generally not significant statistically. The one exception was the relation between alkaline phosphatase activity and primary production, which suggests that production is phosphorus-limited and that periods of high production were followed by periods of phosphorus limitation.

In Peter Lake, changes in cladoceran size following fish manipulations had several significant effects. Increased cladoceran size caused decreases in chlorophyll and primary production, and increases in phosphorus limitation of algae, irradiance at the thermocline, and thermocline depth. These shifts were brief, lasting less than a growing season, but were significant statistically and were large relative to the long-term means for the lake.

In Tuesday Lake, establishment of large cladocerans in 1985 caused significant decreases in chlorophyll, primary production and nitrogen limitation of algae, and significant increases in phosphorus limitation of algae. These shifts were large relative to long-term mean conditions in the lake, and were sustained from 1985 to 1987. Because Tuesday Lake (unlike Peter Lake) has stained water, changes in algal biomass had no effect on physical properties of the water column such as irradiance at the thermocline and thermocline depth. Removal of large daphnids in 1988 had weaker effects than establishment of large daphnids in 1985. Nevertheless, significant increases in chlorophyll concentration and primary production followed the decline of large daphnids. This asymmetric response of Tuesday Lake to opposite food web manipulations is an example of ecosystem hysteresis.

Acknowledgements

We thank M. J. Vanni for comments on a draft.

14 · *Heterotrophic microbial processes*

Michael L. Pace

Introduction

Heterotrophic microorganisms, here defined as bacteria and protozoa, account for a major portion of secondary production and nutrient remineralization in aquatic ecosystems (Stockner & Porter, 1988; Sherr & Sherr, 1991). These organisms are potentially regulated by both their predators and by the supply of organic carbon and inorganic nutrients. In general, prior studies of the trophic cascade have not considered heterotrophic microbial processes (e.g. Carpenter *et al.*, 1987; McQueen *et al.*, 1989; Benndorf, 1990; but see Riemann & Søndergaard, 1986). In previous chapters of this volume, the significance of cascading trophic interactions in determining the structure of zooplankton communities and in explaining shifts in phytoplankton biomass and primary production has been demonstrated. In this chapter, I consider the question, are heterotrophic microbes and the biogeochemical processes mediated by them strongly influenced by these same top-down controls?

One difficulty in answering this question arises from uncertainty about the trophic interactions and recycling processes connecting heterotrophic microbes with phytoplankton, zooplankton and fish. It is an oversimplification to consider the heterotrophic microbial food web as a separate 'loop' wherein bacteria utilize dissolved organic matter and are consumed by protozoa, with energy and nutrients thereby returned to the phytoplankton–zooplankton–fish food chain (Azam *et al.*, 1983). Rather, a more current view of microbial food webs is one of a complex interacting community including phytoplankton, bacteria and protozoans that collectively accounts for primary carbon fixation, nutrient regeneration and production to support metazoans (Sherr & Sherr, 1991). From this perspective, the effect of cascading trophic interactions on heterotrophic microbes will depend on how the cascade changes microbial resources and predation on microbes.

Unfortunately, the key processes limiting the abundance, biomass and productivity of heterotrophic microbes are poorly known. For example, freshwater planktonic bacteria may be limited by the supply of carbon from algae, by phosphorus, or by these two factors plus others such as temperature and possibly predation (Coveney & Wetzel, 1988; Currie, 1991; Toolan, Wehr & Findlay, 1991). Bacteria are consumed by a variety of protozoans as well as certain zooplankton. The relative significance of these consumers depends on plankton community structure (Pace, McManus & Findlay, 1990). Cascading trophic interactions may affect both the potential supply of resources for bacteria (i.e. labile organic carbon, phosphorus) as well as the rate of predation on bacteria. Furthermore, the time scale over which bacteria respond to changes in resources or predators is not known. There may be extremely rapid compensation by bacterial communities to changes in resource supply or predation which might dampen top-down effects.

To begin to evaluate the influence of cascading trophic interactions on microbial processes, the following model was developed. Bacterial communities have enormous potential for growth and should generally be limited by the availability of resources. Resources for these organisms are high-quality organic carbon (e.g. low-molecular-mass carbohydrates, amino acids) and nutrients (e.g. inorganic or organic nitrogen and phosphorus). Both types of resources are typically present at extremely low concentrations in lake waters. If resource limitation is primary then bacterial abundance and production should be related to phytoplankton dynamics, because phytoplankton either respond to the same resources (nutrients) or mediate the resource supply (organic carbon) for bacteria. Increases or decreases in the biomass and productivity of phytoplankton related to top-down effects would lead to similar responses by bacteria (Fig. 14.1). The effects of the trophic cascade on bacteria are, therefore, viewed as **indirect**. This hypothesis suggests that direct predation on bacteria is of secondary importance in terms of regulation.

The effect of cascading trophic interactions on protozoans were proposed to be similar to those on phytoplankton. I hypothesized that shifts in zooplankton community structure to larger consumers would result in increases in direct consumption of protozoans (Fig. 14.1). This hypothesis implies that communities dominated by small zooplankton are less efficient consumers of protozoans. In addition, the hypothesis suggests protozoan abundances are determined more by direct predation from zooplankton rather than indirect effects on protozoan resources.

The goal of this study was to use the food-web manipulations to reveal significant interactions regulating microbial processes in order to evaluate the initial hypotheses proposed above. From 1988 to 1990, I analyzed the response of bacteria and protozoa to the whole-lake experimental manipulations in Peter and Tuesday Lakes relative to the reference system, Paul Lake. A series of enclosure experiments was also performed. Enclosure studies allowed short-term replicated tests of the

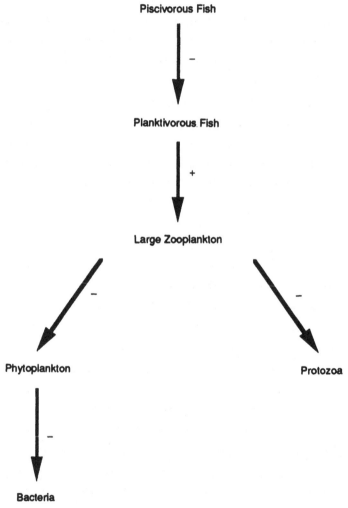

Fig. 14.1. Proposed model of the effects of cascading trophic interactions on bacteria and protozoans. Shifts to large zooplankton are postulated to reduce protozoans in a manner similar to phytoplankton, while bacteria are proposed to be indirectly suppressed by reduction in resource supply from phytoplankton.

influence of changes in zooplankton and nutrients on heterotrophic microbial processes. The whole-lake manipulation studies served as a test of how cascading food-web interactions influence microbial processes at the system scale.

The heterotrophic microbial community

The heterotrophic microbial communities observed in the photic zones of Paul, Peter and Tuesday Lakes are similar to those in other lakes (Stockner & Porter, 1988). These organisms are among the smallest found in plankton. In contrast, *Daphnia pulex* in Paul Lake are typically about 1.5 mm in diameter, and their compound eyes are about 100 μm in width. Bacteria, heterotrophic flagellates, and most ciliates are all considerably smaller than the width of a *Daphnia* eye (Fig. 14.2). Bacteria and heterotrophic flagellates are most often observed and enumerated using epifluorescence microscopy (described below in Methods). As a conse-

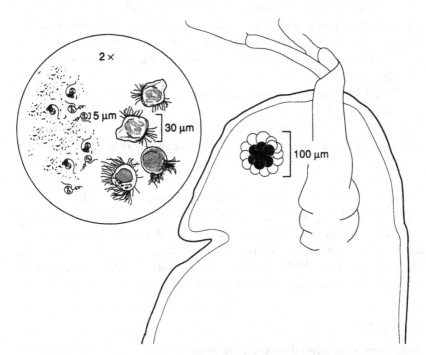

Fig. 14.2. Schematic illustration of the size of heterotrophic microbial plankton relative to *Daphnia pulex*. Note that bacteria, heterotrophic flagellates and ciliates are substantially smaller than the width of the compound eye of *Daphnia*. (Scale for microbes is 2 times the scale for *Daphnia*.)

quence of the preservation and staining procedures used with epifluorescence microscopy, it is not possible to identify these organisms except to broad functional categories. The bacteria in Paul, Peter and Tuesday Lakes are almost all smaller than 1 μm in maximum dimension and consist primarily of cocci, rods and crescents (Fig. 14.2).

Similar to the bacteria, heterotrophic flagellates in Paul, Peter and Tuesday Lakes are small with an average volume of 57 μm³ (equivalent to a diameter of *ca.* 5 μm, assuming a spherical cell). These heterotrophic flagellates are similar in size to the smallest phytoplankton. Heterotrophic flagellates are identified by the absence of chloroplasts. When illuminated with blue light, chloroplasts in algal flagellates are visible as red fluorescence. Heterotrophic flagellates appear pale green with no red fluorescence. Heterotrophic flagellates consume bacteria (Fenchel, 1986) and small phytoplankton (Sherr, Sherr & McDaniel, 1991). These organisms may also acquire nutrition directly from high-molecular-mass dissolved organic carbon (Sherr, 1988). While most flagellates occur as free-swimming forms, a group known as choanoflagellates attach to colonial phytoplankton such as the chrysophyte *Dinobryon*, and the diatom *Asterionella*. Choanoflagellates are ecologically important because they ingest bacteria at rates considerably greater than free forms (Vaqué & Pace, 1992).

Ciliated protozoans are a third important component of the heterotrophic microbial community in the study lakes. Community composition is typical of that found in the pelagic zone of lakes with choreotrichs the dominant group in terms of abundance. Other notable ciliates found in Paul, Peter and Tuesday Lakes are haptorids, peritrichs and tintinnids. Ciliates overlap in size with the most familiar forms of phytoplankton such as diatoms, dinoflagellates, colonial chrysophytes and colonial bluegreen algae (Fig. 14.2). Ciliates consume bacteria, phytoplankton and other protozoans. A number of species also contain functional chloroplasts garnered from ingested algae (Stoecker, Michaels & Davis, 1987). These species are mixotrophic (i.e. acquiring energy from photosynthesis as well as phagotrophy).

Methods

Responses to whole-lake manipulations

Each lake was sampled weekly in conjunction with the routine limnological measurements (Chapter 2). We measured heterotrophic microbial standing stocks and bacterial production at two depths. Water samples from the mixed layer at light levels of 100%, 75% and 50% surface irradiance were pooled to represent the epilimnion. A single sample was taken at the oxygen maximum to represent the metalimnion, a distinct habitat for phytoplankton (Chapter 11) and presumably for other microorganisms. When an oxygen maximum was not present, as often occurs in Tuesday Lake, metalimnetic samples were taken at the lowest metalimnetic depth before the sharp decline in oxygen. Details of our methods for measuring microbial standing stocks and processes with the exception of respiration have been presented elsewhere (Pace, McManus & Findlay, 1990; Pace & Funke, 1991) and are only briefly considered below.

Bacteria and heterotrophic flagellates

Bacteria and heterotrophic flagellates were enumerated using epifluorescence microscopy (Hobbie, Daley & Jasper, 1977; Haas, 1982). We counted heterotrophic bacteria after staining with acridine orange and concentrating the cells on $0.2\,\mu m$ pore size, polycarbonate filters that had been previously stained with irgalan black. Two replicate preparations were made for each depth; over 400 cells were enumerated on each filter. Flagellates were stained with proflavine and subsequently fixed with cold glutaraldehyde (final concentration 1%). Samples were concentrated under low vacuum filtration on $1\,\mu m$ pore size, polycarbonate filters (stained black as above). Two replicate preparations were made for each depth, and at least 40 flagellates were counted on each filter. All preparations for both bacteria and heterotrophic flagellates were made within a few hours of returning from the field.

Bacterial production

Bacterial production was measured by following the incorporation of [³H]thymidine into DNA. The complications and uncertainties associated with this technique have been widely discussed (e.g. Moriarty, 1986; Bell, 1988; Jeffrey & Paul, 1988; Cho & Azam, 1988; Coveney & Wetzel, 1988) and will not be reviewed here. In this study we assume that thymidine incorporation represents a measure of bacterial production comparable among experimental and reference ecosystems. For each depth, six replicate 10 ml samples were taken and amended with $20\,\mu Ci$ of methyl [³H]thymidine (specific activity 80 Ci $mmol^{-1}$). Biological uptake of thymidine in two of the six samples was immediately inhibited by adding either formalin or cold thymidine (see below). Samples were incubated in the dark at *in situ* temperatures for 1 h, filtered on 0.2 μm pore size, polycarbonate filters, and rinsed thoroughly with 5% ice-cold trichloroacetic acid. Filters were frozen and subsequently extracted using acid–base hydrolysis to isolate DNA (Findlay, Meyer & Edwards, 1984). Radioactivity in the DNA fraction was determined by liquid scintillation counting.

To control for nonbiological absorption of the radiolabel and to stop uptake at the end of the 1 h incubations, we used two procedures. In 1988 and 1989, 2 ml of a 5% formaldehyde solution (final concentration 1%) were added to the samples to stop thymidine uptake. Radioactivity in blanks with the 5% formaldehyde solution were generally <10% of sample values, but at times were higher. In 1990 we changed procedures and stopped thymidine uptake by adding 2 ml of a 10 mM solution of unlabeled (cold) thymidine. The two methods were cross-calibrated in 1990 by monthly comparisons in each lake at each sampling depth. There was no difference between cold thymidine and formaldehyde blanks (paired *t* test, $p = 0.65$, $n = 24$). Surprisingly, however, measured thymidine incorporation into DNA was consistently lower in samples where the incubation was stopped with a formaldehyde solution as opposed to cold thymidine. Time course experiments indicated that additions of cold thymidine completely inhibited further uptake of radiolabeled thymidine so that differences between the two methods must be due to some effect of formaldehyde. Apparently, formaldehyde either degraded DNA or lowered the recovery of DNA in the extracted samples. We corrected the 1988 and 1989 data for this effect using a regression equation derived from the cross-calibration comparisons.

Bacterial production data are presented in terms of micrograms of carbon produced per liter per day (μg C l^{-1} d^{-1}). These calculations require conversion of incorporation rates (pmol l^{-1} h^{-1}) into carbon units. We used a factor of 5.2×10^9 cells produced per nanomole of thymidine incorporated into DNA (Smits & Riemann, 1988). This conversion factor is higher than common values of $1-2 \times 10^9$ cells produced per nanomole of thymidine incorporated. The latter conversion factors, however, are most often used in conjunction with thymidine uptake into cold-TCA fractions (as opposed to DNA extracts). Empirical evidence from growth experiments (Smits & Riemann, 1988) suggests that the value we adopted is appropriate for our method. To convert cells produced into carbon, we determined the average size of bacteria in Peter, Paul and Tuesday Lakes (mean ±95% confidence interval $= 0.0213 \pm 0.0021$ μm^3, $n = 959$). The conversion factors of Lee & Fuhrman (1987) and Simon & Azam (1989) are both about 400 fg C μm^{-3} for the smallest bacteria considered in each of these studies. These factors suggest a carbon content of 8 fg C per cell for the average bacterium in the three lakes.

Respiration

Respiration was measured weekly in each lake by determining the consumption of O_2 in 24 h dark bottle incubations. Incubations were conducted *in situ*. Oxygen concentration in initial samples and incubated samples were determined using the Winkler method (Wetzel & Likens, 1991). The analytical limit to detecting a difference between two samples was 1.6 μM.

Enclosure studies

Manipulations of herbivores and nutrients
We performed a series of experiments in Paul and Peter Lakes to address the effects of *Daphnia* and nutrient additions on heterotrophic microbes. These studies were conducted in 45 l enclosures, described by Elser *et al.* (1988), which were open to the surface. Enclosure experiments were all conducted *in situ* over a four-day period. We measured a series of parameters in these studies to assess the response of various components of the planktonic community including chlorophyll *a*, bacterial production, and the abundance of bacteria, heterotrophic flagellates and ciliates. Details of the methods used to measure these parameters are either described above or detailed in Pace & Funke (1991).

Two experimental designs were used. In the first set of experiments, triplicate enclosures with and without additions of *Daphnia* were compared. Experiments were conducted simultaneously in Peter and Paul Lakes in June 1988. A *t* test assuming equal variances between treatments was used to compare the means. *Daphnia* were primarily *D. pulex* with some *D. rosea*. Final densities of *Daphnia* in the enclosures were *ca.* 40 animals l^{-1}. In the second set of experiments there were four treatments: controls, *Daphnia* additions, nutrient additions, and additions of *Daphnia* plus nutrients. Experiments were conducted in August 1988 in Peter Lake and August 1989 in Paul Lake. Final concentrations of *Daphnia* were 11 animals l^{-1} in the Peter experiment and 16 animals l^{-1} in the Paul experiment. The nutrients NH_4Cl and KH_2PO_4 were added to final concentrations of 10 μmol l^{-1} N and 1 μmol l^{-1} P. A two-factor ANOVA was used to compare treatments.

Effects of organic carbon, phosphorus and nitrogen on bacteria
Enclosure experiments were also conducted to test the effects of potentially limiting resources on bacterial communities. Organic carbon, inorganic nitrogen and inorganic phosphorus were added to enclosures in a factorial design to test which of these resources alone or in combination would promote the growth of bacteria. In addition, changes in phytoplankton as measured by chlorophyll were assessed to determine if increases in phytoplankton drive changes in bacterial communities.

Two experiments were conducted in Paul Lake during 1989 using the enclosures described above. In the initial experiment we added glucose, NH_4Cl and KH_2PO_4 to final concentrations of 180 μmol C l^{-1}, 23 μmol N l^{-1}, 0.65 μmol P l^{-1} representing a C:N:P ratio (by atoms) of 106:16:1. These nutrients were added singly and in all possible combinations to enclosures. Each treatment was run in triplicate, and no additions were made to three enclosures (controls). Enclosures were filled with surface lake water, and 250 ml were removed to provide a sample of initial conditions. Treatments were applied randomly to each enclosure. Enclosures were sampled again after 48 h. A second experiment was conducted in June 1989 following the same design and protocol except that nutrient additions were reduced $4 \times$ (44 μmol C l^{-1}, 5.7 μmol N l^{-1}, 0.16 μmol P l^{-1}).

In each enclosure for the initial and final samples, duplicate preparations were made for bacterial counts. Chlorophyll was measured following the methods described in Chapter 13. In the June experiment bacterial size was determined for 100 cells from each enclosure using an image analyzer as described in Findlay *et al.* (1991). A factorial analysis of variance was used to compare treatments.

Results

Enclosure experiments

Manipulations of Daphnia *and nutrients*
Results of the enclosure studies manipulating *Daphnia* and nutrients have been presented elsewhere (Pace & Funke, 1991). These results are briefly reviewed here in the context of the model effects of the trophic cascade on the heterotrophic microbial community proposed in Fig. 14.1. The primary contrast between Peter and Paul Lakes during the two single

factor (*Daphnia* additions vs. controls) experiments was that *Daphnia* populations were near normal densities in Paul Lake (9 animals l^{-1}) but at low densities in Peter Lake (< 0.1 animals l^{-1}). In the Peter Lake experiment, therefore, the *Daphnia* addition represented a shift of the zooplankton community to larger herbivores. In both experiments, bacterial abundance and production were very similar in treatments with and without *Daphnia*. In the Peter Lake experiment reductions of chlorophyll (28%) were observed, and this result was consistent with the hypothesis that a shift in zooplankton community structure to large herbivores would result in direct reduction of phytoplankton biomass. The results of the Peter Lake experiment, however, were not consistent with the hypothesis that changes in bacteria would be related to changes in phytoplankton as no effect on either bacterial abundance or production was detected.

When the factorial experiments were conducted, *Daphnia* was relatively abundant in Peter Lake (15 animals l^{-1}) and present at nominal densities in Paul Lake (2.5 animals l^{-1}). Again, *Daphnia* had little or no effect on bacterial abundance and production with one exception. In the Paul Lake experiment additions of *Daphnia* had a positive effect on bacterial growth. The experiment increased the density of *Daphnia* about eight times over the background condition in the lake. The regeneration of organic carbon and/or nutrients by *Daphnia* may explain the stimulation of production in the Paul Lake experiment.

Nutrients had a strong positive impact on bacterial abundance and production in the factorial experiments. For example, there were about twice as many bacteria in treatments with nutrient additions (Fig. 14.3). Although we did not document cell size, bacteria in the nutrient addition treatments were much larger indicating bacterial biomass was also significantly stimulated by nutrient additions.

Phytoplankton did not respond to nutrient additions in the same manner as bacteria. In the combined *Daphnia*–nutrient treatment the increase in chlorophyll was less than in the nutrient addition treatment. Bacteria were not affected by the presence of *Daphnia* as illustrated by plots of chlorophyll versus bacteria on day 4 in the Peter and Paul experiments (Fig. 14.3). In terms of our initial hypothesis, the results of these experiments are consistent with the idea of resource limitation of bacterial communities. Nutrients rather than the availability of algal material (i.e. release of dissolved organics plus the production of detritus), however, appeared to be the primary factor determining bacterial abundance and production.

Daphnia additions had a strong effect on protozoans. For example, in all four experiments there were control treatments (no additions of either *Daphnia* or nutrients) and *Daphnia* addition treatments. In the *Daphnia* addition treatments protozoan densities were reduced relative to controls in three of the four experiments (Fig. 14.4). In each case, ciliates were more strongly reduced relative to controls than were heterotrophic flagellates (Fig. 14.4). Either grazing was more intense on ciliates, or flagellates were able to compensate for losses with increased growth. The negative effect of *Daphnia* on protozoans supports our initial hypothesis that shifts to larger zooplankton communities would

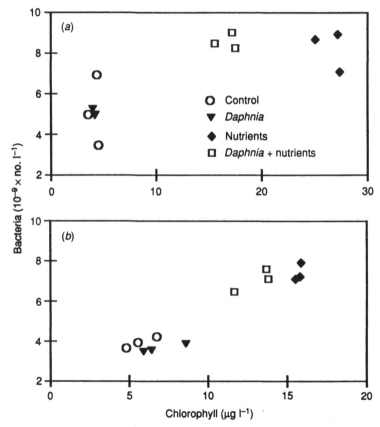

Fig. 14.3. Chlorophyll versus bacteria in the factorial experiments conducted in (a) Peter Lake and (b) Paul Lake. There were three replicates for each treatment; two of the *Daphnia* addition points overlap. Note the similarity of bacteria in the two treatments with nutrients, but how these treatments are separated in terms of chlorophyll concentration. Nutrient treatments with *Daphnia* had consistently lower chlorophyll.

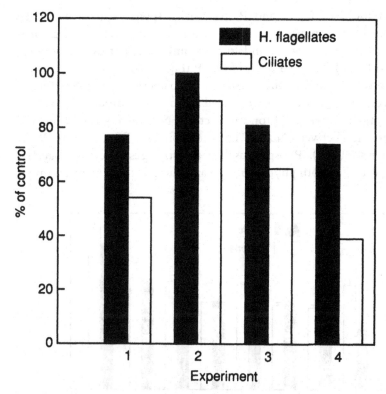

Fig. 14.4. Percentage reduction of heterotrophic flagellates and ciliates in *Daphnia* addition treatments relative to controls in two single-factor experiments (1 = Peter, June 1988; 2 = Paul, June 1988) and two factorial experiments (3 = Peter, August 1988; 4 = Paul, August 1989).

result in reductions of protozoan communities as a consequence of increases in predation.

Nutrient additions also resulted in increased abundances of protozoans in the factorial experiments. The positive effect of nutrients on protozoans might have been direct through the mechanism of nutrient uptake or indirect in that food resources including bacteria and algae were in greater supply.

Manipulations of organic carbon and nutrients
Experiments with additions of organic carbon and the inorganic nutrients nitrogen and phosphorus were designed to test whether the response of bacterial communities to limiting resources could be predicted from changes in phytoplankton alone or whether bacterial resource limitation was distinct from phytoplankton. In the first experiment, with relatively

high additions of C, N and P, bacterial response largely paralleled phytoplankton response (Fig. 14.5). Highest concentrations of chlorophyll and bacteria were found in the combination treatments which included N and P as well as C, N and P (Fig. 14.5). Chlorophyll alone explained over 84% of the variability in bacteria among treatments. However, analysis of variance revealed an important difference between the response of phytoplankton and bacteria. Nitrogen, phosphorus, and the interaction between N and P had the strongest effect on phytoplankton (all $p < 0.0001$). Phosphorus had the strongest effect on bacteria (Table 14.1). For both phytoplankton and bacteria there were a number

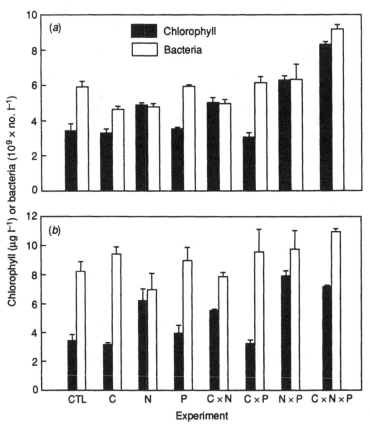

Fig. 14.5. Final chlorophyll concentration and bacterial abundance in Paul Lake enclosures with various combinations of nutrient additions. (*a*) The May experiment, with high nutrient additions; (*b*) the June experiment, with lower nutrient additions. Abbreviations: CTL, control; C, carbon; N, nitrogen; P, phosphorus.

Table 14.1. *Values of p from factorial analysis of variance of chlorophyll and bacterial responses to high and low additions of glucose (C), phosphorus (P) and nitrogen (N)*

Experiment	Treatment	Chlorophyll	Bacteria
high	C	0.038	0.086
	N	0.0001	0.028
	P	0.0001	0.0001
	C × N	0.0010	0.0016
	C × P	0.039	0.0015
	N × P	0.0001	0.0013
	C × N × P	0.0050	0.29
low	C	0.054	0.16
	N	0.0001	0.81
	P	0.004	0.021
	C × N	0.72	0.91
	C × P	0.67	0.92
	N × P	0.030	0.080
	C × N × P	0.76	0.74

of significant interactions with carbon. This was surprising in phytoplankton, where only a nutrient response was expected. Either some of the phytoplankton were able to use carbon heterotrophically or possibly there was some facilitation of phytoplankton by bacteria when carbon was added. One possibility is that organic carbon additions stimulated consumption of bacteria by those phytoplankton with phagotrophic capabilities (Sanders, 1991). Facilitation could also be based on the production by bacteria of vitamins, nutrients, or other constituents necessary to support algal growth.

The first experiment suggested that much of the net increase in bacteria could be explained by changes in phytoplankton alone and that bacteria also responded directly to phosphorus additions. However, phytoplankton response was a joint function of N and P. We also observed but did not quantify shifts in bacterial size especially in the C, N, P combination treatment where there were many large bacteria. We conducted a second experiment with fourfold lower carbon and nutrient additions and recorded changes in both bacterial abundance and size.

As in the first experiment, phytoplankton responded to the addition of N and P, and there was a positive interaction between N and P (Fig. 14.5). Organic carbon additions in this experiment appeared to have a slight negative effect on phytoplankton of marginal significance (Table

14.1). As in the first experiment, phosphorus was the most important factor determining bacterial abundance (Fig. 14.5, Table 14.1). There was a possible interaction between N and P ($p = 0.08$) that can be attributed to the lower number of bacteria observed in the nitrogen addition treatment (Fig. 14.5). Differences in bacterial size among treatments explained less than 2% of the overall variability for the 1200 bacteria we measured. Although there were some statistically significant differences among treatments (e.g. phosphorus $p = 0.045$), the magnitude of these differences was not ecologically interesting.

Changes in chlorophyll did not predict changes in bacteria in the second experiment (regression $r^2 = 0.01$, $p = 0.59$). Chlorophyll changes were a function of N and P whereas bacteria responded to phosphorus but not nitrogen. This result indicates that phytoplankton and bacteria are limited by nutrients, but that this limitation is distinct for the two groups of organisms. Organic carbon produced by phytoplankton may not be the principal factor limiting bacterial abundance.

Whole-lake experiments

Paul Lake

Temporal patterns of bacterial abundance and production varied among years in the reference ecosystem (Fig. 14.6), but there were no strong trends in these patterns with one exception. Bacterial production in the epilimnion (Fig. 14.6*b*) and metalimnion was higher in 1988 than in 1989 or 1990. The overall mean (± 95% confidence interval) of epilimnetic production in 1988 was $12.5 \pm 2.7\,\mu g\,C\,l^{-1}\,d^{-1}$ compared with 4.4 ± 0.9 and $3.8 \pm 1.1\,\mu g\,C\,l^{-1}\,d^{-1}$ for the following two years. Higher production cannot be attributed to greater phytoplankton biomass, higher primary production, or higher phosphorus concentrations in 1988 (Chapter 13). Temperature is a possible explanation. In 1988 the lake heated rapidly with epilimnetic temperatures exceeding 20 °C on day 153, substantially earlier than in 1989 or 1990. Furthermore, average epilimnetic temperature over the summer season (days 140–260) was 22.1 °C, higher than average temperatures for all other study years (1984–90).

Bacterial abundances were similar among years. Abundances ranged from 2 to 6 and 2 to 8×10^9 cells l^{-1} in the epilimnion (Fig. 14.6*a*) and metalimnion, respectively. The abundance of heterotrophic flagellates varied from 1.9 to 24×10^5 cells l^{-1} in the epilimnion (Fig. 14.6*c*); similar abundances were observed in the metalimnion. There were no clear

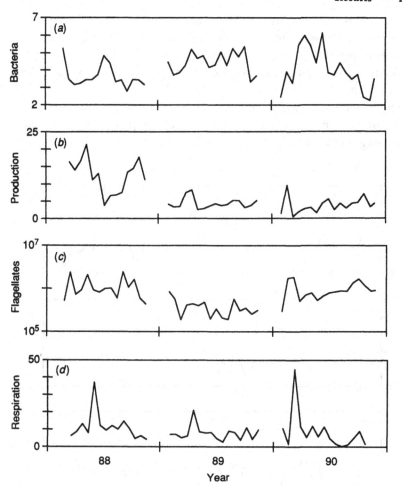

Fig. 14.6. Epilimnetic time series in Paul Lake during the summer stratified seasons of 1988–90. (*a*) Bacterial abundance ($10^9 \times$ no. 1^{-1}); (*b*) bacterial production (μg C 1^{-1} d^{-1}); (*c*) heterotrophic flagellate abundance (no. 1^{-1}); (*d*) respiration (μmol O$_2$ 1^{-1} d^{-1}).

seasonal patterns. Flagellate abundances were lower in 1989 than in the other two years.

Respiration, which is a measure of the metabolism of the planktonic community, was similar across years, with peak rates of 22–44 μmol O$_2$ 1^{-1} d^{-1} occurring briefly in June or early July each year (Fig. 14.6*d*). These peaks were not associated with high rates of primary production or with high concentrations of chlorophyll. Respiration appeared to be stimulated by the waning of specific phytoplankton populations (e.g.

dinoflagellates, chrysophytes). Throughout the remainder of the season respiration rates were relatively stable. As with bacterial production, average respiration was greater in 1988 ($11.3 \pm 4.7 \, \mu mol \, O_2 \, l^{-1} \, d^{-1}$) but 95% confidence intervals across years overlapped (1989: 7.8 ± 2.3; 1990: $7.9 \pm 5.5 \, \mu mol \, O_2 \, l^{-1} \, d^{-1}$).

There were no significant correlations between time series of microbial parameters and chlorophyll, primary production, or cladoceran length in Paul Lake.

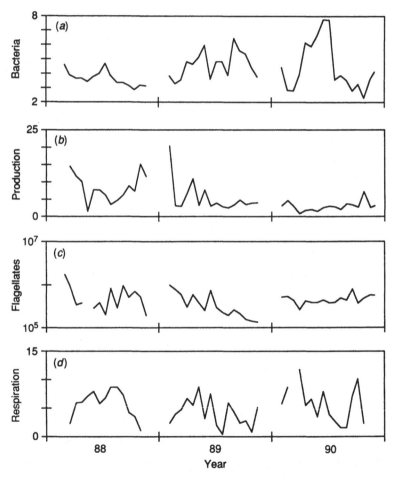

Fig. 14.7. Epilimnetic time series in Peter Lake during the summer stratified seasons of 1988–90. (*a*) Bacterial abundance; (*b*) bacterial production; (*c*) heterotrophic flagellate abundance; (*d*) respiration. (All units of measurement as in Fig. 14.6.)

Peter Lake

The time series of bacterial abundance, bacterial production and heterotrophic flagellates in Peter Lake were generally similar to those of Paul Lake. For example, production was higher in 1988 relative to the following two years (Fig. 14.7b). Seasonal patterns of bacterial and flagellate abundance also varied among years in a manner similar to Paul Lake (Fig. 14.6 and 14.7). Respiration rates were similar across years although more variable in 1989 and 1990 (Fig. 14.7d).

The experimental manipulations in Peter Lake produced a series of short-term shifts in the food web, which resulted in changes in the mean length of cladocerans and subsequent cascading effects on chlorophyll and primary production (Chapter 13). For bacteria, there was a positive correlation between epilimnetic bacterial production and chlorophyll (but not primary production) at a time lag of zero (Fig. 14.8a,b). The filtered time series of bacterial production in the epilimnion can be modeled with an autoregressive term of lag 1 and epilimnetic chlorophyll ($t = 1.83$, d.f. $= 46$, $p \approx 0.1$). While chlorophyll is of marginal statistical significance, this analysis provides some support for the hypothesis proposed initially that changes in chlorophyll lead to corresponding changes in bacterial production. We cannot attribute the apparent positive effect of chlorophyll on bacteria directly to cascading trophic interactions. Chlorophyll responded to shifts in cladoceran length at zero lag (Chapter 13), but there was no similar effect of changes in cladoceran length on bacterial production (Fig. 14.8c). Instead, bacterial production was negatively related to cladoceran length at a time lag of eight weeks. This long gap between shifts in herbivore grazing as represented by cladoceran length and bacterial production is not easily explained by cascading effects through the phytoplankton. Other microbial parameters including bacterial and heterotrophic flagellate abundance, metalimnetic bacterial production, and respiration were unrelated to chlorophyll, primary production, or cladoceran length.

Tuesday Lake

In late 1988 the second manipulation of the food web in Tuesday Lake occurred. As a consequence of the recovery of minnow populations, the zooplankton community shifted from dominance by *Daphnia pulex* to a community of smaller species (Chapters 4 and 8). Associated with this change was a modest increase in phytoplankton biomass and primary production (Chapter 13). Given these changes, I expected increases in

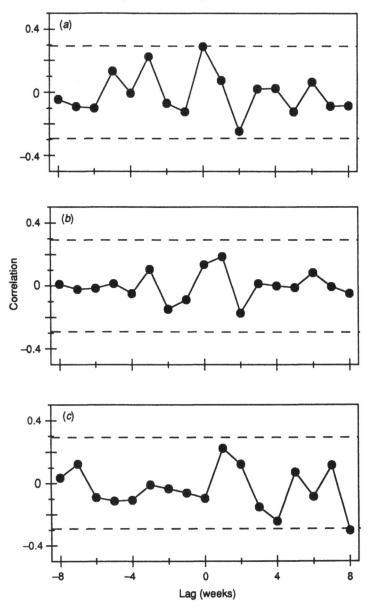

Fig. 14.8. Cross correlations between filtered time series for selected variables in Peter Lake hypothesized to partly determine rates of bacterial production. (*a*) Epilimnetic bacterial production and chlorophyll; (*b*) epilimnetic bacterial production and primary production; (*c*) epilimnetic bacterial production and cladoceran length. Dotted lines denote correlations significant at $\alpha = 0.05$.

heterotrophic flagellates, bacterial abundance and bacterial production, especially in the epilimnion of Tuesday Lake where the increase in chlorophyll and primary production was most pronounced. This expectation was tested using randomized intervention analysis (Carpenter *et al.*, 1989; Chapter 3).

The dynamics of bacterial abundance and production in Tuesday Lake were similar to those observed in Paul Lake. As in Paul Lake, bacterial production was highest and bacterial abundance lowest in 1988 relative to the other two years (Fig. 14.9*a,b*). Randomized intervention analysis (RIA) indicated that the bacterial production in the epilimnion of Tuesday Lake was distinct from Paul Lake before and after the second manipulation. Both lakes had relatively high production in 1988, perhaps owing to environmental factors. Production, however, declined more dramatically in Paul Lake (and in Peter Lake) than in Tuesday Lake in the subsequent two years (Fig. 14.6*b* and 14.9*b*). The ability of Tuesday Lake to sustain higher productivity relative to Paul Lake after the second manipulation is consistent with expectations based on the increase in chlorophyll and primary production. As with chlorophyll and primary production, the shift in bacterial production was small. Epilimnetic bacterial abundance increased in Tuesday Lake relative to Paul Lake in a manner consistent with bacterial production, but this change was not significant (Table 14.2).

Other parameters did not show any distinct shift as a result of the manipulation with the possibility of metalimnetic heterotrophic flagellates (Table 14.2). Populations of these flagellates were initially substantially higher in Paul Lake relative to Tuesday Lake and became more similar after the manipulation. This is the response that would have been predicted from a relaxation in predation on flagellates as a consequence of the removal of large cladocerans; however, this apparent shift was of marginal significance ($p = 0.069$). Furthermore, epilimnetic heterotrophic flagellates either did not change or declined in the epilimnion relative to Paul Lake (Figs 14.6*c*, 14.9*c*; Table 14.2). There is then little evidence for changes in heterotrophic flagellates in concert with food web shifts observed in Tuesday Lake.

Respiration in Tuesday Lake was, on average, higher than in the other two lakes (Fig. 14.9*d*). Tuesday Lake typically supports a higher algal biomass (Chapter 11) and is surrounded by a bog (Chapter 2). These two factors probably account for the greater respiration in Tuesday Lake. As in Paul Lake, we observed higher respiration during the warm year of 1988 than in the subsequent two years (Fig. 14.9*d*). RIA indicates that respiration rates did not change in response to the shift in the food web in late 1988 (Table 14.2).

Discussion

The overall effect of the whole-lake manipulations over the period of 1988–90 was to produce short-term perturbations in the food web of Peter Lake and longer-term shifts in Tuesday Lake. While changes in

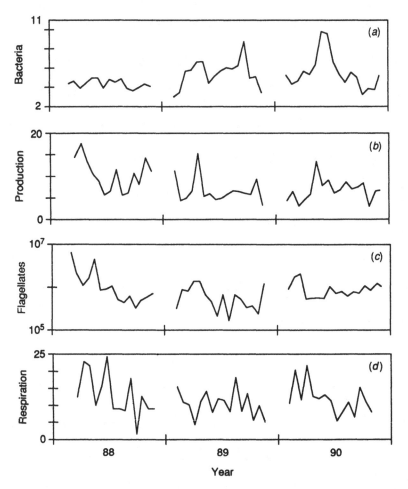

Fig. 14.9. Epilimnetic time series in Tuesday Lake during the summer stratified seasons of 1988–90. (*a*) Bacterial abundance ($10^9 \times$ no. 1^{-1}); (*b*) bacterial production (μg C 1^{-1} d^{-1}); (*c*) heterotrophic flagellate abundance (no. 1^{-1}); (*d*) respiration (μmol 1^{-1} d^{-1}).

Table 14.2. *Summary of randomized intervention analysis for the second manipulation in Tuesday Lake*

Differences in heterotrophic microbial parameters between the experimental (Tuesday) and reference (Paul) lakes were compared for time series before (D_{pre}) and after (D_{post}) the manipulation. $T = (D_{post} - D_{pre})$. P is the two-tailed probability that T is significantly different from 0. Abbreviations: B, bacterial; H, heterotrophic; epi, epilimnion; meta, metalimnion.

Depth	Parameter	D_{pre}	D_{post}	T	P
epi	B. abundance	0.686	1.203	0.52	0.19
epi	B. production	− 1.806	2.157	3.96	0.003
epi	H. flagellates	5.040	0.824	− 4.22	0.23
epi	respiration	0.029	0.101	0.72	0.49
meta	B. abundance	0.612	0.531	− 0.08	0.85
meta	B. production	1.619	0.117	− 1.50	0.12
meta	H. flagellates	− 4.855	− 0.136	4.72	0.069

some components of the food webs were dramatic (e.g. zooplankton), there was relatively little change in the abundance of heterotrophic microbes, bacterial production or epilimnetic respiration. I attribute the limited responses by the heterotrophic microbial community to a lack of strong, sustained shifts in the phytoplankton communities from 1988 through 1990. Given the shorter time series of observations for microbial parameters, it is more difficult to provide unequivocal conclusions about the impact of the trophic cascade on these organisms.

Nevertheless, patterns of bacterial production in the epilimnion of both Peter and Tuesday Lakes were consistent with the original model. I proposed that bacterial communities would covary with phytoplankton, and that cascading trophic effects on bacteria would be mediated by the phytoplankton. In Peter Lake the time series of bacterial productivity was related to chlorophyll supporting the first point of the proposed model. The dynamics of chlorophyll in Peter Lake were determined in part by the trophic cascade initiated by the experimental manipulations as illustrated in Chapter 13. The linkage between cladocerans and chlorophyll in Peter Lake, and in turn, between chlorophyll and bacterial production supports the thesis of the original model that shifts in zooplankton size structure would have an indirect effect on bacteria. I could not, however, detect the effect of changes in cladoceran length on bacterial production in the Peter Lake experiment except at long time lags. Analysis of these time series is limited by the number of observations (Chapter 3). Given the transitory effects of the manipulations during 1988–90 and the proposed two-step linkage between bacteria and zooplankton, it is not surprising that changes in zooplankton cannot be directly related to bacterial production. The Peter Lake results support the hypothesis, but a longer period of analysis or additional manipulations are required to arrive at stronger conclusions.

The decline of large cladocerans and the recovery of the original food web in Tuesday Lake provides an easier case for testing the hypothesis. Using RIA to test for a step change related to the manipulation, bacterial production increased in Tuesday Lake relative to the reference ecosystem Paul Lake. This change was directly parallel to the increases in epilimnetic chlorophyll and primary production linked to the food-web shifts (Chapter 13). Results of the whole-lake manipulations in Tuesday Lake support the hypothesis that cascading trophic interactions from changes in fish community structure are manifested even at the level of heterotrophic microorganisms.

In the whole-lake experiments, bacterial production was a more sensitive measure of bacterial community response than bacterial abundance. This result is not surprising, for two reasons. First, bacterial productivity varies considerably both within and among systems while even over large trophic gradients the number of bacteria only changes by about an order of magnitude (Cole, Findlay & Pace, 1988). Second, bacterial communities appear to consist of both highly active and relatively inactive populations (e.g. reviews by Karl, 1986; Newell, Fallon & Tabor, 1986). Total counts of bacteria include both active and inactive cells. The proportions of the two categories of cells can change independently of large changes in total numbers (Newell et al., 1986). Bacterial abundance as determined by conventional epifluorescence microscopy methods is probably a poor indicator of community conditions.

While the whole-lake experiments suggest a strong linkage between bacterial production and phytoplankton, the enclosure experiments revealed distinctions between the factors regulating these groups. Whole-lake experiments represent food-web manipulations independent of changes in the nutrient status of these lakes. In the enclosure experiments, shifts in nutrients led to rapid and large changes in bacterial abundance. Bacteria were unaffected by dramatic increases in the biomass of consumers, and independent studies indicate that cladocerans, not protozoans, are the primary consumers of bacteria in these lakes (Vaqué & Pace, 1992). Phytoplankton, on the other hand, were clearly influenced by both nutrient and grazer additions. A further distinction in terms of the type of nutrient limitation was observed in the C, N and P addition experiments. Bacteria responded primarily to phosphorus while phytoplankton responded most strongly to N plus P additions. If bacteria are strongly P limited while phytoplankton are colimited by N and P, the dynamics of these two groups may be independent, at least over shorter time scales, as was the case in the second C, N and P addition experiment (Fig. 14.5). Limitation of bacteria by phosphorus has been observed in other studies (Coveney & Wetzel, 1988; Currie, 1991; Toolan et al., 1991), and bacteria are effective competitors for phosphorus (Currie & Kalff, 1984). These observations suggest that the capacity of the bacterial community to sequester nutrients (particularly phosphorus) may be an important mechanism regulating nutrient availability to other components of the food web. Variability in the uptake and release of phosphorus by bacteria may influence phytoplankton and zooplankton dynamics.

Heterotrophic flagellates were generally unaffected by changes in food web structure in both Peter and Tuesday Lakes. I had initially postulated that shifts from small to large zooplankton communities would result in decreases in protozoans. This idea was supported in the enclosure experiments but not in the whole-lake manipulations. The heterotrophic flagellate communities were apparently able to compensate for shifts in predation by changes either in growth rates or in community structure. In the longer-term, whole-lake studies, compensatory responses appear to result in no change in the overall abundance of these flagellates.

One factor which could uncouple heterotrophic microbial processes from the trophic cascade would be large carbon inputs from the littoral zone or allochthonous sources. In humic Scandinavian lakes, Hessen, Andersen & Lyche (1990) have inferred that substantial carbon flows from allochthonously derived DOC through planktonic bacteria. During long-term ^{14}C tracer experiments in a humic lake, bacterial production exceeded primary production and the pelagic system was heterotrophic (respiration > primary production). The lake studied by Hessen et al. (1990) had a high concentration of DOC (7.5 mg C l^{-1}), and 75% of the particulate carbon was detrital. Paul, Peter and Tuesday Lakes also have relatively high concentrations of DOC (1990 averages: Paul 4.5, Peter 3.7, Tuesday 7.9 mg C l^{-1}). Further, surface water respiration is typically greater than or of the same magnitude as epilimnetic primary production. In these lakes, however, bacterial production appears to be a much lower proportion of phytoplankton production (range for all lakes 4–11%).

These comparisons suggest that in Paul, Peter and Tuesday Lakes there is a significant flow of carbon through heterotrophic bacteria from allochthonous or littoral sources. Bacterial respiration of allochthonous carbon would explain the high rates of respiration relative to production. Such a large carbon flow would only be possible at the low bacterial productivities observed if bacteria use the allochthonous material with a low efficiency (i.e. high respiration per unit production). Bacterial utilization of carbon from outside sources in this way would probably dampen microbial responses to manipulations of pelagic food-web structure.

Summary

Heterotrophic microbial processes were followed in Paul, Peter and Tuesday Lakes in the years 1988–90. Bacterial abundance, heterotrophic flagellate abundance and microbial respiration were similar across the three years in Paul Lake. Bacterial production, especially in the epilimnion, was higher in all three lakes during 1988, perhaps owing to higher water temperatures observed in that year.

Despite large changes in the zooplankton of Peter Lake in response to fish additions during 1988–90 (Chapter 10), there were few detectable shifts in microbial standing stocks, productivity or respiration. There was a positive correlation between filtered time series of epilimnetic bacterial production and chlorophyll, suggesting that bacterial productivity was closely coupled to phytoplankton dynamics. However, conclusions regarding the effect of the food-web shifts on heterotrophic microbes in Peter Lake are currently limited by the short period of measurements (< 50 observations).

In Tuesday Lake recovery of the minnow populations resulted in the disappearance of *Daphnia pulex* in late 1988 (Chapter 10). The modest increase in phytoplankton biomass and primary production that followed (Chapter 13) was paralleled by a shift in epilimnetic bacterial production. Bacterial production increased in Tuesday Lake relative to the reference system, Paul Lake. No other significant changes in the microbial community were observed in response to the manipulation in Tuesday Lake.

Enclosure experiments conducted in Paul and Peter Lakes generally supported the results of the whole-lake experiments. Bacteria responded to changes in phytoplankton and increases in nutrients, but not to changes in zooplankton. Bacteria were strongly phosphorus-limited, and their responses to nutrient additions were distinct from those of the phytoplankton. Increases in bacteria were most directly related to increases in phosphorus, while phytoplankton were colimited by phosphorus and nitrogen. In contrast with the whole-lake experiments, protozoans were strongly affected by *Daphnia*. Predation by *Daphnia* may be an important regulator of protozoans.

These studies support the view that cascading trophic interactions have indirect effects on bacteria as mediated by the phytoplankton but direct effects on protozoans as a consequence of zooplankton predation. Microbial communities, however, appear to be fairly conservative in their overall response to food-web changes. The conservative response probably reflects a high degree of functional redundancy in these communities and the availability of alternative resources (such as allochthonous carbon) that partly uncouple microbial processes from the planktonic food web.

Acknowledgements

I thank Elizabeth Funke, David Lints and Gail Steinhart for technical assistance, and Jonathan Cole for reviewing the manuscript. Steve Carpenter and Jim Kitchell provided the opportunity to work on these problems in experimental lakes. Marty Berg, Ron Hellenthal, John Morrice and Pat Soranno facilitated the field work. Financial support was provided by NSF grants BSR–880593 and BSR–9019873. This is a contribution to the program of the Institute of Ecosystem Studies, The New York Botanical Garden.

15 · *Annual fossil records of food-web manipulation*

Peter R. Leavitt, Patricia R. Sanford, Stephen R. Carpenter, James F. Kitchell and David Benkowski

Introduction

Limnologists are interested in why lakes vary from year to year. Studies of temporal variability under baseline conditions are needed to quantify the relative importance of mechanisms regulating production, to interpret ecosystem experiments and to help solve lake management problems (McQueen *et al.*, 1986; Benndorf, 1987; Carpenter, 1988*a*).

Long-term studies, ecosystem experiments and paleolimnology provide information on interannual variation in lakes. Long-term studies potentially span many short and intermediate-length processes (10^{-4}–10^1 y) (Edmondson & Litt, 1982; Goldman, Jassby & Powell, 1989; Schindler *et al.*, 1990) but are rare and sometimes purely descriptive or site-specific. Results of ecosystem experiments may apply more broadly (Carpenter, 1991), but are also costly and rare. Further, many are too brief ($<10^1$ y) to detect the long-term responses of lakes to perturbation. Paleolimnology is relatively inexpensive and can yield records that are otherwise unobtainable.

Paleolimnology is the study of lake ecosystem structure and function using the historical record in sediments. Lake sediments accumulate through time and integrate material from the lake, its basin and catchment, and atmospheric sources (Frey, 1969; Binford, Deevey & Crisman, 1983; Battarbee *et al.*, 1990). Development of high resolution sampling techniques (Glew, 1988; Davidson, 1988; Leavitt *et al.*, 1989), well-defined taxonomy and autecology (e.g. Walker, 1987; Kingston & Birks, 1990) and automated analysis of some fossils (Mantoura & Llewellyn, 1983) allows paleoecological analysis on time scales relevant to population interactions in lakes and watersheds (10^{-1}–10^4 y).

Paleolimnology can address a wide range of environmental issues. Morphological remains of terrestrial and aquatic plants (Charles & Smol,

1988; Battarbee *et al.*, 1990; Liu, 1990; Brenner, Leyden & Binford, 1990), invertebrate exoskeletons (Frey, 1959; Walker & Mathewes, 1989) and bacteria (Nilsson & Renberg, 1990) have been used to document anthropogenic perturbations including deforestation, agriculture, acidification, eutrophication, pollution and climate change (Frey, 1969; Binford *et al.*, 1983; Charles & Smol, 1988; Davis, 1989; Battarbee *et al.*, 1990; Smol, 1990; Walker *et al.*, 1991). Analysis of sediment morphology, fine structure, mineralogy and inorganic geochemistry quantifies watershed processes such as erosion (Sandman, Lichu & Simola, 1990), fire (Hickman, Schweger & Klarer, 1990), fossil fuel combustion (Renberg & Wik, 1985) volcanic activity (Reasoner & Healy, 1986) and pollution (Tolonen, Alasaarela & Liehu, 1988) as well as changes in land use patterns (Züllig, 1989; Brenner *et al.*, 1990). Recently, food-web structure, competition and predation have been investigated with fossil analyses (Kerfoot, 1974, 1981; Kitchell & Kitchell, 1980; Kitchell & Carpenter, 1987; Leavitt *et al.*, 1989; Uutala, 1990).

Reconstruction of food-web processes (production, competition, herbivory and predation) from lake sediments would be simpler if key populations all left fossils. Fish remains are uncommon, and inferences about size-selective predation by fish rest on deductions from fossil zooplankton communities (Hrbáček, 1969; Kerfoot, 1974; Kitchell & Kitchell, 1980). Several important invertebrates in aquatic food webs leave fossils that can be identified, occasionally to species: chironomids (Walker, 1987), chaoborids (Uutala, 1990), rotifers and copepods (as eggs) (J. F. Kitchell & P. R. Sanford, unpublished data), ostracods (Carbonel *et al.*, 1988), protozoa (Douglas & Smol, 1987), chydorids (Frey, 1988) and the Cladocera (Hrbáček, 1969; Kitchell & Kitchell, 1980; Kerfoot, 1981). Algal microfossils include diatom frustules, scales and cysts of chrysophytes, dinoflagellate cysts, and occasionally cell wall remains or reproductive cells (Smol, 1987, 1990). However, many important algal taxa leave only biochemical fossils such as fatty acids or pigments (Brown, 1969; Züllig, 1989).

Fossil pigments may be particularly good indicators of food-web change. Two groups of plant pigments – chlorophylls and carotenoids – are well preserved in lake sediments (Brown, 1969; Brown, McIntosh & Smol, 1984; Sanger, 1988). Carotenoids are especially promising paleolimnological indicators because they include pigments that discriminate between all major groups of algae (Table 15.1). Bacterial carotenoids are particularly diagnostic, even to the genus level (Brown, 1969; Brown *et*

Table 15.1. *Pigments studied, their taxonomic specificity, and location of their principal sources*

P, pelagic or planktonic; L, littoral, benthic or attached; T, terrestrial. Upper case letters indicate quantitatively more important sources than lower case.

Pigment	Taxa	Source
alloxanthin	Cryptophyta	P
isorenieratene	Chlorobiaceae (brown varieties)	P
α-carotene	Cryptophyta, Chlorophyta, Tracheophyta	P l t
fucoxanthin	Chrysophyta	P L
lutein–zeaxanthin	Chlorophyta, Euglenophyta, Cyanophyta, Tracheophyta	P L t
β-carotene	Plantae, some bacteria	P L t
chlorophyll *a*	Plantae	P l t
chlorophyll *b*	Chlorophyta, Euglenophyta, Tracheophyta	P l t
chlorophyll *c*	Chrysophyta, Pyrrophyta	P l
pheophorbide *a*	Chl *a* derivative (grazing)	P l
pheophytin *a*	Chl *a* derivative (general)	P L t
pheophytin *b*	Chl *b* derivative (general)	P L t
pheophytin *c*	Chl *c* derivative (general)	P L

Source: Leavitt *et al.* (1989); Carpenter & Leavitt (1991).

al., 1984; Züllig, 1989). Carotenoids are the sole fossils of cryptophyte algae and form the main record of green and bluegreen algae (Table 15.1). Unique derivatives of carotenoids and chlorophylls are formed by herbivores and can be used to monitor community grazing pressure (Daley, 1973; Repeta & Gagosian, 1982; Carpenter & Bergquist, 1985). Pigments are contributed from a wide variety of sources from exclusively planktonic (alloxanthin) to strongly terrestrial (pheophytin *a*) and can differentiate between aquatic and terrestrial processes (e.g. Gorham & Sanger, 1975).

Interpretation of fossil records can be difficult and ambiguous. The effects of selective deposition or preservation on fossil records are poorly known (Brown, 1969). Preferential deposition (Haberyan, 1985, 1990), resuspension (Davis, Moeller & Ford, 1984), preservation (Johnson, 1974) and postdepositional disturbance (Robbins *et al.*, 1989) can complicate records of algal and animal microfossils. Plant pigments are sensitive to changes in light, pH, enzymes, oxygen and temperature (Davies, 1976) and may be especially affected by selective losses (Hurley & Armstrong, 1991). Factors that influence formation of the animal microfossil record are only rarely documented (Kerfoot, 1981). Accurate reconstruction of community interactions from the paleoecological record requires either unbiased fossil deposits or clear understanding of mechanisms for selective preservation or degradation.

This chapter calibrates the use of fossil pigments and zooplankton in whole-lake manipulations of food-web structure. In the first section, experiments show that grazing by herbivores and pigment photo-oxidation interact to regulate carotenoid and chlorophyll flux to the sediments. Sediment trap studies show that fossils from planktonic Cladocera document the species composition, size structure and herbivory of the pelagic zooplankton community. In the second section, we verify paleolimnological signals with contemporaneous limnological data. Fossil pigment and zooplankton analyses in Peter and Paul Lakes demonstrate that changes in piscivory can reconfigure food webs down to the microbial level and that these events are accurately recorded in lake sediments. These interactions are modified by perturbations of the abiotic environment and may take decades to stabilize. In Tuesday Lake, we show how the outcome of food-web manipulation depends on the strength of interaction between piscivorous and planktivorous fish. We also examine how interannual variation in zooplanktivory by fish sets the rhythm of primary production and pigment deposition.

Methods

Site histories

This study took advantage of ongoing research and 30 y of experimental manipulation history in Peter, Paul and Tuesday Lakes to calibrate pigments as paleoecological indicators. These lakes lie in protected watersheds and have recently spanned wide ranges of chlorophyll concentration and productivity, and have harbored contrasting plankton communities (Chapters 11, 13). The lake sediments are annually laminated (varved) at all deepwater sites (Leavitt et al., 1989; Carpenter & Leavitt, 1991) and yield high-resolution paleoecological records. Profundal sediments in these lakes appear undisturbed by mixing or bioturbation (Leavitt & Carpenter, 1989; Leavitt et al., 1989; Carpenter & Leavitt, 1991).

Paul Lake is ideal for study of food-web manipulation because it has a well documented history and has contained four distinct fish assemblages in the past 50 years (Leavitt et al., 1989). In 1951 Arthur Hasler divided the basin of Paul and Peter Lake with an earthen dike, removed the native fishes (largemouth bass, yellow perch) and stocked both lakes with rainbow trout. These fish were added frequently between 1951 and 1961; however, the trout did not reproduce and by 1967 had been replaced by a cyprinid assemblage (redbelly and finescale dace). Minnows dominated the fish community until 1975 when largemouth bass were stocked. Bass are efficient predators and by 1980 had removed the minnow population. Currently, Paul Lake is a virtual monoculture of largemouth bass (Chapter 4) and has a zooplankton assemblage dominated by large-bodied *Daphnia* and *Diaptomus* (Chapter 8).

Peter Lake underwent the same changes in fish community structure as Paul Lake. Peter Lake also received additions of hydrated lime (1951–3, 1955–7, 1959, 1962, 1969) or $CaCO_3$ (1976) (Elser *et al.*, 1986*b*; Leavitt *et al.*, 1989).

Winterkill simplifies the food web in Tuesday Lake (Carpenter & Leavitt, 1991). The natural fish community was dominated by redbelly and finescale dace, both highly planktivorous minnows, while small cladocera (e.g. *Bosmina longirostris*) were common herbivores (Chapter 8).

Tuesday Lake has had three experimental manipulations: artificial destratification by deep-water aeration (1956) (Schmitz, 1958); liming and rainbow trout stocking (1961) (UNDA, 1988); and largemouth bass addition and minnow removal (1985) (Carpenter *et al.*, 1987; Chapter 2). An additional disturbance occurred in 1951–1952 when a road was built through both the catchment forest and a *Sphagnum* bog mat that encircles the lake (UNDA, 1988; Leavitt *et al.*, 1989; Carpenter & Leavitt, 1991). Comparison of the effects of food-web and abiotic perturbations allows direct contrast of trophic and nutrient control of ecosystem structure.

Experimental studies

The central question of paleolimnology is whether sedimentation of fossils accurately monitors changes in the abundance of populations in the water column. Both phytoplankton and zooplankton fossils are subject to selective deposition or degradation. Accurate interpretation of the fossil record requires a quantitative understanding of such potential sources of bias. Therefore, we conducted several experimental studies of processes that determine the sedimentary signature of water column communities.

All carotenoids and chlorophylls were quantified by reversed-phase high pressure liquid chromatography, (RP-HPLC) (Leavitt *et al.*, 1989;

Leavitt & Carpenter, 1990a). Automated RP-HPLC allows rapid, simultaneous analysis of all colored pigments and derivatives (Mantoura & Llewellyn, 1983). Pigments were identified by visible light absorption spectra, pigment polarity, and simple chemical identification of oxygenated functional groups (Leavitt et al., 1989; Leavitt & Carpenter, 1990a).

Epilimnetic water samples from Peter, Paul and Tuesday Lakes were analyzed weekly in 1986 for algal community composition and carotenoid content. Standing stocks were compared with material collected in sediment traps suspended in the lakes at two depths, equivalent to 1–3% incident irradiance (shallow) and aphotic waters (deep) (Leavitt & Carpenter, 1990b). Algal biovolume was estimated as described in Chapter 11.

Pigment losses during algal decomposition were estimated by incubating hypolimnetic water in situ in opaque containers for either 17 weeks (1986) or 4 weeks (1987) (Leavitt & Carpenter, 1990b). These conditions mimic those in which detrital particles sink and sediment traps are commonly suspended (cold, dark, stagnant). Decay constants (d^{-1}) were used to correct for pigment loss in sediment traps.

Photo-oxidation decay constants $(m^{-2} E^{-1})$ (1 Einstein = 1 mol photons) were obtained from slopes of linear regressions of incident irradiance versus pigment loss in a series of in situ bottle experiments in 1986 (Leavitt & Carpenter, 1990b). The half-life (d) for pigment bleaching was compared with that required for algae to sink out of light. Algal sinking rates were calculated using observed cell size distribution and estimated density under different scenarios of turbulence (Leavitt & Carpenter, 1990b). Pigments may be selectively preserved despite rapid instantaneous rates of loss if they are contained in algae that sink rapidly.

Rates of ingestion of algae by zooplankton were calculated under two extreme scenarios which bracket reality (Leavitt & Carpenter, 1990b). First, ingestion was assumed to be proportional to both herbivore size and algal cell size: a conservative or exclusive measure (Burns, 1968). Alternatively, mesocosm experiments in these lakes suggest that all cells < 30 μm long could be eaten: a lenient or inclusive index (Bergquist et al., 1985; Bergquist & Carpenter, 1986). Inclusive measures of herbivory were insensitive to increases in the upper limit for ingestible cell size to 50 μm (Leavitt & Carpenter, 1990b).

Photic zone mass balance budgets were calculated for a-phorbins (in units of μmol m^{-2}) by direct measurement of chlorophyll production (P), changes in standing stock (D), photo-oxidation (X), and sedimentation (S) for either Chl a alone (subscript chl), pheophytin and pheophorbide (cdp) or all three (tot). Nonplanktonic inputs were calculated by difference (Carpenter et al., 1986; 1988; Leavitt & Carpenter, 1990b):

$$N_{tot} = D_{tot} - P_{chl} + X_{cdp} + S_{tot}.$$

Mass budgets allow direct comparison of the relative importance of individual processes that regulate pigment sedimentation, but have wide confidence estimates on nonplanktonic inputs (Carpenter et al., 1986, 1988).

The formation of the animal microfossil record was studied by comparing pelagic zooplankton records (Chapter 8) with the cladoceran remains in sediment traps in Peter, Paul and Tuesday Lakes, 1985–9. Remains from traps suspended in hypolimnetic waters were collected biweekly (Carpenter et al., 1986, 1988), subsampled and chemically treated (Frey, 1986) before microscopic enumeration. Deposition rates (fossils m^{-2} d^{-1}) were calculated from three two-week trap periods corresponding to early, mid and late summer samples in Paul and Tuesday Lakes. Daphnia carapaces and postabdominal claws, and Bosmina carapaces (ca. 50 of each) were measured and converted to mean animal lengths (where appropriate) by regression analysis (Kitchell & Kitchell, 1980).

Sediment analyses

Repeated change in the fishes of Paul, Peter and Tuesday Lakes over about 40 years allows independent assessment of effects of both effective zooplanktivores (minnows) and efficient piscivores (largemouth bass). Addition of hydrated lime to Peter Lake is an abiotic perturbation that influenced food-web interactions (Kitchell & Kitchell, 1980). Extensive historical records are available for most experiments (Elser et al., 1986b; Carpenter et al., 1987; UNDA, 1988; Leavitt et al., 1989) and were used to test fossil signals.

Annually laminated sediments were collected with a freeze corer from the deepest part of each lake (Leavitt et al., 1989; Carpenter & Leavitt, 1991). Freeze-coring produces the least disturbance of fossil deposits (O'Sullivan, 1983). Varves or annual-equivalent intervals were isolated and organic content and pigment concentration determined for 38 pigments between 1940 and 1986, except Tuesday Lake (1889–1986). Approximately 20–30 additional compounds were resolved by our chromatograms but remain unidentified.

Postdepositional disturbance of sediments may reduce the resolution of paleoecological records. Annual profiles and mean concentrations of pigments from a transect of four cores across the basin of Paul Lake were compared 1956–86. Sediments in Paul Lake are varved in the central portion of the basin, but visibly mixed in shallow-water deposits (Leavitt & Carpenter, 1989). In addition, the three deep water sites (15 m, 9 m, 7 m) were below the maximum depth of light penetration, while the shallow water location (4 m) was within the photic zone and the epilimnion. Comparison of cores quantified the effect of sediment mixing and benthic algal production on fossil pigment profiles.

Spectral analysis of pigment stratigraphies was conducted for sediment from Tuesday Lake (1889–1982). Concentration profiles were detrended using first-difference to remove the interfering effects of long-term trends in pigment concentration (Chatfield, 1980). Normalized, cumulative plots of variance versus period (in years) were tested for significant difference from a white-noise signal (Carpenter & Leavitt, 1991). If variance ordinates fell outside a 95% confidence interval, signals were considered significant. The range of periods (in years) over which signals were significant was recorded for each pigment.

Fossil cladoceran communities in Peter and Paul Lakes were analyzed as the abundance of *Bosmina* remains relative to those of *Daphnia* (% *Bosmina*). This designation is equivalent to percentage small-bodied zooplankton and reflects changes in zooplanktivory by size-selective fishes (Hrbáček, 1969). Many other animal fossils were noted in sediments during our analyses (e.g. *Chaoborus* mandibles, rotifer resting eggs, chironomids, other cladocerans).

Abundance of cladoceran fossils from Tuesday Lake was expressed as concentration (individuals g^{-1} organic matter). Prey morphology was used to infer changes in predation by invertebrate predators. *Bosmina longirostris* that have longer morphological features (antennules, mucrones) are less sensitive to predation by invertebrate predators (Kerfoot, 1974, 1981, 1987b; Sprules, Carter & Ramcharan, 1984). *Bosmina longirostris* antennule and mucrone lengths were measured to the nearest micrometer following Kerfoot (1975). Antennules were also differentiated into short *cornuta* (curved, *sensu* Hutchinson, 1967) and longer straight forms.

Results and discussion

What determines pigment sedimentation?

The main inputs to pigment standing crops in the water column include *in situ* production, contributions from terrestrial or allochthonous sources (Gorham & Sanger, 1975) and benthic production and resuspension of shallow water sediments (Swain, 1985; Bianchi & Findlay, 1990; Hickman & Schweger, 1991) (Fig. 15.1). Pigments in suspended algae may be ingested by herbivores and digested (Leavitt & Brown, 1988), incorporated (Green, 1957), modified (Repeta & Gagosian, 1982; Carpenter & Bergquist, 1985) or passed through the gut unharmed (Klein, Gieskes & Kraay, 1986; Leavitt & Brown, 1988; Barlow, Burkill & Mantoura, 1988). Further pigment loss occurs during algal sinking when pigments are bleached by sunlight (Welschmeyer & Lorenzen, 1985*a*; Carpenter *et al.*, 1986, 1988) or oxidized by chemical or microbial processes (Leavitt, 1988). Postdepositional degradation or derivative formation can also occur (Leavitt, 1988; Hurley & Armstrong, 1991). Pigments that pass the gauntlet of degradation are incorporated into the sediments and form the fossil record (Fig. 15.1).

Carotenoids monitor different features of algal populations than do biovolume estimates of algal standing crops (Fig. 15.2). Pigment concentrations in the epilimnion were altered by changes in abundance of individual species, succession of taxa, ambient light intensity and contributions from nonplanktonic sources (Leavitt & Carpenter, 1990*b*). Because pigment and biovolume concentrations are different (yet equally valid) units for estimating phytoplankton composition, the most appropriate comparisons are between changes in carotenoid standing stock and patterns of pigment sedimentation.

Pigment flux to traps did not quantitatively represent algal abundance in the water column (Fig. 15.2). Sediment traps did not record either weekly changes or relative abundance of epilimnetic carotenoids. Loss of resolution is expected because traps were sampled less frequently than the water column. However, sediment-trap material overrepresented both alloxanthin and pheophorbide *a* at the expense of lutein–zeaxanthin and Chl *a* – pheophytin *a*, respectively. Bias increased with depth in the water column and culminated in surficial sediments where alloxanthin is the dominant carotenoid (Leavitt *et al.*, 1989; P. R. Leavitt & S. R. Carpenter, unpublished data), a common finding in both marine and fresh waters (Repeta & Gagosian, 1987; Hurley & Armstrong, 1991). Overrepresentation of alloxanthin and pheophorbide *a* jeopardizes the utility of fossil pigments in paleolimnology unless the mechanisms that produce biases are accounted for.

We compared pigment degradation in lakes with contrasting food webs to determine the main mechanisms controlling pigment sedimentation. Experimental studies suggested that the principal route of pig-

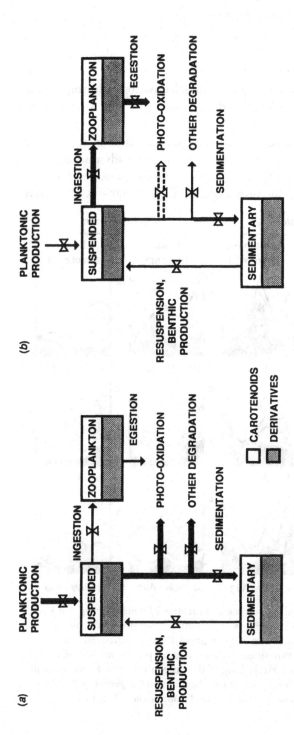

Fig. 15.1. Main fluxes of undegraded pigments (open box) or their colored derivatives (stippled box) in food webs dominated by (a) planktivores and (b) piscivores. Arrow width indicates importance of process. Boxes represent pools of pigments.

ment sedimentation is dependent on species composition within the food web. When piscivores are absent or planktivores predominate, zooplankton are small and relatively ineffective (Fig. 15.1a). In this case, pigments are deposited in accordance with selective losses during sinking due to photo-oxidation and algal decay, processes which tend to be taxon-specific.

In general, photo-oxidation constants were greatest in transparent Peter Lake and least in Tuesday Lake where humic acids absorb potentially damaging wavelengths of light. Alloxanthin (oxidation half-life, $T_{1/2} = 0.47 - 1.45$ d) and lutein-zeaxanthin ($T_{1/2} = 0.60 - 0.92$ d) were

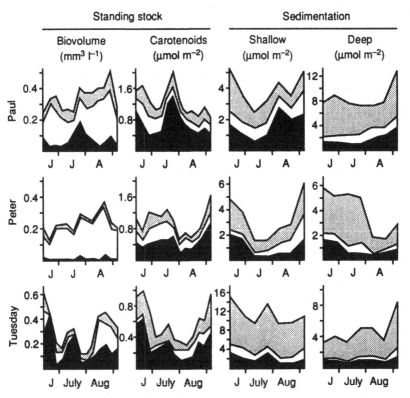

Fig. 15.2. Cumulative algal biovolume and carotenoid standing crops for Paul, Peter and Tuesday Lakes in 1986. Black, fucoxanthin (diatoms, chrysophytes); white, lutein and zeaxanthin (green and bluegreen algae); hatched, alloxanthin (cryptophytes). Pigment composition in shallow (third panel) and deep sediment traps (far right) represent material sinking to depths equivalent to 1–3% incident light and aphotic waters, respectively. (From Leavitt and Carpenter (1990b). Copyright © *Canadian Journal of Fisheries and Aquatic Sciences.* Used by permission.)

bleached in less time than it takes detrital algae to sink into darkness (minimum $T_{1/2} = 2.71 - 5.00$ d and $0.70 - 1.55$ d, respectively) (Leavitt & Carpenter, 1990b).

Pigment losses during aphotic algal decomposition were regulated by the oxygen content of, and length of exposure to, hypolimnetic waters (Leavitt & Carpenter, 1990a). The rate of pigment decay was further affected by both algal community composition and microfaunal processing of detritus. Decay rates differed among carotenoids (to -0.0870 d^{-1}), with β-carotene the most refractory pigment (to -0.0167 d^{-1}). In contrast, Chl a degraded rapidly and variably (to -0.1226 d^{-1}) and its losses were not compensated by stoichiometric production of recognizable derivatives.

The importance of photo-oxidation and hypolimnetic decay decreases in piscivore-dominated lakes (Fig. 15.1b). When piscivores are abundant or planktivores rare, large zooplankton are common. These herbivores graze algae and transport pigments of edible algae to the sediments in their feces. Photo-oxidation is potentially high because water clarity is great. However, fecal transportation bypasses much of the bleaching. On the other hand, clear water also allows deep blooms to form and transport pigments to sediments via cell sinking. These distinctive blooms deposit a characteristic subset of pigments (Leavitt et al., 1989; Leavitt & Carpenter, 1990b).

Herbivory is a potent process in pelagic food webs and can affect pigment deposition. Incorporation of algae into fecal material can increase sinking rates up to 1000-fold, depending on the species of plankton (Welschmeyer & Lorenzen, 1985a,b; Kitchell & Carpenter, 1987). Herbivory reduces pigment abundance, alters relative proportions and modifies the chemical structure of pigments (Repeta & Gagosian, 1982; Carpenter & Bergquist, 1985; Leavitt & Brown, 1988; Barlow et al., 1988). However, pigments are less degraded than nonpigmented organic material (Daley, 1973; Klein et al., 1986; Leavitt & Brown, 1988). Therefore, ingestion of algae by herbivores bypasses some major pigment loss processes.

Grazing by large Cladocera increased preservation of alloxanthin in traps and sediments. Cryptophytes were the most easily ingested phytoplankton in Peter, Paul and Tuesday Lakes (Fig. 15.3). Zooplankton consumed 25–75% of alloxanthin standing stock per week. Inclusive estimates of grazing were 4-fold greater than those from the conservative model, but showed similar differences among algal groups (Leavitt & Carpenter, 1990b). Therefore, high ingestion combined with low digestion selectively transported alloxanthin to deepwater traps or sediments.

Fecal vectors have less effect on pigments that are less edible or more easily digested (Leavitt & Brown, 1988). Pigments in inedible cells may bleach more slowly but are exposed to light longer and suffer greater total losses. Slow sinking also increases pigment losses during algal decomposition by exposing moribund cells to elevated oxygen concentration or microbe effects.

Chlorophyll mass budgets also revealed that large Cladocera regulate pigment sedimentation in concert with photo-oxidation (Carpenter *et al.*, 1986, 1988; Leavitt & Carpenter, 1990*b*). Pigment sedimentation was unrelated to primary production (Carpenter *et al.*, 1986, 1988). However, when large *Daphnia pulex* replaced small crustaceans as the major herbivore in Tuesday Lake in August 1985 (Chapter 8), pheophorbide deposition increased 2–3-fold over early summer and 1984 levels (Carpenter *et al.*, 1988). Total summer phorbin flux also increased almost 3-fold and was positively correlated ($p < 0.05$) with mean grazer length. Continued reduction in standing crop by grazing further reduced total pigment losses to photo-oxidation in Tuesday Lake in 1986 (Leavitt & Carpenter, 1990*b*). Comparisons with Paul and Peter Lakes indicated that herbivory and photo-oxidation could control sedimentation either

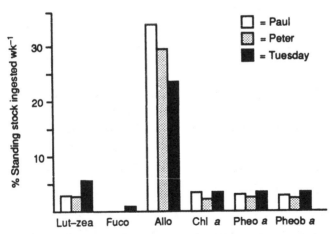

Fig. 15.3. Ingestion of pigment standing crop (as % per week) calculated assuming all algae < 30 μm GALD are edible (exclusive model). Lut-zea, lutein-zeaxanthin; Fuco, fucoxanthin; Allo, alloxanthin; Pheo, pheophytin; Pheob, pheophorbide. (From Leavitt and Carpenter (1990*b*). Copyright © *Canadian Journal of Fisheries and Aquatic Sciences*. Used by permission.)

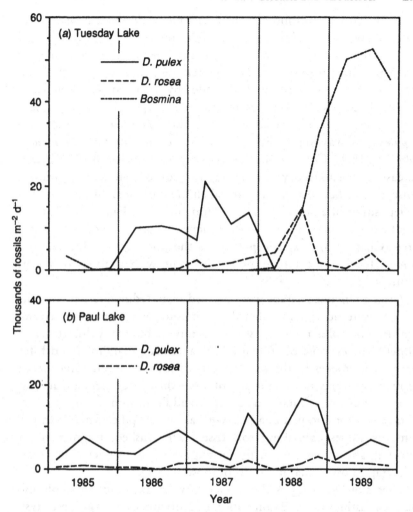

Fig. 15.4. Animal microfossil deposition in (*a*) Tuesday Lake and (*b*) Paul Lake for sediment traps set from early, mid- and late summer in 1985–9.

independently or in concert, in a similar way to marine ecosystems (Welschmeyer & Lorenzen, 1985*a*,*b*).

What determines animal deposition?

Sediment trap catches of animal remains accurately represented changes in the abundance of pelagic populations in Tuesday Lake (Fig. 15.4*a*). Removal of dace and stocking with largemouth bass in 1985 reduced zooplanktivory by fish and allowed *Chaoborus* to eliminate *Bosmina* (Fig. 15.4a; Chapter 8). *Daphnia pulex* postabdominal claws were scarce until late summer consistent with contemporaneous plankton records (Chapter 8 and 10). Low levels of vertebrate planktivory in 1987 (no bass, few minnows) were clearly recorded by sustained high abundance of both large *Daphnia* species in sediment traps. However, as dace populations increased, vertebrate zooplanktivory intensified and large *D. pulex* gave way to mid-sized *D. rosea* and eventually small *Bosmina* in 1988 (Fig. 15.4a; Chapter 8). The recovery of minnows in 1989 further increased zooplanktivory and was clearly recorded in traps by continued dominance of *Bosmina* and the virtual elimination of *Daphnia*.

Deposition of animal remains in unmanipulated Paul Lake changed little through time (Fig. 15.4b). Sediment trap and water column data agree (Chapter 8). *Daphnia pulex* was the dominant cladoceran and *Bosmina* fossils were uncommon, consistent with analyses of recent sediments (Leavitt *et al.*, 1989).

Daphnia remains in traps were derived from sedimentation of molts and senescent individuals rather than deposited as a result of predation. *Daphnia* was represented by few whole animals, but many disarticulated helmets, valves, postabdominal claws, antennae, mandibles and filter combs. Differences in the state of preservation between valves from sediment traps and those in the guts of fish in these lakes suggest that most empty valves and helmets were senescent adults or molts. Empty valves in traps were still joined as a pair with an intact spine, whereas in fish stomachs valves containing soft tissues are crushed, fragmented or distorted, and spines are usually missing. Analysis of overwinter sediment traps showed that molts decompose rapidly.

Sources of *Daphnia* postabdominal claws in traps and sediments also could be distinguished. Examination of individuals undergoing ecdysis demonstrated that molted claws are lightly colored, while claws from live zooplankton are dark-colored. Virtually all claws from traps were dark, and therefore from dying daphnids. Thin molts presumably decay rapidly and are not well preserved in sediments (but see Kerfoot, 1981). Analysis of claws showed that the mean size of *D. pulex* in sediment traps was consistently larger than that of individuals in the zooplankton (trap size/plankton size $= 1.22 \pm 0.01$). Senescent individuals are the oldest and largest. Similarly, animals in traps were generally smaller than individuals recovered from fish guts (Hodgson & Kitchell, 1987), except when mean *Daphnia* length was at its annual maximum. Therefore, most animals in traps are deposited through direct sedimentation, rather than in fish feces. Fecal transport is presumably more important when there are large changes in vertebrate zooplanktivory (e.g. transition from bass to minnows). In many cases, fecal transportation produces the same fossil signal as direct deposition through sedimentation.

Sediment traps accurately collected rare zooplankton during the period of study. *Diaphanosoma* has minute postabdominal claws that were detected in samples from Tuesday Lake in 1985 and 1989 after enumeration at 200 × magnification. Claws from 1984 were present at 3–6% of the fossil sum, consistent with plankton analyses. *Holopedium* was represented by postabdominal claws, second antennae, and gelatinous sheaths. *Holopedium* claws were also present in the most recent postmanipulation sediments. Similarly, *Polyphemus* was only recovered from postmanipulation sediment traps and sediments.

Fossil record of food web interactions

The experiments and sediment-trap studies summarized above provide essential clues for the interpretation of sediment stratigraphies. The acid test of paleolimnology, however, is the capacity of sediment records to portray long-term events in the water column. In this section, we compare sediment records from Paul, Peter and Tuesday Lakes with limnological data from 1950 to 1986. We hypothesized that changes in fish community composition have distinctive effects on the relative abundance of large daphnids in the zooplankton, algal standing crop, and water transparency: the principal determinants of pigment sedimentation. We found that shifts between piscivore- and planktivore-dominated food webs were clearly recorded in the sediments. We then used sediment records from Tuesday Lake to test the variance transmission hypothesis (Carpenter, 1988*b*): that algal sedimentation rate should exhibit periodicities synchronous with the life cycle of the dominant fish in the food web.

Paul Lake

The effects of fish community change are clearly recorded in the paleoecological record of fossil pigments and zooplankton. Loss of trout due to reproductive failure and winterkill, coupled with the invasion of minnows, resulted in increased size-selective zooplanktivory and replacement of large *Daphnia* by small *Bosmina* (Fig. 15.5*a*). These herbivores were unable to effectively graze algal standing crops, such that water transparency declined and deep water blooms of chrysophytes were lost. Transition from a cyprinid community to one of largemouth bass resulted in a reciprocal shift in zooplankton: large-bodied *Daphnia* replaced smaller *Bosmina*. These large herbivores reduced algal standing crop, increased water clarity and allowed a deep bloom of photosynthetic bacteria (Fig. 15.5a). In all respects the paleo-ecological record of fossil pigments and zooplankton proved remarkably

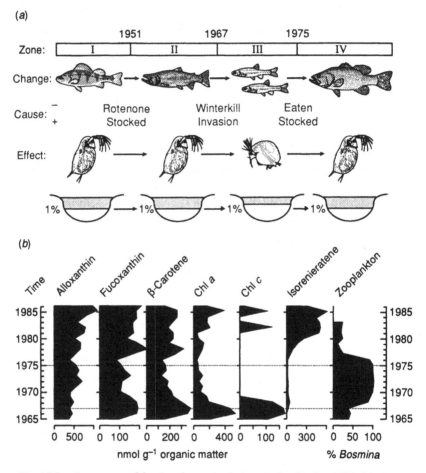

Fig. 15.5. Summary of food web manipulations in Paul Lake, 1939–86. (*a*) Schematic diagram of changes in the fish assemblage, zooplankton and water column during summer: 1% delimits photic zone. (See text for details.) (*b*) Sedimentary record of selected fossil pigments and zooplankton for Paul Lake, 1965–86. %*Bosmina* = *Bosmina*/(*Bosmina* + *Daphnia*). (From Leavitt *et al.* (1989). Copyright © by American Society of Limnology and Oceanography. Used by permission.)

sensitive to known changes in the plankton during the period 1951–86 (Leavitt et al., 1989).

The phytoplankton community in 1965 was dominated by metalimnetic blooms of the chrysophytes *Dinobryon sertularia*, *Mallomonas* sp. and *Chrysosphaerella* (Malueg, 1963). This algal community was clearly shown in the sediments by high concentrations of fucoxanthin, β-carotene, Chl *a* and Chl *c*, together characteristic of Chrysophyta (Fig. 15.5*b*). Algal microfossil analysis verified that chrysophytes, but not diatoms, dominated the plankton (J. P. Smol, Queen's University, personal communication). Fossil zooplankton community structure was 60% *Daphnia* (>90% *D. pulex*), consistent with contemporaneous records (UNDA, 1988).

The first important change in fish community structure occurred by 1967, when the failing rainbow trout population was replaced by a cyprinid community (Fig. 15.5*a*). Redbelly dace exhibit rapid population growth and diel migration behaviors that maximize their effect as zooplanktivores (Naud & Magnan, 1988). Coincident with minnow colonization, the fossil zooplankton assemblage shifted from predominantly large-bodied *Daphnia* to a community composed entirely of small *Bosmina longirostris* (Fig. 15.5*b*). This shift should decrease water transparency, as large grazers are replaced by smaller, less effective herbivores (Mazumder et al., 1988; Chapter 13). Reduction in transparency was signalled by the decline in the chrysophyte bloom indicators fucoxanthin, β-carotene, Chl *a* and Chl *c* (Fig. 15.5*b*).

Lake conditions remained stable until 1975 when largemouth bass were stocked. With addition of this piscivore, size-selective zooplanktivory was reduced and small *Bosmina* were replaced by large-bodied *Daphnia* (Fig. 15.5*b*). Introduction of effective *Daphnia* should increase sedimentary flux and concentration of pigments (Carpenter et al., 1988). As expected, increased herbivory rapidly reduced algal standing crops, leading to increased sedimentary pigment concentrations in 1977–9.

Reduced algal standing crop improved water clarity and was signalled by a dramatic rise in the carotenoid isorenieratene (Fig. 15.5*b*). Isorenieratene is characteristic of the green sulfur photosynthetic bacteria Chlorobiaceae: obligate anaerobes that require both light and hydrogen sulfide for photosynthesis (Schmidt, 1978). High levels of isorenieratene confirmed neolimnological records of the photosynthetic bacteria (Parkin & Brock, 1980) and indicated that light penetrated to the anoxic layer of the lake. Further evidence for increased transparency came from coeval limnological records which showed that the light penetration in Paul

Lake (>5.5 m) was the greatest since 1951 (Chapter 13; Leavitt *et al.*, 1989).

The fossil pigment record of food-web alterations withstood moderate mixing of sediments (Leavitt & Carpenter, 1989). Three major pigment signals remained ecologically interpretable in an unlaminated core from 9 m of water: (1) elevated chrysophyte pigment concentrations characteristic of deepwater blooms (1964–70); (2) transient, *Daphnia*-mediated increases in algal deposition (1976–80); and (3) sharp increases in bacterial abundance (isorenieratene 1975–86) (Fig. 15.5*b*).

Effects of sediment mixing were inversely related to fossil signal strength (defined as C_P/C_B, the ratio of peak to baseline concentrations): weak signals ($C_P/C_B < 1.5$) were rapidly blended into background while strong signals ($C_P/C_B > 5$) maintained resolution despite homogenization to depths > 1 cm (> 5 y accumulation). Pigment profiles from shallow water cores (7 m and 4 m depth) were unrelated to historical events and were dominated by benthic production by diatoms or complete mixing of sediments (Leavitt & Carpenter, 1989).

Peter Lake

Piscivore manipulation in Peter Lake again altered the food web down to the microbial level, but the outcome was modified by chemical perturbation of the physical environment (pH, oxygen concentration, water clarity). Complex interactions between biotic and abiotic control of algae took several decades to be resolved. Evaluation of ecosystem experiments benefitted from the long-term perspective provided by accurate fossil analyses.

Transitions in fish community structure had strong effects on zooplankton and were clearly recorded in the sediments (Fig. 15.6*a*). Replacement of rainbow trout by minnows resulted in a shift from *Daphnia* to *Bosmina* by 1970. Introduction of bass reversed these trends and returned *Daphnia pulex* to dominance (1978; Fig. 15.6*a,b*). In Peter Lake, early lime additions (Hasler, Brynildson & Helm, 1951) increased transparency and allowed rainbow trout to eliminate *D. pulex* in favor of smaller *D. rosea* (identified as *D. longispina* by Johnson & Hasler, 1954). Fossil zooplankton records document that *Daphnia* species replacement required 15 y (Leavitt *et al.*, 1989). Apparently cyprinids are more effective zooplanktivores than trout, which required multiple lime additions to effect changes in herbivore composition.

Addition of 4500 kg of hydrated lime to Peter Lake in 1969 altered the food web (Fig. 15.6*a*) and was seen in the fossil pigment record (Fig. 15.6*b*). Areal algal production increased owing to lime addition, as

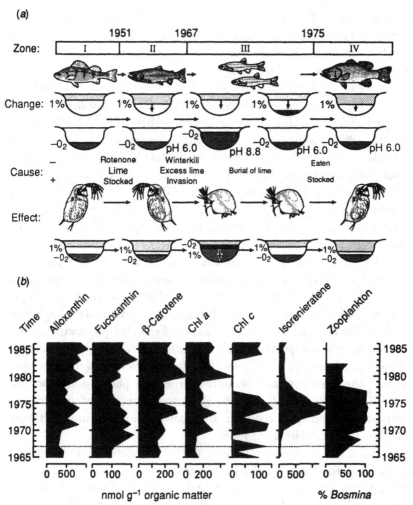

Fig. 15.6. Summary of food web manipulations in Peter Lake, 1939–86. (*a*) Schematic diagram of changes in the fish assemblage, zooplankton and water column during summer: 1% delimits photic zone, $-O_2$ delimits anoxic waters. (See text for details.) (*b*) Sedimentary record of selected fossil pigments and zooplankton for Peter Lake, 1965–86. %*Bosmina* = *Bosmina*/(*Bosmina* + *Daphnia*). (From Leavitt *et al.* (1989). Copyright © by American Society of Limnology and Oceanography. Used by permission.)

shown by elevated concentrations of all pigments between 1970 and 1976. Previous lime additions also removed humic acids, increased the photic zone (Stross, 1958; Stross & Hasler, 1960) and allowed deep blooms of chrysophytes (Leavitt *et al.*, 1989).

Green sulfur bacteria populations bloomed because of lime addition (isorenieratene in Fig. 15.6b). As in Paul Lake, Chlorobiaceae blooms indicate that light penetrated into anoxic waters. However, the bloom in Peter Lake monitored a reduction in deep water oxygen concentrations rather than clear water (Fig. 15.6a).

Photosynthetic bacteria were rare prior to 1969 because light did not penetrate into anoxic waters (UNDA, 1988). Addition of lime and loss of *Daphnia* increased algal production and further reduced water clarity (UNDA, 1988). However, lime addition also increased the pH of water above the sediments from <6 to *ca.* 8.8 (UNDA, 1988). Increased pH stimulated bacterial decomposition of organic matter and increased hypolimnetic oxygen consumption (e.g. Waters, 1957). Therefore, as the volume of anoxic waters increased, hydrogen sulfide moved into the photic zone and provided a habitat for photosynthetic bacteria (Fig. 15.6).

Food-web structure and lime interacted to limit bacterial populations after 1974, as shown both by reductions in isorenieratene concentrations (Fig. 15.6b) and by direct observations (Parkin & Brock, 1980). High algal production initially caused by liming was maintained through low herbivory by *Bosmina*. However, because of elevated production, sedimentary lime was rapidly buried; its buffering capacity reduced, pH declined and decomposition processes slowed. With reduced oxygen demand, the anoxic waters withdrew from the photic zone and the bacterial bloom declined (Fig. 15.6a,b). Historical records of water color, transparency, and pH confirm this paleoecological interpretation (UNDA, 1988; Leavitt *et al.*, 1989).

Return of oxygen to hypolimnetic waters stimulated pigment degradation, as predicted from bottle experiments (Leavitt & Carpenter, 1990a). Concentrations of all sedimentary pigments declined parallel to the bacterial bloom and remained low until the bass–induced transition in zooplankton community structure (Fig. 15.6b). Replacement of *Bosmina* with large-bodied *Daphnia* resulted in reduced algal standing crops and sharply increased pigment concentrations in lake sediments (1979–81). Reduced algal standing crop increased water clarity and was marked by reestablishment of deep chrysophyte blooms (Elser *et al.*, 1986b; Chapter 12) and high Chl *c* concentration in sediments (Fig. 15.6b). Photosynthetic bacteria (as isorenieratene) did not respond, consistent with the presence of oxygen in deep waters.

High concentrations of pigments in surface sediments are common in these lakes and may also reflect incomplete degradation (Hurley & Armstrong, 1991). Incomplete fossil record formation, combined with a mild manipulation and brief effects in 1985, produced little indication of recent fish experiments, consistent with mass balance studies.

Tuesday Lake
Food-web interpretation

Introduction of rainbow trout (1961) and largemouth bass (1985) showed that additions of top carnivores have different outcomes depending on the species added. Final food-web structure depended on the strength of interaction between piscivorous and planktivorous fish (Fig. 15.7) (P. R. Leavitt et al., unpublished manuscript). In both cases, the trophic cascade was mediated by invertebrate predators (Chaoborus punctipennis). Effective predation by bass removed minnows and allowed large herbivores (Daphnia pulex, Holopedium gibberum) and Chaoborus to dominate. Chaoborus eliminated small cladoceran grazers (Bosmina longirostris, Diaphanosoma birgei). In contrast, moderate predation by rainbow trout only reduced the population of minnows; consequently, intermediate-size herbivores (D. rosea) increased in abundance and Bosmina persisted. While changes in piscivory affected herbivores, increased algal abundance resulting from allochthonous nutrients (1951–2) did not change fossil zooplankton species composition or community size structure. Increased primary production enhanced Chaoborus survivorship and these predators suppressed Bosmina populations (Fig. 15.7).

Road construction (1951–2) probably increased nutrient loading to Tuesday Lake (cf. Likens et al., 1970) and resulted in higher biomass of edible green algae (Fig. 15.8). Concentrations of characteristic pigments (lutein–zeaxanthin, α-carotene, β-carotene and Chl b) increased 2- to 3-fold after the road cut and were paralleled by enhanced herbivory, as detected by the grazing indicator pheophorbide a. Similar results were seen in Peter and Paul Lakes when the road was constructed (Leavitt et al., 1989). Because concentrations of all cladoceran fossils decline or remain stable (Fig. 15.8), the increase in the grazing indicator suggests either that other, unrecorded herbivores benefit from the nutrients (e.g. copepods or rotifers) or that the chlorophytes were especially easy to graze (P. R. Leavitt et al., unpublished manuscript).

Increased primary production enhanced survival of Chaoborus punctipennis and led to a decline in the main cladoceran herbivore, Bosmina longirostris (Fig. 15.7) (P. R. Leavitt et al., unpublished manuscript). Survivorship of omnivorous early-instar Chaoborus is increased by algal blooms (Yan et al., 1991). Bosmina populations decline until 1957 and individuals with cornuta-form (curved) antennules are reduced to 50% of

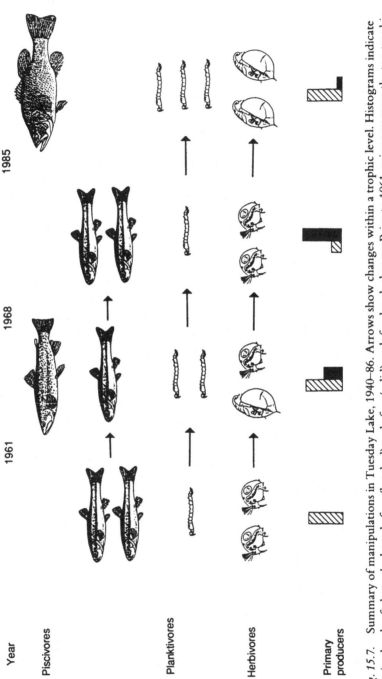

Fig. 15.7. Summary of manipulations in Tuesday Lake, 1940–86. Arrows show changes within a trophic level. Histograms indicate relative levels of phytoplankton before (hatched) and after (solid) each food web change. Prior to 1961, minnows were the top trophic level. Rainbow trout were the top trophic level in 1961–8. Minnows were again the top trophic level in 1968–84. Largemouth bass were the top trophic level in 1985–6.

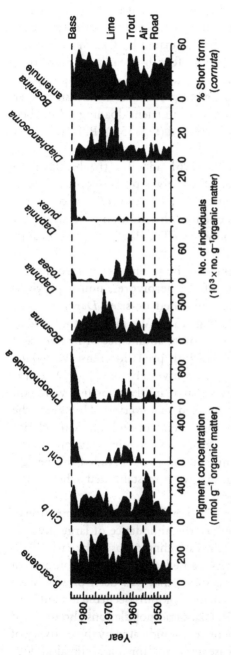

Fig. 15.8. Sedimentary record of selected fossil pigments and zooplankton for Tuesday Lake, 1945–85. (From P. R. Leavitt *et al.*, unpublished manuscript).

previous values. *Bosmina* morphology is sensitive to changes in predation by invertebrates (Kerfoot, 1974, 1981). Animals with curved antennules are preferentially ingested by raptorial invertebrate predators (Kerfoot, 1987*b*). *Bosmina* were not completely eliminated by *Chaoborus* because zooplanktivory by minnows prevented fourth-instar predators from achieving sufficient population size (P. R. Leavitt *et al.*, unpublished manuscript). All invertebrates returned to predisturbance densities after hypolimnetic aeration reduced internal nutrient loading, disturbed forests recovered, and algal blooms subsided (Fig. 15.8) (P. R. Leavitt *et al.*, unpublished manuscript).

Introduction of rainbow trout in 1961 added a trophic level to Tuesday Lake. However, trout are themselves planktivores (Scott & Crossman, 1973) that are only moderately effective predators on cyprinids (Hodgson *et al.*, 1991). Consequently, the addition of trout caused an intermediate response in the food web of Tuesday Lake (P. R. Leavitt *et al.*, unpublished manuscript).

Moderate reduction in zooplanktivory by minnows changed the size structure and species composition of the herbivores, unlike previous nutrient inputs. Abundance of intermediate-size *Daphnia rosea* increased following trout stocking (Fig. 15.8). We infer that *Chaoborus* predation increased, reducing both absolute abundance of *Bosmina* and the relative abundance of individuals with predation-sensitive *cornuta* morphology. The fact that the largest invertebrates (e.g. *D. pulex, Holopedium*) did not increase shows that planktivorous fish were still present (P. R. Leavitt *et al.*, unpublished manuscript). However, the effectiveness of trout as predators was dependent on high water clarity caused by lime addition (1961) and natural increases in light penetration (1965–70) (Fig. 15.8). Consequently, minnows reinvaded the pelagic zone and premanipulation conditions were reestablished when water transparency declined (e.g. 1964–5).

Trout were winterkilled before 1970 (von Ende, 1979). Elimination of piscivory allowed minnows to reduce the abundance of large invertebrate grazers, leading to increased algal standing crops in the early 1970s (Fig. 15.7). Declining water clarity caused a loss of deep blooms of chrysophytes (as Chl *c* and fucoxanthin) that had been present since treatment with lime. Similar losses of deep blooms occurred in Paul Lake after minnows invaded (Fig. 15.5). *Chaoborus* populations also declined when planktivory by minnows increased, as indicated by the recovery of *Bosmina* populations and the increase in *cornuta*-form individuals. Food-web structure was relatively constant until the addition of largemouth bass in 1985.

Introduction of an effective piscivore caused massive changes in invertebrate communities (Carpenter *et al.*, 1987; Chapter 8) and was accurately recorded in the fossil record (P. R. Leavitt *et al.*, unpublished manuscript). Large herbivores (*Daphnia pulex*, *Holopedium gibberum*) dominated the Cladocera for the first time in 40 years, although the response of mid-size *D. rosea* was less than that during trout addition (Fig. 15.8). In contrast, small Cladocera (*Bosmina longirostris* and *Diaphanosoma birgei*) were virtually eliminated by a *Chaoborus* population that doubled after minnow extirpation (Elser *et al.*, 1987b; Chapter 4). This outcome contrasts the moderate effect of trout in 1961 and shows that the final food-web configuration depends on the intensity of predation by piscivorous fish.

Spectral analysis
In lakes with constant food-web structure, the variance transmission hypothesis states that pigments in edible algae should be deposited with a periodicity set by lifespan of the dominant predator in the food web (Fig. 15.9a–d) (Carpenter, 1988b; Carpenter & Leavitt, 1991). For a fish population to propagate itself, there must be a successful cohort at least once per lifespan. If density-dependent limitations are strong, however, there will be a successful cohort no more than once per lifespan (Carpenter, 1988b). Zooplanktivory by this cohort selectively removes the largest herbivores. Because pigment sedimentation varies directly with grazing by large, nonselective zooplankton (Carpenter *et al.*, 1988), pigment deposition should be entrained to zooplanktivory by fish and reflect variability in fish predation. Pigments in less edible algae or with nonplanktonic sources should show weaker linkage to fish. In Tuesday Lake, cyprinids live 3–5 y (Stasiak, 1978; He, 1986) and the hypothesis predicts that sedimentation of pigments in edible algae should vary with that period.

Pigments in edible algae were deposited with a periodicity set by the lifespan of minnows in Tuesday Lake. Plots of normalized variance versus period showed that both alloxanthin and the grazing indicator pheophorbide *a* had variance maxima at a period of 3–4 y, as predicted (Fig. 15.9e). The strength of the alloxanthin peak is especially interesting because it is the carotenoid most strongly affected by zooplankton grazing (Fig. 15.3). Alpha-carotene, another cryptophyte pigment, was also deposited with a period of 3–5 y (Fig. 15.9f). In contrast, pigments with terrestrial or littoral sources (pheophytins *a*, *b* and *c*) show no significant signal while those associated with less edible particles (lutein–zeaxanthin, fucoxanthin) had relatively flat, weak spectra with peaks

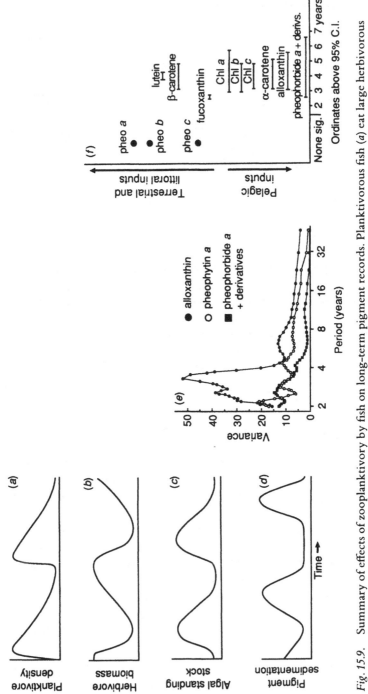

Fig. 15.9. Summary of effects of zooplanktivory by fish on long-term pigment records. Planktivorous fish (*a*) eat large herbivorous zooplankton (*b*) that in turn graze phytoplankton (*c*) and regulate pigment sedimentation (*d*). (*e*) Variance spectra of pigment concentrations in Tuesday Lake sediments (1889–1984). (*f*) Variance spectra significantly different from a white-noise signal arranged by pigment source. (From Carpenter & Leavitt (1991). Copyright © by the Ecological Society of America. Used by permission.)

near 2 y (Fig. 15.9*f*). Therefore, changes in zooplanktivory by fish during constant food-web structure were most clearly recorded by deposition of pigments in edible algae.

Paleolimnology can differentiate between sources of temporal variation in lakes. Zooplanktivory by minnows decreased pigment sedimentation in Tuesday Lake by reducing the abundance of the dominant grazer, *Bosmina longirostris*. Other sources of variation can be revealed by removing predator effects with running averages that are scaled by predator lifespan (e.g. 3–5 y). In Tuesday Lake, such smoothed time series showed cycles of 15–20 y for pigments derived from terrestrial sources (Carpenter & Leavitt, 1991).

Synthesis

Paleolimnology provides information that is often otherwise unavailable. Rapid, inexpensive analyses can create time series that lack artifacts that may arise from methodology and personnel changes in other long term studies. Paleolimnological records are biased representations of ecosystem processes. However, quantification of processes that regulate formation of the fossil record identifies this bias and can allow unambiguous analysis.

Results of our experimental and paleoecological studies agree. During periods of intense piscivory, large herbivores dominate and their feces transport pigments of edible algae to the sediments. Photo-oxidation is potentially high because water clarity is great. However, fecal transport bypasses much of the bleaching. Increased light penetration allows deep blooms to form and transport pigments to sediments via cell sinking. Metalimnetic blooms often deposit a characteristic subset of pigments.

When piscivores are absent or planktivores predominate, zooplankton are small and relatively ineffective. Pigment deposition is regulated by photo-oxidation and hypolimnetic decay, which cause selective, taxon-specific losses. Herbivores eat only the smallest algal cells, so fecal transportation is minimal and total photo-oxidation is high. Most pigments sink directly because of compensatory increases in algal standing stock and because water clarity is low during periods of reduced herbivory.

Because the principal mechanism of pigment sedimentation (grazing vs. sinking) differs with food-web structure, the strongest fossil signals occur during transition between food-web configurations (planktivore- or piscivore-dominated). Deposition of pigments in algae that are only edible by large herbivores should be particularly sensitive to fish manipulation. Similarly, ratios of large *Daphnia* to small *Bosmina* represent zooplankton community size structure and record changes in zooplanktivory by fish. Our study contrasts with other paleolimnological approaches, which typically use static species assemblages to indicate limnological conditions. The use of fossil pigments complements other studies and provides unique and valuable information on food-web processes in lakes.

Most fossil pigment signals were interpretable without historical verification, although analyses benefit from simultaneous study of zooplankton remains. Major fossil records of food-web change remained despite some mixing of sediments (Leavitt & Carpenter, 1989). Deep water blooms of photosynthetic organisms are marked by dramatic increases in concentrations of taxon-specific pigments in lake sediments. Chlorophylls degrade rapidly; undegraded Chl is therefore a sensitive indicator of deep algal populations. Ratios of β-carotene to Chl *a* may differentiate deep blooms (low ratio) from overall high production (high ratio). Transient (2–3 y), nonselective increases in sedimentary pigment concentrations accompany introductions of large-bodied herbivores and are useful indicators of changes in total herbivory. Future comparisons of zooplankton microfossil and pigment profiles in lakes with historical food-web manipulations will be informative.

Some aspects of fossil pigment interpretation remain difficult. Concentrations of sedimentary pigments are weakly related to primary production within a year, but may show better correspondence over a wider range of nutrient input (e.g. Züllig, 1989). Selective degradation limits the usefulness of pigments in sediments as indicators of relative algal abundance. Peredinin (from dinoflagellates) was particularly degraded during algal decomposition (Leavitt & Carpenter, 1989). Further, the relative importance of degradative processes may also change with trophic state and water column depth (relative penetration of light and oxygen). Additional experimental studies are needed to resolve these issues.

Fossil zooplankton should accurately monitor changes in pelagic Cladocera. Comparison of *Daphnia* in sediment traps, plankton and fish guts showed that remains mostly arise from senescent individuals that sink directly, and not from molts or victims of fish predation. Close correspondence of zooplankton community composition between plankton samples and traps from 1985 to 1989 also suggest that sedimentary deposits will accurately monitor changes in herbivore abundance and species composition.

Fish community structure can be deduced from analyses of fossil cladoceran communities. The size structure of the fossil community is a functional measure of selective zooplanktivory. In our lakes, shifts in species composition and size structure were highly correlated.

Dominance by large *Daphnia pulex* (as postabdominal claws) indicated very low levels of planktivory by fish, suggesting a piscivore-dominated lake. In contrast, fishless conditions may be identified by the presence of mandibular fossils from *Chaoborus americanus*, a predatory insect characteristic of fishless lentic habitats (Uutala, 1990). Mixtures of moderate-sized *Daphnia* (e.g. *D. rosea*) were characteristic of assemblages in which piscivores and planktivores coexist. In such mixed communities, size-selective planktivory is important in shaping pelagic zooplankton assemblages. Absence of daphnid remains combined with abundant fossils from small cladocerans (*Bosmina, Diaphanosoma*) indicated the highest levels of planktivory.

Interactions between vertebrate and invertebrate predators affect food-web structure (Chapters 7, 8 and 10) and require further paleoecological study. Invertebrate predators (*Chaoborus*, copepods) regulate abundance of smaller zooplankton (Chapter 8; Kitchell & Kitchell, 1980; Leavitt *et al.*, 1989) and leave corporeal remains (Uutala, 1990) as well as indirect fossil records (i.e. changes in prey morphology) (P. R. Sanford *et al.*, unpublished data). In Tuesday Lake, the impact of *Chaoborus* was dependent on the strength of interaction between piscivores and planktivorous fish. Effective piscivory released *Chaoborus* from control by planktivorous fish and allowed phantom midges to eliminate small Cladocera. Interactions between lake productivity and invertebrate predation are also possible (e.g. Neill & Peacock, 1980).

Summary

Experimental and paleolimnological techniques were used to calibrate an important suite of sedimentary indicators of phytoplankton, the carotenoids. Comparison of carotenoid and algal biovolume standing stocks revealed that pigment concentrations were influenced by species replacements, nonplanktonic inputs and changes in algal abundance. Pigment sedimentation was not closely related to algal production over the relatively narrow range observed in these lakes.

Carotenoid sedimentation was regulated by photo-oxidation and herbivore grazing. Sediment traps were biased towards pigments that were incorporated into zooplankton feces, sank rapidly and bypassed the photo-oxidation that otherwise destroyed pigment standing stocks.

Large-bodied herbivores produced the most effective fecal vectors for pigments in these lakes. Edible phytoplankton such as cryptophytes were selectively preserved in the fossil record. Additionally, hypolimnetic decay caused underestimation of dinoflagellate abundance and biased estimates of total sedimentation based on chl *a*. Beta-carotene was a more reliable estimator of total algal deposition.

Comparison of seasonal sediment trap contents with changes in zooplankton communities (1985–9) showed that animal fossil deposition faithfully recorded all changes in size-selective planktivory resulting from fish manipulations. Most *Daphnia* in traps were derived from deaths of older, larger individuals rather than from ecdysis.

Annual records of fossil pigments and zooplankton were compared with contemporaneous records from Peter and Paul Lakes (1940–86). Paleolimnological profiles recorded all major plankton dynamics resulting from fish community changes and liming including: (1) transitions in cladoceran size structure and species composition; (2) changes in water clarity that affected vertical zonation of primary producers; (3) changes in the absolute abundance of all algal divisions except dinoflagellates; and (4) periods of increased herbivory. Undegraded chl *a* and *c* and specific carotenoids monitored metalimnetic (chrysophyte) and hypolimnetic (bacterial) populations. Major sedimentary signals withstood partial mixing of sediments but were decoupled from pelagic events in shallow waters owing to complete mixing and benthic algal production.

In Tuesday Lake, multiple manipulations of fish between 1950 and 1985 showed that the outcome of predation, and final food-web structure, depended on the intensity of piscivory. Largemouth bass removed minnows and allowed large herbivores (*Daphnia pulex*) and invertebrate predators (*Chaoborus punctipennis*) to dominate. Small grazers (*Bosmina longirostris, Diaphanosoma birgei*) were eliminated by invertebrate predators. Under moderate predation by rainbow trout, minnows remained, only intermediate-size herbivores (*Daphnia rosea*) increased in abundance, and *Bosmina* persisted. Increased algal abundance resulting from road construction did not affect the zooplankton community because invertebrate predators (*Chaoborus*) prevented growth of herbivore populations.

Spectral analysis of sedimentary carotenoids in Tuesday Lake confirmed that food-web interactions scaled interannual variability as predicted by the variance transmission model (Chapter 16). Annually resolved fossil pigment profiles (1889–1982) documented that variability in zooplankton grazing was entrained to fluctuations in planktivore populations. Strong year classes of minnows reduced the abundance of herbivores and hence sedimentation of easily ingested algae. Deposition of pigments from other sources (inedible algae, nonplanktonic sources) was unrelated to changes in planktivory or herbivory.

Acknowledgements

We thank S. Barta, J. P. Smol, M. S. V. Douglas, M. B. Berg for aid with analyses. D. W. Schindler and T. Kratz provided helpful criticism of the manuscript. Preparation of this chapter was supported by a Natural Sciences and Engineering Research Council of Canada (NSERC) Operating Grant to D. W. Schindler and an NSERC postdoctoral fellowship to PRL.

16 · *Simulation models of the trophic cascade: predictions and evaluations*

Stephen R. Carpenter and James F. Kitchell

Introduction

Previous chapters have detailed the responses of Peter and Tuesday Lakes to fish manipulations. Some of the changes were anticipated, while others were surprises. Our research was guided by models of the trophic cascade in lakes. To what extent did these models forecast the experimental results? The purpose of this chapter is to assess how our predictions fared, and how our view of the trophic cascade has been modified by the experimental outcome. First, we must explain why we developed simulation models of the trophic cascade, how the models were structured, and the predictions that derived from the models.

The trophic cascade is, in essence, a simple idea. In the complexity of real lakes, however, it involves the collective outcome of life history, predator–prey, and physical–chemical processes that cannot be adequately represented by simple verbal, graphical or mathematical models. Computer simulations are one way of integrating these complex interactions. They elaborate the conceptual framework and develop specific, testable predictions. We began simulation studies of the trophic cascade in 1981, three years before initiating the ecosystem experiments. Many of the predictions of those models can now be evaluated.

This chapter has four parts. First, we review three simulation models that produced the hypotheses we tested in the field. Second, we address predictions specific to the outcome of our ecosystem experiments. Third, we turn to more general expectations that should apply to trophic cascades in many lakes. Finally, we evaluate the status of modeling and theory germane to trophic cascades in lakes.

Models

Three simulation models were developed in relation to our ecosystem experiments. The models were designed for different purposes. Consequently, they have different structures (Fig. 16.1). More importantly, they address questions across a gradient of time scales, ranging from a few days to a few centuries.

The **size-structured model** of plankton interactions (Fig. 16.1a) was developed to study the ecosystem consequences of well-known allometric relationships (Carpenter & Kitchell, 1984). Size-selective predation molds zooplankton size distributions (Brooks & Dodson, 1965). Zooplankter size affects the range of particle sizes that can be ingested (Burns, 1968), feeding rate (Peters & Downing, 1984), and nutrient excretion rate (Peters, 1983). Rates of nutrient uptake, growth, and sinking loss by phytoplankton are related to cell or colony size (Malone, 1980; Walsby & Reynolds, 1980). These empirical relations were used to estimate parameters for dynamics of 11 size classes of zooplankton feeding on 11 size classes of phytoplankton, with phosphorus as the limiting nutrient (Fig. 16.1a). The zooplankton size distribution was assumed to be fixed for each run of the model. The model was run for time periods of less than six days. On this scale, zooplankton could be assumed static, and analysis focused on dynamics of phytoplankton and phosphorus. An exhaustive analysis of parameter uncertainty for this model was presented by Bartell, Brenkert & Carpenter (1988).

Model output included phytoplankton size distributions, biomass and primary production as functions of initial zooplankton biomass and mean size (Carpenter & Kitchell, 1984). Increased biomass or size of grazers favored larger phytoplankters and lower phytoplankton biomass. As zooplankter size increased, primary production decreased. Zooplankton biomass was unimodally related to primary production. At high zooplankton biomass, production was low, owing to heavy grazing. At low zooplankton biomass, production was low, owing to low rates of nutrient resupply by excretion. At intermediate zooplankton biomass, stimulatory effects of nutrient regeneration more than compensated for modest grazing losses, and production was maximized. Since algal biomass declined monotonically as zooplankton biomass increased, primary production was maximal at intermediate levels of algal biomass (see below, Fig. 16.2).

The **seasonal dynamics model** (Fig. 16.1b) was built to study sources of variance within and among growing seasons (Carpenter &

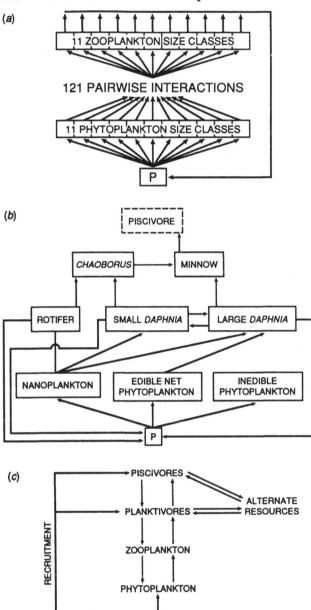

Fig. 16.1. Flow charts of simulation models described in the text. (*a*) Size-structured plankton model (Carpenter & Kitchell, 1984); (*b*) seasonal dynamics model (Carpenter & Kitchell, 1987); (*c*) variance transmission model (Carpenter, 1988*b*, 1989).

Kitchell, 1987). The program simulated summer stratified seasons of 100–150 days. Zooplankton and the invertebrate predator *Chaoborus* were dynamic variables. Predation by fishes changed through the season as a forcing function.

Development of the seasonal dynamics model was motivated by the observation that about half of the variance among lakes in primary production could not be explained by nutrient inputs (Schindler, 1978). Model analyses showed that this unexplained variance was similar in magnitude to that produced by fluctuations in fish predation (Carpenter & Kitchell, 1987). We proposed that maximum interannual variability in primary production should occur at intermediate fish stocks, owing to maximized variation in recruitment and the pulses of predation by fish year classes on zooplankton. Analyses of the seasonal dynamics model also showed that correlations among adjacent trophic levels depended upon the time scale over which data were averaged (Carpenter & Kitchell, 1987).

The **variance transmission model** (Fig. 16.1*c*) was written to study interannual variability over time spans of decades to centuries (Carpenter, 1988*b*). This model was also used to determine the consequences of manipulation strength and duration for statistical detection of ecosystem response (Carpenter, 1989). Parameters of the model were based on dynamics of Paul, Peter and Tuesday Lakes during 1984–6. This model extended to multiyear time scales a prediction noted above: correlations among adjacent trophic levels depended upon the time scale over which data were averaged (Carpenter, 1988*b*). The model predicted that variance in primary production would be maximal at time scales corresponding to the lifespan of the top predator in the food web (Carpenter, 1988*b*). Model analyses showed that changes due to fluctuations in top predator biomass were attenuated as they passed down the food chain (Carpenter, 1989). Consequently, strong and sustained manipulations of fishes are needed to produce changes in primary production that are detectable in short time series of only a few years (Carpenter, 1989).

Specific predictions

We developed site- and year-specific predictions for responses of Peter and Tuesday Lakes to our manipulations. Despite the importance of prediction for hypothesis testing, the ecological literature usually does not provide outlets for testable predictions when they are system-specific. Our predictions, largely derived from modeling studies,

Table 16.1. *Specific predictions made for Peter Lake*

An outcome of 'yes' indicates that results corroborated predictions. An outcome of 'no' indicates that results contradicted predictions.

Prediction	Outcome
(a) Response to piscivore removal and planktivore addition in 1985	
1. Bass growth rates increase	yes
2. Planktivory increases	yes
3. No change in *Chaoborus*	yes
4. *Daphnia* decreases; small zooplankters increase	yes
5. More recycling, less nutrient loss	no
6. Smaller phytoplankton become more dominant	no
7. Phytoplankton biomass and production increase	yes
(b) Consequences of 1985 manipulation in later years	
1. Responses persist for at least two years	no
2. 1985 bass year class has maximum piscivory in 1988	no
3. Planktivory declines smoothly from 1985 to 1988	no
4. Herbivore biomass rises smoothly from 1985 to 1988	no
5. Chlorophyll and primary production decline smoothly from 1985 to 1988	no
(c) Response to removal of piscivores and addition of planktivores in 1989–90	
1. Planktivory becomes more variable	?
2. Herbivore biomass becomes more variable	yes
3. Chorophyll and primary production become more variable	yes

appeared in grant proposals written in 1982 and 1985. Specific predictions for Peter Lake are collected in Table 16.1 and those for Tuesday Lake in Table 16.2. No hypotheses were developed for Paul Lake, which was intended to represent the baseline variability of an undisturbed food web, and serve as a reference system for effects of factors that affected all three lakes.

Peter Lake

In 1985, minnows were added to Peter Lake while an equivalent biomass of piscivorous bass was removed (Chapter 2). Predicted responses included increases in growth rates of remaining bass, planktivory rates, and phytoplankton biomass and primary production (Table 16.1a). Large-bodied herbivores (*Daphnia*) were predicted to decrease in biomass, while biomass of smaller bodied zooplankters increased. All of

Table 16.2. *Specific predictions made for Tuesday Lake*

An outcome of 'yes' indicates that results corroborated predictions. An outcome of 'no' indicates that results contradicted predictions.

Prediction	Outcome
(a) *Effects of piscivore addition and planktivore removal*	
1. Rapid decrease in planktivory by fishes	yes
2. No change in *Chaoborus*	no
3. Large herbivores increase	yes
4. Total zooplankton biomass increases	no
5. Nutrient regeneration by zooplankton increases	no
6. Nutrient loss rate from epilimnion increases	yes
7. Total epilimnetic nutrient concentration decreases	no
8. Larger phytoplankton become more dominant	no
9. Phytoplankton biomass and production decrease	yes
10. Bass winterkill in 1985–6	no
(b) *Effects of piscivore removal and planktivore reintroduction*	
1. Minnow populations increase	yes
2. Large zooplankton eaten; small zooplankton return to dominance	yes
3. Nutrient concentrations in epilimnion increase	no
4. Smaller phytoplankton become more dominant	yes
5. Phytoplankton biomass and production increase	yes
6. Herbivory increases smoothly as minnows recover	no
7. Chlorophyll and primary production increase smoothly as minnows recover	no

these predictions proved correct. However, we had expected that minnows, rather than young-of-the-year largemouth bass, would be the principal cause of increased planktivory. We were surprised by the rapid elimination of minnows from the pelagic zone and by the size of the bass cohort recruited in 1985 (Chapter 4).

Two predictions about the short-term results of the 1985 manipulation were not correct. As a consequence of the shift in herbivore size, we had expected more rapid recycling of nutrients within the water column, and reduced losses to sedimentation. Shifts between nutrient limitation by N or P, explainable by changes in grazer size, did occur (Elser *et al.*, 1988; Chapter 13). However, no change in nutrient concentrations was detected in the water column (Chapter 13) and no change was detectable in rates of sedimentation (S. R. Carpenter, unpublished data).

We had expected smaller phytoplankters to replace larger ones after the decline in herbivore size. The reverse occurred in 1985: large gelatinous colonies bloomed in Peter Lake after the shift in grazer size (Carpenter *et al.*, 1987; Chapter 11).

Predictions about the state of Peter Lake after 1985 stemmed from the expectation that minnows would become established in the lake. Since minnows were rapidly eliminated, none of these predictions was supported by the data (Table 16.1*b*). Maximum piscivory by the 1985 bass year class occurred in 1989, close to the prediction (1988). We had expected planktivory to decline smoothly as the bass year class aged and their diets shifted to benthos and smaller fishes. Planktivory did decline from 1985 to 1987, but the changes were more abrupt than expected, owing to rapid elimination of the minnows and rapid diet shifts by bass as they grew (Chapter 6).

Once the 1985 year class of largemouth bass attained maximal piscivory rates, our research plan called for steady removal of the bass. At the same time, a planktivore component was established in the lake by stocking rainbow trout and golden shiners (Chapter 2). The intent was to simulate the effects of exploitation pressure exerted by sport fisheries on natural piscivore populations. We expected exploitation to increase the temporal variance in biomass and process rates at all trophic levels (Beddington & May, 1977; Carpenter & Kitchell, 1987). Herbivore biomass, chlorophyll and primary production did become more variable among weeks in 1989–90 (Table 16.1*c*). Data are not sufficient to evaluate the variability in planktivory during that period.

Tuesday Lake

In 1985, minnows were removed from Tuesday Lake while a nearly equivalent biomass of piscivorous bass was added (Chapter 2). We predicted that planktivory by fishes would decline rapidly, that biomass of large herbivores (*Daphnia*) would increase, that nutrient loss rates from the epilimnion would increase and that phytoplankton biomass and primary production would decrease (Table 16.2*a*). All of these forecasts proved correct (Chapters 6, 8, 10 and 13).

Our conclusion that nutrient loss rates from the epilimnion increased derives from the fact that sedimentation rates increased for a few months after *Daphnia* attained dominance of the zooplankton (Carpenter *et al.*, 1988; Leavitt & Carpenter, 1990*b*). These papers reported sedimentation rates of algal pigments, which are directly related to those of nutrients. Phosphorus sedimentation rates (averaged for June, July and August) were 1.1 mg P m^{-2} d^{-1} in 1984, 1.6 mg P m^{-2} d^{-1} in 1985, and 1.0 mg P m^{-2} d^{-1} in 1986. Thus, any increase in phosphorus loss rates were confined to 1985. No change could be detected in total nutrient concentrations in the water column (Chapter 13), contrary to our prediction. Changes in nutrient loss rates were short-lived and did not cause sustained changes in nutrient concentrations.

Several other predictions were not borne out by the data. Because *Chaoborus* is cryptic and migratory, we had expected that it would not respond to the manipulations. *Chaoborus* biomass increased in 1985 because prey were abundant early in the summer (Chapter 7). *Chaoborus* predation on rotifers and copepodites contributed to the shift toward dominance by larger grazers, which are less vulnerable to *Chaoborus* (Hall *et al.*, 1976). Through 1986–90, *Chaoborus* biomass declined, probably because of low concentrations of its preferred prey, rotifers and juvenile copepods (Chapter 7).

Total zooplankton biomass was expected to increase. However, zooplankton biomass decreased only slightly, despite the massive community shift toward large grazers (Chapters 8, 10). These changes in the zooplankton community caused substantial changes in the phytoplankton (Chapters 11, 13). Pace (1984) argued that changes in zooplankton size, rather than biomass, were responsible for variations in chlorophyll concentration that could not be explained by phosphorus levels. Our results support his point.

Nutrient regeneration by zooplankton was expected to increase. Since zooplankton biomass decreased slightly while the size distribution shifted toward larger individuals, nutrient regeneration rates probably declined. Changes in nutrient regeneration were sufficient to alter nutrient limitation of the phytoplankton (Elser *et al.*, 1988; Chapter 13).

The size-structured model predicted that the larger zooplankton in Tuesday Lake would promote dominance by larger phytoplankters (Carpenter & Kitchell, 1984). The weighted mean length of phytoplankters responded positively to increases in cladoceran length at scales of weeks (Chapter 11). Similar short-term shifts in phytoplankton size distribution occurred in enclosure experiments in Tuesday Lake (Bergquist *et al.*, 1985). At time scales of years, however, the manipulation caused drastic reductions in the larger species of phytoplankton, including three *Peridinium* species, *Chrysosphaerella longispina* and *Microcystis aeruginosa.* Overall, the evidence does not support the prediction that larger zooplankters foster dominance of larger phytoplankters.

The most important surprise from the 1985 manipulation in Tuesday Lake was the overwinter survival of the bass. We had expected winterkill to eliminate the bass, leading to rapid reversal of manipulation effects at all trophic levels. However, bass survival rates were high and effects of the manipulation were sustained for several years.

We had expected that piscivore removal and minnow reintroduction in 1987 would rapidly reverse effects of the 1985 manipulation (Table 16.2*b*). While several of our predictions were borne out, the rate of ecosystem recovery was slower than we had expected. Increasing minnow populations eventually eliminated *Daphnia*, and small zooplankters returned to dominance (Chapter 8). However, the *Daphnia* were surprisingly resistant to predation. Diel vertical migration was initiated as a predator avoidance behavior shortly after minnows were introduced (Chapter 9) and *Daphnia* were relatively abundant until August 1988 (Chapter 8). Chlorophyll and primary production increased after the *Daphnia* population collapsed, but did not attain the levels observed in 1984. Several phytoplankton species that had dominated the premanipulation community, most notably the large dinoflagellate *Peridinium limbatum*, did not regain their former densities. That absence accounts for the incomplete recovery of chlorophyll and primary production. Smaller phytoplankton remained dominant.

Several predictions were contradicted by our findings (Table 16.2*b*). We had expected more efficient nutrient recycling to increase nutrient concentrations, but no changes in nutrient levels were detected (Chapter 13). Relatively smooth transitions in herbivory, chlorophyll and primary production were expected as the minnow population recovered. In contrast, changes in lower trophic levels were rather abrupt, especially at the time of the *Daphnia* collapse in early August 1988 (Chapter 13).

General predictions

Many of the specific predictions discussed above were based on ideas that are general, in the sense that they should apply to a wide range of lakes (Carpenter & Kitchell, 1984; Carpenter *et al.*, 1985; Carpenter & Kitchell, 1987; Carpenter, 1988*a*; Kitchell *et al.*, 1988; Carpenter, 1989). In the following sections, we have classified these general expectations and, in most cases, evaluated them in the light of results presented in earlier chapters and in Tables 16.1 and 16.2.

Inverse biomass trends

Biomasses at adjacent trophic levels are expected to be inversely related, while biomasses at trophic levels separated by an intermediate trophic level are expected to be directly related (Carpenter et al., 1985). The trophic cascade shares this expectation with many other concepts of community and ecosystem organization (Hairston et al., 1960; Oksanen et al., 1981; Menge & Sutherland, 1987; Persson et al., 1988). The idea of inverse biomass trends is no more than a useful heuristic simplification. Theoretical analyses have shown that inverse biomass trends should be observed only under certain conditions. For example, they depend upon the time span over which the trends are measured, whether an equilibrium or nonequilibrium model applies, and the number of trophic levels in relation to potential productivity of the ecosystem (Smith, 1969; Oksanen et al., 1981; Carpenter & Kitchell, 1987; Walters et al., 1987; Persson et al., 1992). Therefore, acceptance or rejection of the idea is not at issue. Progress occurs when we are better able to resolve the conditions under which inverse biomass trends are or are not observed.

Compensatory population dynamics or behavior of certain species can determine whether inverse biomass trends occur. Processes at the population level can accelerate or dampen ecosystem responses to perturbation. Several clear examples are evident in previous chapters.

(1) Predator avoidance behavior of small fishes caused planktivory to decline extremely rapidly in the presence of piscivores in Peter and Tuesday Lakes in 1985 (Chapters 4–6). Behavior at the population level accelerated and intensified ecosystem responses to piscivores.

(2) Another predator avoidance behavior (diel vertical migration) enabled *Daphnia* populations to persist despite fairly high planktivory by fishes in Tuesday Lake in 1987, and in Peter Lake in 1988–90 (Chapter 9). In this case, behavior at the population level mitigated ecosystem responses to planktivores.

(3) Compensation at the community level was evident in Peter Lake in 1989–90. After trout eliminated *Chaoborus*, a smaller daphnid (*Daphnia dubia*) established dense populations in coexistence with the larger, migratory *D. pulex* (Chapter 8). Species replacement increased grazer biomass despite substantial planktivory by fishes.

In sum, predator avoidance behavior and species replacement processes often buffer the response of prey biomass to predation. However, trophic cascades from fish to phytoplankton do not necessarily require changes in zooplankton biomass. Changes of the keystone grazer *Daph-*

nia pulex, which has a broad diet and rapid numerical response, have more significance for phytoplankton than do changes in overall zooplankton biomass.

Herbivore effects on primary producers

In pelagic ecosystems, interactions between grazers and primary producers involve a compensatory mechanism not found at higher trophic levels: nutrient excretion by the herbivores stimulates and supports growth of the plants. During stratification in lakes, recycling by herbivores can be a major source of nutrients for the algae (Lehman, 1980; Sterner, 1986; Elser *et al.*, 1988). Consequently, herbivory can actually stimulate net algal growth rate when nutrients are limiting and grazer biomass is low (Fig. 16.2). The key rates that determine the relative height of the curves in Fig. 16.2 depend strongly on the sizes of the interacting zooplankters and phytoplankters. Analysis of the size-structured model of zooplankton–phytoplankton interactions led to the hypotheses evaluated below (Carpenter & Kitchell, 1984).

Increases in zooplankter size were predicted to cause declines in phytoplankton biomass and primary production. This prediction is unequivocally supported by the data, at time scales ranging from one or two weeks to one or two years.

Increases in zooplankter biomass were predicted to cause declines in phytoplankton biomass. This hypothesis was not generally supported by our data, although increases in zooplankter biomass cause declines in phytoplankton biomass at certain time scales. The hypothesis cannot hold where predator–prey oscillations of zooplankton and phytoplankton arise (Carpenter & Kitchell, 1987, 1988; Murdoch & McCauley, 1985; Lathrop & Carpenter, 1992). The oscillations involve both overgrazing of algae by zooplankton and stimulation of zooplankton by algae. The bouts of heavy grazing are likely to shift algal community structure toward grazing-resistant morphologies, thus ending the cycle (Lathrop & Carpenter, 1992). The net result is short-term (days to weeks) depressions of algal biomass by increased grazer biomass. Multilake comparative studies have also shown negative relations between grazer biomass and algal biomass when the effects of nutrients were removed statistically (Carpenter *et al.*, 1991). However, the effects of grazer biomass are not as consistent as those of grazer size (Pace, 1984; Carpenter *et al.*, 1991).

Primary production was predicted to be a unimodal function of

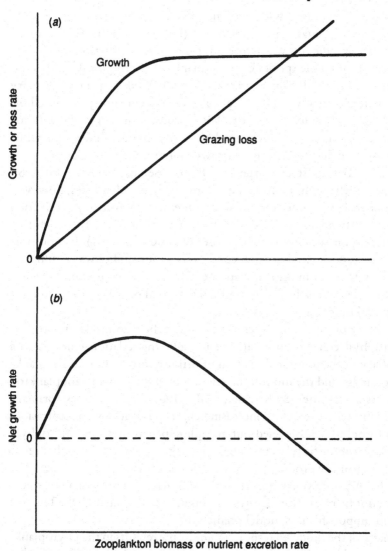

Fig. 16.2. Explanation of the unimodel response of phytoplankton production to zooplankton biomass. Excretion by zooplankton is assumed to be the major source of nutrients for growth. As zooplankton biomass increases, gross growth rate increases hyperbolically and grazing loss increases linearly (*a*). The difference between the curves is the net growth rate (*b*).

zooplankton biomass, with maximal primary production occurring at intermediate biomasses of zooplankton (Fig. 16.2). In enclosure experiments, this prediction was borne out for edible size fractions of chlorophyll and for edible species of phytoplankton (Bergquist & Carpenter, 1986; Elser et al., 1987a; St. Amand, 1990). With respect to primary production or growth rate of all phytoplankton (not just the edible fraction), our results are mixed. Total production or growth rate has increased or remained unchanged as grazer biomass increases, while biomass-specific production or growth rate increase as grazer biomass increases (Bergquist & Carpenter, 1986; Elser & MacKay, 1989; St. Amand, 1990; epilimnetic data in Chapter 12). In other lakes, however, researchers have found unimodal responses in total primary production (Sager & Richman, 1991) or positive, neutral and negative responses depending on lake trophic state (Elser & Goldman, 1991). It is possible that the range of experimental conditions in our studies was insufficient to generate a full unimodal response. More experimentation is apparently needed to resolve the hypothesis that total primary production is a unimodal function of grazer biomass.

Primary production was also predicted to be a unimodal function of chlorophyll concentration. In the models, this relation emerges as a corollary of the unimodal relation of primary production to zooplankton biomass, and the monotonic relation of chlorophyll to zooplankton biomass (Carpenter & Kitchell, 1984, 1987). In the lakes, primary production increased with increasing chlorophyll at low concentrations of chlorophyll, and leveled out at high concentrations of chlorophyll. Similar asymptotic relations appear in multilake comparisons that relate integral primary production to algal biomass (Brylinsky & Mann, 1973; Smith, 1979). This curvilinearity probably arises from self-shading of the algae, not from the size-class transitions invoked in the models. Thus we see no support for this model prediction.

Many of the modeled responses hinge on size shifts in the phytoplankton induced by shifts in size or biomass of the zooplankton. Specifically, the model predicts that increases in either biomass or size of zooplankters will shift the phytoplankton toward larger cells or colonies. Consequences for primary production derive from the dependency of algal metabolic rates on cell or colony size (Carpenter & Kitchell, 1984). Evidence for these size dynamics is equivocal. Larger phytoplankters are more resistant to grazing, and size shifts in short-term enclosure experiments are similar to those predicted by the model (Bergquist et al., 1985; Bergquist & Carpenter, 1986). In the whole-lake experiments, however, shifts that agreed with model predictions were no more common than were exceptions. For example, shifts to small zooplankters in Peter Lake were followed by blooms of large colonial phytoplankters, consistent with model expectations. In contrast, the shift to large zooplankters in Tuesday Lake eliminated several large phytoplankters, which did not reappear when the large zooplankters were removed by planktivory. It seems likely that failures of the size-structured model are due to life history and ecophysiological properties of the algae that cannot be represented by size alone.

Hysteresis

When a piscivore manipulation is made and then reversed, 'the sequence of ecosystem states and the rate of transition among states in the reverse pathway will differ from those of the forward pathway' (Carpenter et al., 1985). The dynamics of Tuesday Lake during 1984–90 are an example of ecosystem hysteresis (Fig. 16.3). Piscivore enhancement rapidly changed Tuesday Lake to a condition similar to that of Paul and Peter Lakes. Ecosystem responses to piscivore additions in 1985 were complete by 1986. Responses to the reverse manipulation, however, took several years. The lag resulted from the slow recovery rate of the minnow population, the predator avoidance behavior of the Daphnia, and the continued recruitment failure of dinoflagellates.

Correlations depend on time scale

Ecosystem comparisons using regression (or similar static, linear statistical models) have a long history of use in ecosystem ecology (Cole, Lovett & Findlay, 1991). In limnology, regression analyses have made notable contributions to research, for example on nutrient and contaminant loading (Reckhow & Chapra, 1983) and the dependency of ecosystem process rates on organism size (Peters, 1983).

We investigated the possibility of using regression to analyze trophic cascades. Simulation models were used to generate data, which were then subjected to correlation and regression analysis. If the analyses could reliably determine the functional relations built into the simulation models, credibility would be lent to regression analyses of relations among trophic levels in lakes. Unfortunately, the regression analyses of the simulation models depended strongly upon the time scale at which the simulation output was sampled (Carpenter & Kitchell, 1987, 1988;

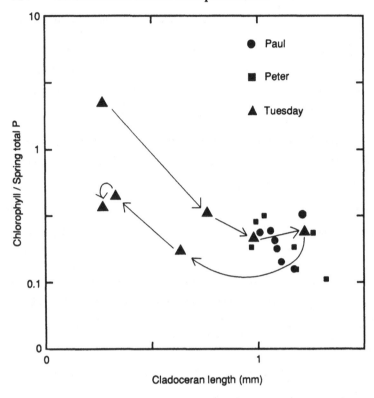

Fig. 16.3. Ratio of mean epilimnetic chlorophyll *a* concentration during summer to mean total P concentration in spring (mg chl mg^{-1} P; note log scale) versus mean length of cladoceran grazers during summer for Paul, Peter and Tuesday Lakes in each of the seven years of this study. For Tuesday Lake, arrows show trajectory of the lake from year to year during 1984–90. Total P concentration in spring (the mean of all epilimnetic measurements made in May) is a surrogate for phosphorus loading (Carpenter *et al.*, 1991). Thus the vertical axis represents the biomass of algae generated per unit nutrient supplied. The horizontal axis is an index of grazing intensity.

Carpenter, 1988*b*). Adjacent trophic levels could be positively or negatively correlated, depending upon the sampling interval! Consequently, important restrictions apply to the interpretation of regressions relating adjacent trophic levels in lakes (Carpenter *et al.*, 1991). Regression models can work reasonably well to explain static differences among lakes, but they are not reliable for analyzing dynamics within lakes. These findings prompted us to adopt a dynamic statistical approach that accounted explicitly for time lags and serial dependencies (Chapter 2). The time-dependency of correlations found in the simulation models is corroborated by the time series analyses presented in this book (e.g. Chapter 13).

Temporal variance

The seasonal dynamics model (Fig. 16.1b) indicates that variance in primary production that is not explainable by nutrient supply is about the same as that induced by fluctuations in fish predation (Carpenter & Kitchell, 1987). This hypothesis has proven hard to test. Variance cannot be unequivocally assigned to nutrient versus fish sources, because the linear models necessary for such calculations are usually not appropriate. One data set that meets the assumptions of multiple linear regression suggests that nutrient and fish effects on variance are similar (Carpenter et al., 1991).

The responses of the experimental lakes to manipulation were weaker than the responses observed in the seasonal dynamics model (Fig. 16.1b) to similar manipulations. Compensatory population dynamics and behavioral shifts which were not included in the model attenuated the effects of fishes in the lakes. The variance transmission model (Fig. 16.1c) was calibrated using results of the field experiments and therefore yields fish effects more similar to those observed in the lakes (Carpenter, 1988b, 1989). In this model, the consequences of compensation are mimicked by fitting stochastic processes to the data, rather than through mechanistic understanding.

Our data show that variance in primary production that is not explainable by nutrients is partly, but not entirely, explainable by fish predation. The important question of how much variance is due to fish predation remains unresolved at this time.

On the basis of model results, we argued that intermediate fish stock densities should maximize the variability in biomass and production of lower trophic levels (Carpenter & Kitchell, 1987). Exploitation of piscivore stocks increases the variance of stock size (Beddington & May 1977). At intermediate stock densities, variance in recruitment may be maximal (Carpenter & Kitchell, 1987). As diets shift during the ontogeny of young fishes, variability in predation translates into fluctuations in zooplankton and phytoplankton. The variability of Peter Lake created by fishing and stocking during 1989–90 corroborates this hypothesis. However, by its very nature, a rigorous test of the hypothesis requires many more years of data from lakes experiencing a range of fish stock densities.

The variance transmission model predicts that variance in primary production is maximal at time scales corresponding to the lifespan of the top predator in the food web (Carpenter, 1988b). Spectral analyses of annually resolved data from the Tuesday Lake core support the hypothesis (Chapter 15; Carpenter & Leavitt, 1991). Thus we have a single confirmatory test from a lake with relatively short-lived fishes. Further tests are needed in fishless lakes and in lakes dominated by longer-lived piscivores.

Manipulation strength

Ecosystems are variable experimental units. For that reason, substantial and sustained manipulations are necessary to obtain interpretable results (Kitchell et al., 1988). Quantitative comparisons of experimental designs using the variance transmission model indicate that piscivory must be altered by five- to tenfold for periods of at least two or three years to detect changes in the phytoplankton using intervention analyses (Carpenter, 1989).

Paul, Peter and Tuesday Lakes represent a gradient of manipulation strength. Paul Lake was not manipulated. Peter Lake received a series of relatively brief manipulations of intermediate strength. Tuesday Lake underwent massive manipulations for sustained periods of time. How does our perception of cascade effects differ among the three lakes?

The temporal development of evidence for trophic cascades in the three lakes can be compared using Bayesian statistics. In Bayesian statistics, the degree of credibility of alternative models is compared using probabilities calculated for each model. These probabilities are updated over time as additional data become available. If the probabilities of the models remain about equal, then learning (in the sense of discriminating alternative models) is not occurring. If, on the other hand, the probabilities of the models diverge and approach unity for one model and zero for the others, then learning is taking place. In the Bayesian scheme, the best experimental program is that which causes the most rapid learning rate, where learning rate is proportional to the rate of discrimination of alternative models.

Our goal is to estimate the probability of the cascade model against reasonable alternatives in the three lakes. The most parsimonious Bayesian approach involves two models. The 'cascade' model predicts ecosystem response on the basis of food web structure. The 'physical–chemical' model predicts ecosystem response on the basis of physical and/or chemical characteristics of the water column. We want to compare how the probabilities of the two models change over time in each lake.

Our specific approach to the modeling and Bayesian calculations was as follows. Integrated volumetric primary production (Chapter 13) was the dependent variable for both models. The models were transfer functions (Chapter 3). The 'physical–chemical' (P–C) model was the best-fitting transfer function involving physical or chemical inputs. The 'cascade' model was the best-fitting transfer function using zooplankton as inputs. At the end of each year, both models were fitted using cumulative data (e.g. the fit for 1987 used data from 1984–7). The models were compared using Bayes' formula:

$$P_C(t) = L_C(t)P_C(t-1)/[L_C(t)P_C(t-1) + L_{PC}(t)P_{PC}(t-1)],$$

where $P_C(t)$ is the posterior probability of the cascade model at the end of year t. $P_C(t-1)$ and $P_{PC}(t-1)$ are the prior probabilities of the cascade and P–C models, respectively, at the end of year $t-1$. Since the probabilities of the two models are assumed to sum to one, $P_{PC}(t) = 1 - P_C(t)$. The initial prior probabilities of the two models were set at 0.5. $L_C(t)$ and $L_{PC}(t)$ are the likelihoods of the cascade and P–C models, respectively, given all of the data available through the end of year t. The likelihoods depend on the standard deviation (S) of the residuals from each model, the degrees of freedom (DF) of S, and the number (N) of residuals:

$$L = \exp(-DF/2)/[(2\pi S^2)^{0.5}]^N.$$

This formula for the likelihoods assumes that the residuals are centered around zero and have approximately normal distributions (Walters, 1986).

In Paul Lake, the P–C and cascade models were never clearly discriminated (Fig. 16.4). Neither model attained a probability less than 0.1, and dominance switched between the two models in 1987 and 1990. In Peter Lake, the probability of the P–C model has declined slowly but steadily throughout the study. By 1990, the probability of the P–C model was less than 0.0001, and the probability of the cascade model exceeded 0.9999. In Tuesday Lake, dominance of the cascade model was established most rapidly. The probability of the P–C model had declined to about 10^{-9} by 1986, and continued to decline slightly in later years. The cascade model emerged as the dominant model in Peter and Tuesday Lakes because the food webs were manipulated. Had we manipulated physical or chemical forcing variables, the P–C model might have become dominant.

The effect of manipulation strength on learning rate (Walters, 1986) is clear. Strong, sustained manipulation in Tuesday Lake rapidly produced unequivocal changes that clearly discriminated the alternative models. In Paul Lake, seven years of monitoring without manipulation could not distinguish between the two models. Peter Lake is an intermediate case, where moderate pulse manipulations gradually produced changes that discriminated between the models.

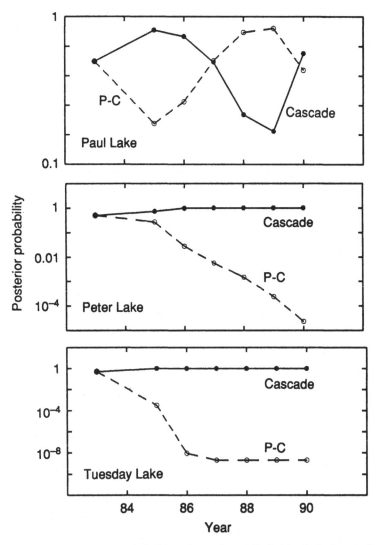

Fig. 16.4. Posterior probabilities of cascade and P–C (physical–chemical) models for primary production in Paul, Peter and Tuesday Lakes, 1984–90.

Status of modeling for trophic cascades

Ecosystem models are simplifications; they are incomplete and yield uncertain predictions. They are useful because they organize concepts and develop logical consequences in ways that may be tested by empiricism and experimentation. Models also have value for communicating essential concepts in relatively compact, comprehensible form (Walters, 1986; Fagerstrom, 1987). Because the models are known *a priori* to be incomplete and uncertain, disproving them is a trivial exercise. It is much more challenging to find the deficiencies in ecosystem models that are important for improving concepts or reducing uncertainty in predictions (Beck, 1983). Understanding why a model has failed and discovering alternatives are often the key steps in the creative process.

At this point, serious deficiencies are evident in the size-structured model (Fig. 16.1a; Carpenter & Kitchell, 1984) and the seasonal dynamics model (Fig. 16.1b; Carpenter & Kitchell, 1987). Two points are particularly devastating:

(1) phytoplankton community dynamics deviate in important ways from predictions of the size-structured model;
(2) food web responses to fish manipulations, while significant, are smaller in magnitude than predicted by the seasonal dynamics model.

The variance transmission model (Fig. 16.1c; Carpenter, 1988b), on the other hand, is corroborated by results to date. Among our models, the variance transmission model is hardest to invalidate because of its high empirical content and stochastic nature. Nevertheless, it does account for compensation (the tendency of fish effects to attenuate at lower trophic levels; cf. point (2) above), the observed variance spectra of primary production (Carpenter & Leavitt, 1991), and the need for strong and sustained manipulations of fishes to cause detectable effects at lower trophic levels (Fig. 16.4).

How can future efforts improve on the size-structured model and the seasonal dynamics model? One rather specific point is that precision in phytoplankton–zooplankton interactions cannot be achieved simply by adding more size classes to the model. Too many species-level responses of phytoplankton depend on biological idiosyncrasies that are poorly represented by size. Several modeling strategies are promising alternatives to the approach we used in the size-structured and seasonal dynamics models. If the phytoplankton are to be represented by a fairly large number of state variables, then modeling several ecophysiology and life history strategies (Bartell *et al.*, 1988) or functional groups (Vanni *et al.*, 1992*b*) is possible. If a more aggregated model is needed, simply modeling the phytoplankton as a single-state variable (as was done in the variance transmission model) is the parsimonious approach.

Compensatory mechanisms are a far broader and more difficult challenge to the modeling process. These dampen the response of lower trophic levels to fish manipulations. Species shifts and behavioral changes were the prominent compensatory mechanisms in our ecosystem experiments. Microbial (Chapter 14) and littoral zone processes (Lodge *et al.*, 1988; Gulati *et al.*, 1990) may also comprise important compensatory mechanisms. Compensations were not anticipated on the basis of models with a modest number of state variables, such as the seasonal dynamics model (Fig. 16.1*b*). These models generally predicted stronger effects of fishes on lower trophic levels than those we observed.

In many food webs, interactions among a small number of species, guilds or functional groups have a disproportionately large influence on system dynamics. R. T. Paine has shown that community dynamics often derive from a fairly small number of keystone species (1966) and strong interactions (1980). For modelers, this is an attractive simplification and may account for the fact that fairly good simulations of lake ecosystems can be obtained using models with only a few state variables. Unfortunately, experimental manipulations or natural perturbations can change ecosystems so that once reliable models are no longer appropriate (Beck, 1983; Walters, 1986). Different state variables and/or parameters are then necessary to model the altered state.

Failed models can of course be revised and improved. For example, the seasonal dynamics model fitted data from Peter Lake reasonably well in 1984 (Carpenter & Kitchell, 1987) but did not adequately predict the lake's responses to manipulations over 1985–90. The model failed because it did not include mechanisms that proved critically important, such as habitat shifts in fishes (Chapter 4) and the appearance of the cryptic, epilimnetic *Daphnia dubia* (Chapter 8). Such mechanisms could easily be added, and it is likely that the improved model would fit the observed dynamics. However, *post hoc* tinkering raises an important question (Beck, 1983). The resulting model may be little more than curve-fitting and may have no effect on the model's capacity to predict the outcome of any future manipulations. On the other hand, the modifications may have added an important process that substantially improves the predictive capacity of the model. Further experimental tests will be needed to determine which of these possibilities pertains.

Compensatory mechanisms create both a danger and an opportunity for lake ecosystem modeling. The danger is that models will become elaborate and unwieldy as we attempt to keep pace with an increasing number of special cases. Overly complex models are often hard to understand and test, and apply only to very specific systems or circumstances (Beck, 1983; Walters, 1986). Opportunity derives from the compensatory mechanisms that are among the unexpected responses that follow large-scale perturbations. Ecosystem perturbations can create surprising changes (Edmondson, 1979; Kitchell *et al.*, 1988), and by definition surprises aren't expected on the basis of extant models. Without expectations, there are no surprises. Well-posed models that develop testable expectations will be essential to learning from future ecosystem manipulations.

To what extent can food-web dynamics explain the variance in lake productivity that cannot be accounted for by nutrients? Although it is clear that changes in fish communities have substantial effects on primary production, this fundamental question remains unanswered. In principle, it could be resolved by a factorial experiment involving four lakes, with baseline and enriched nutrient levels crossed with planktivore- and piscivore-dominated fish communities. We are now conducting such an experiment.

While independent effects of nutrients and fishes can be generated experimentally, these factors probably do not operate independently in lake ecosystems. We have explained trophic cascades in the terms of community ecology and trophic interactions, but they must also be reconciled with nutrient budgets and nutrient cycles. In terms of the phosphorus cycle, food-web manipulation represents a reallocation of P among components of the ecosystem (Chapter 17). Since fishes represent a large compartment with relatively slow turnover (Kitchell *et al.* 1979), they strongly influence the rate at which the P cycle re-equilibrates after perturbations (DeAngelis, Bartell & Brenkert, 1989; Carpenter *et al.*, 1992*b*). Analysis of trophic cascades in the context of whole-lake nutrient cycles (through models, empiricism, and ecosystem experiments) will be an important step toward integrating nutrient and food web effects on lake ecosystems.

Acknowledgements

Very helpful reviews of this chapter were provided by Kathy Cottingham, Xi He, Mike Pace and Daniel Schindler.

17 · Synthesis and new directions

James F. Kitchell and Stephen R. Carpenter

Introduction

Preceding chapters provide the theoretical, analytical and empirical background for food-web interactions in an ecosystem context. In this chapter, we summarize what we consider to be the major accomplishments of our work and our interpretation of certain important, unexpected results. We also provide our view of the next generation of research issues involving the interactions of food-web structure and nutrient status in lakes, and speculate about the generality of trophic cascades in terrestrial and aquatic ecosystems.

Our primary goal in designing these experiments was to evaluate the role of food-web interactions in regulating primary production rates of planktonic algae. Regressions based on data from many lakes revealed that nutrient loading rates could account for only about half of the observed variance in primary production; roughly an order of magnitude of variability among lakes remained unexplained (Carpenter & Kitchell, 1984; Carpenter et al., 1985). We reasoned that a substantial share of that was due to differences in trophic interactions and developed a set of experiments designed to test that idea. Manipulations of fish populations in Peter and Tuesday Lakes were intended to yield maximum contrast in food web structure while the reference system, Paul Lake, remained as a monitor of interannual variance due to other sources.

We found that piscivores had rapid, massive effects on planktivores (Chapters 4–6). Predator avoidance behavior exhibited by small fishes caused planktivory in the pelagic zone to decrease much more rapidly than it would have done through piscivory alone (Chapter 5). Changes in planktivory had sustained effects on herbivore body size, but not on herbivore biomass (Chapter 10). Compensatory responses in the zooplankton community were complex; they included indirect effects due to

Table 17.1. *Summary of manipulation effects on primary production and related variates in Peter and Tuesday Lakes, based on time series analyses presented in Chapter 13*

In Tuesday Lake, responses were sustained for three years after the *Daphnia* onset of August 1985, or for two years after the *Daphnia* collapse of August 1988. In Peter Lake, responses were calculated from transfer functions for a 1 mm increase in cladoceran length. NS indicates no significant response at the 5% level.

Response variate	Tuesday Lake		Peter Lake: 1mm increase in Cladoceran length
	Daphnia onset	*Daphnia* collapse	
epilimnion chlorophyll	$-5.8\,\mu g\,l^{-1}$	$+2.3\,\mu g\,l^{-1}$	$-1.7\,\mu g\,l^{-1}$
primary production	-33%	$+31\%$	-37%
alkaline phosphatase activity/chlorophyll (P limitation index)	$+62\%$	-27%	$+119\%$
ammonium enhancement response (N limitation index)	-93%	$+45\%$	NS
irradiance at thermocline	NS	NS	$+82\%$
thermocline depth	NS	NS	1.5 m deeper

changes in the intensity of invertebrate and vertebrate predation, as well as responses to changes in the algal community, and sometimes contradicted the general expectations of the cascade hypothesis. Herbivore body size proved the better indicator of both fish predation and effects on phytoplankton. Large *Daphnia* caused decreases in chlorophyll and primary production, and amplified limitation of phytoplankton by phosphorus rather than nitrogen (Table 17.1). These changes increased transparency and deepened the thermocline in Peter Lake, but had no effect on water-column physics in stained Tuesday Lake.

Our experimental treatments induced a range of zooplankton community structures and chlorophyll levels that approaches the breadth found in the Great Lakes region of North America (Fig. 17.1). Only fishless lakes have zooplankton communities that exhibit larger average individual sizes than those encountered in our experiments (Hall et al., 1976; Kerfoot & Sih, 1987). Only culturally eutrophic lakes in agricultural basins of the region attain higher chlorophyll levels (Kitchell, 1992).

At the annual scale represented in Fig. 17.1, weekly dynamics of plankton responses are obscured. The widest fluctuations in chlorophyll are evident in Tuesday Lake. Despite the wide range of zooplankton lengths that occurred in Peter Lake, variations in annual mean chlorophyll are no greater than occurred in Paul Lake, the reference lake. As discussed in preceding chapters, the strongest grazing effects in Peter Lake occurred at shorter-than-annual time scales. The overall pattern in Fig. 17.1 suggests that chlorophyll takes on a wide range of values, depending on factors other than grazing, when zooplankton mean length is less than about 0.25 mm. When zooplankton mean length exceeds 0.25 mm, grazing constrains chlorophyll concentration to less than about 8 $\mu g\ l^{-1}$.

These results show clearly that some of the unexplained variance in lake productivity can be attributed to food-web effects. In this respect,

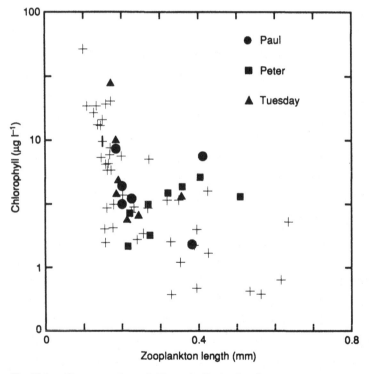

Fig. 17.1. Concentration of chlorophyll *a* in the photic zone (note log scale) versus mean length of all zooplankton during summer stratified seasons in 25 lakes of the Great Lakes region of North America (Carpenter *et al.*, 1991). Each lake is represented by at least two years on the plot. Coordinates for Paul, Peter and Tuesday Lakes during the seven years of this study are indicated. Overall correlation statistics: $r = -0.516$, $n = 81$, $p < 0.001$.

our central hypothesis was confirmed. However, some of the unexplained variance remains as such. Compensatory responses at the species, population and community scales alter the direct, complete transmission of trophic interactions from predators to primary producers (McQueen et al., 1986, 1989). Nutrient limitations, behavioral responses to changes in predation intensity, the numerical response capacity of plankton, and species replacement processes are among the mechanisms that can attenuate the transmission of variance from piscivores to phytoplankton (Carpenter, 1988a, 1989). That these compensatory mechanisms are incomplete adds to the difficulty of predicting ecosystem responses and invites the attention of future research efforts (Strong, 1992).

A simplistic view of trophic interactions (Is it top-down or bottom-up?) is insufficient and misleading. Variability in primary production rates includes both the donor-controlled response due to nutrient loading and the recipient-controlled effects of competition and selective predation. The response to changes in either of these forces is complex. The interactions of those rates and their expression in the behavior of lake ecosystem variables remains open to investigation (Carpenter et al., 1991; Carpenter & Kitchell, 1992; DeMelo et al., 1992; McQueen, 1990).

Trophic cascades and phosphorus pools

In our study systems, as in most lakes, phosphorus is typically the essential and limiting currency of trophic dynamics. Given that, we attempted a general analysis of ecosystem responses by expressing major state variables in terms of phosphorus standing stocks (Fig. 17.2).

Total system phosphorus in unmanipulated Paul Lake varied by roughly 20% over the seven years of our studies. Least variable of the component standing stocks were the major predators Chaoborus and piscivorous bass (Fig. 17.2), each of which accounted for approximately 20–25% of total phosphorus. In other words, slightly less than half of the phosphorus in the pelagic trophic system was located in predators. The only major change in zooplankton was a substantial reduction in 1986 which corresponded with the highest observed biomass of Chaoborus and the greatest value for total seston. Food-web effects are implicated therein but any effect was modest relative to those in the manipulated systems. A more cogent observation is that the majority of the key nutrient in these systems is sited in the biota. It should be no surprise that the effects of selective predation and changes in prey resources would be manifest at the level of ecosystem processes.

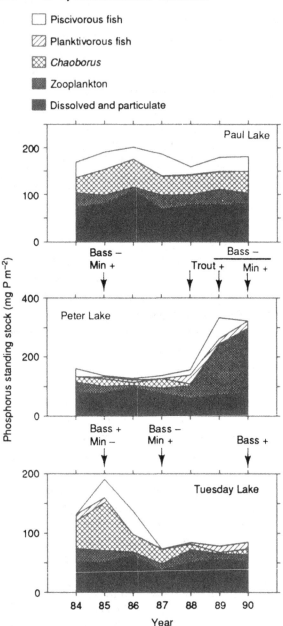

Fig. 17.2. Distribution of phosphorus among major components of the food webs in Paul, Peter and Tuesday Lakes. Standing stocks were calculated for the pelagic photic zones of the lakes. The dissolved and particulate pool consists mainly of seston (phytoplankton plus detritus) and is referred to as seston in the text.

Our initial expectation was that Peter Lake would undergo the most dramatic changes in community structure and ecosystem processes. The extensive literature on size-selective predation by fishes led us to believe that removing the piscivores and adding a substantial population of zooplanktivorous fishes to Peter Lake would yield an unequivocal result (Hrbacek *et al.*, 1961; Brooks & Dodson, 1965; Hall *et al.*, 1976). Large *Daphnia* would disappear, small zooplankton would increase and phytoplankton would bloom in the absence of an effective, non-selective grazer. The reciprocal manipulation of Tuesday Lake offered less promise because we were not sure that we could effectively deplete the local stocks of minnows and we were concerned about the lags involved in establishing populations of large *Daphnia*. As discussed in Chapter 16, we were wrong in both expectations.

A major event in the chronology of Peter Lake was the immense year class of largemouth bass formed in 1985. That cohort demonstrated strong density-dependent reduction in growth and intense competitive interactions as it progressed through trophic ontogeny and suppressed subsequent year classes through cannibalism (Chapter 4). Through the gradual changes in trophic position, the ecosystem-scale effects of this year class contributed to the highly variable behavior of zooplankton and phytoplankton. The trajectory of this cohort during 1985–90 simulated effects of year-class development in lakes experiencing high levels of fishery exploitation. That is the case for most lakes in the north-temperate region of the world; it may explain why the variability of basic production dynamics prompted our study, and it serves as evidence of the relevance of our results to those charged with the management of lentic resources.

While the reciprocal fish exchange of 1985 had negligible effect on phosphorus pools in Peter Lake, the trout additions of 1988–9 substantially increased phosphorus standing stocks (Fig. 17.2). No comparable changes occurred in Paul Lake, suggesting that the phosphorus increment cannot be explained by inputs from the watershed. The most likely explanation is that the lake was fertilized by the direct addition of fish phosphorus and the changes in trophic interactions that resulted as trout intensified the linkage between the pelagic system and the benthos through their predation on benthic invertebrates (Chapter 6). Contrary to our expectations, the additional phosphorus accumulated in the zooplankton, an ecosystem consequence of community interactions and rapid nutrient recycling.

Rainbow trout predation in Peter Lake reduced *Chaoborus* populations and their predation on small *Daphnia* (Chapters 7 and 8). The net effect was a huge increase in zooplankton biomass, which was dominated by the small to intermediate sizes of *Daphnia* species (Chapters 8 and 10). Large, grazing-resistant algae flourished intermittently as a result (Chapter 11). By 1990, total phosphorus standing stock had doubled over that of the pretreatment year (1984) and the years when zooplanktivory was high (1985–8). *Chaoborus* populations were strongly reduced and small *Daphnia* accounted for 66% of the total phosphorus present in the food web. The 1988–9 plankton community in Peter Lake is typical of lakes experiencing substantial fishing pressure.

Contrary to our initial expectations, Tuesday Lake exhibited a rapid and sustained response to the initial manipulation. Colonization by large *Daphnia* and their effect as a dominant grazer were apparent within one season. Winterkill did not remove the bass population. During the summer of 1986, the Tuesday Lake ecosystem resembled that of Paul Lake where bass remained abundant, planktivorous fishes were rare, large *Daphnia* were abundant, and both chlorophyll concentrations and primary production rates were low.

The reciprocal fish exchange of 1985 caused a modest increase in phosphorus standing stock in Tuesday Lake (Fig. 17.2). This increase cannot be explained by the bass addition, since a nearly equivalent biomass of minnows was removed (Chapter 2). The phosphorus increment resulted from growth by bass supported by feeding on littoral and/or benthic invertebrates and other nonpelagic prey (Carpenter *et al.*, 1992*b*). Removal of bass between the summers of 1986 and 1987, however, caused a substantial decline in phosphorus standing stock (Fig. 17.2). The small amount of minnow biomass added in 1987 did not compensate for the bass biomass removed (Chapter 2).

Tuesday Lake recovered slowly when the manipulation was reversed. As in Peter Lake, sestonic phosphorus showed no dramatic response (Fig. 17.2). Total phosphorus distribution among the consumers in Tuesday Lake's food web was a mirror image of that in Peter Lake. Highest standing stock occurred in 1985 when predaceous bass and *Chaoborus* were most abundant and, as in Paul Lake, accounted for roughly half of the total. Total system phosphorus was greatest in 1985–6 when the biomass of piscivores was at its peak and bass diets included virtually everything larger than small zooplankton. Total system phosphorus was lowest in 1987 after the bass were removed and a modest population of adult minnows was reestablished. *Chaoborus* declined during 1987–90 as the minnows recovered to premanipulation levels.

The large-scale dynamics of total system phosphorus generally correspond with additions or removals of fishes (Fig. 17.2). In their role as predators, fishes directly govern the amount, sizes and species composition of the plankton. Acting indirectly through predation on the key invertebrate predator (*Chaoborus*), as well as through their direct effects as size-selective predators of large *Daphnia* and sources of remineralized phosphorus (Carpenter *et al.*, 1992*b*; Schindler, 1992), fishes effected nearly twofold changes in total system phosphorus. Addition of rainbow trout to Peter Lake provided both an enhancement of nutrients in fish biomass and a series of trophic interactions that doubled total system phosphorus. Trout preferentially preyed upon *Chaoborus* and benthic invertebrates, thereby contributing to increased zooplankton biomass. Interestingly, sestonic phosphorus showed no dramatic changes.

Total seston phosphorus in Tuesday Lake also showed strong compensatory responses. However, phosphorus pools in zooplankton, *Chaoborus*, planktivorous fishes and piscivores each changed dramatically in concert with food web manipulations. By 1990, the fish and seston phosphorus pools in Tuesday Lake resembled those of the pretreatment year; however, the middle of the trophic system (zooplankton and *Chaoborus*) had not recovered to previous levels. The phytoplankton community also exhibited significant differences from that of 1984. Both low phosphorus standing stocks and poor recruitment of dinoflagellates from the sediments may contribute to these lags (Chapter 11).

Phosphorus distribution in Paul Lake probably represents steady-state conditions. In the absence of food-web perturbation, such as exploitation of fish populations, the Paul Lake ecosystem exhibited relatively modest changes. Steady-state assumptions may be safely proffered for lakes that experience little or no fishery exploitation. However, such lakes are rare.

Trophic cascades and ecosystem variability

One of the main ideas examined in our research was that increased variability in the plankton would follow from greater variability in fish recruitment. Variability in fish populations is expected as the events regulating early survival are strongly influenced by weather- and density-dependent ecological interactions (Cushing, 1982; DeAngelis, 1988; Carpenter & Kitchell, 1987). Variance in year-class strength is greatest when adult fish densities are reduced by exploitation. To promote maximum sustainable yields, most fish populations are managed and/or sustained at intermediate population levels that would

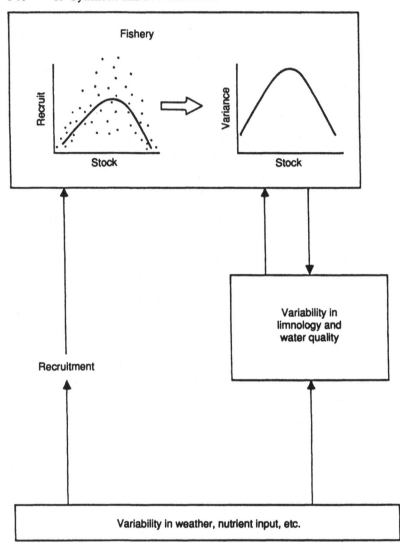

Fig. 17.3. Variability in limnology and water quality derives from direct effects of weather, nutrient input, and other physical–chemical variables, and from indirect effects of fish recruitment through the trophic cascade. Exploitation and management of fish stocks tend to sustain intermediate stock levels where recruitment, its variance, and variability in limnological variates are maximal.

assure highest average recruitment as well as maximum variation in recruitment (Fig. 17.3). The feedback of recruitment events yields strong or weak cohorts which exhibit trophic ontogeny that passes through food webs to yield maximum variability in limnological variables (Carpenter & Kitchell, 1987). The magnitude and the lags in those events are dictated by life-history characteristics of the dominant fishes in the system (Carpenter, 1988b). The results from Peter and Tuesday Lakes offer general support to those ideas and to the argument that fishing contributes to ecosystem variance through its effects on food-web interactions (Carpenter & Leavitt, 1991). Lessons from Lake Washington (Edmondson, 1991) and Lake Mendota (Kitchell, 1992) offer evidence of similar mechanisms and magnitude, thereby attesting to the generality of the argument. Where fishing occurs, we should expect that food-web effects will be amplified and that variability in ecosystem responses will be substantial. History tells us that fisheries are among the first and most common human effects imposed on lake ecosystems. Limnological studies are belated, if they occur at all. The global database deriving from limnological studies is likely to include a substantial and generally unknown set of food-web effects deriving from fishery exploitation.

Our experiments were designed, in part, to test for lags and hysteresis in response rates and their expression in ecosystem parameters (Carpenter et al., 1985). This plan involved some concession to the unexpected because we knew (from paleolimnological studies) that strong interactions (Paine, 1980) prohibited some species from the current community. Nevertheless, we were often surprised by changes in species composition and behavior, and their capacity to dampen or exaggerate ecosystem response.

Some of these surprises were easy to understand, at least in retrospect. We now appreciate the capacity of predator avoidance behaviors to accelerate (Chapters 4 and 5) or delay (Chapters 8 and 9) responses of prey, which have important consequences for ecosystem processes. The magnitude and timing of ecosystem response to food-web manipulation can be altered by changes in the risk of predation (Werner & Gilliam, 1984). Other surprises remain poorly understood. Large phytoplankton such as the dinoflagellates (*Peridinium* spp.) in Tuesday Lake were eliminated as large *Daphnia* came to dominance, but did not recover when the grazing pressure there was reduced (Chapter 11). It is possible that the seed bank of encysted forms (De Stasio, 1990) was exhausted by three years of *Daphnia* dominance in Tuesday Lake and that the populations of large *Peridinium* can only recover their former dominance after an extended period of gradual population growth. Alternatively, the low standing stock of phosphorus in Tuesday Lake in 1989–90 may have been insufficient to support *Peridinium* (Fig. 17.2).

The converse of the *Peridinium* mystery is the rapid response of zooplankton in Tuesday Lake during 1985. Given the long history of intensive planktivory there (Chapter 2), we expected that the ephippial reserve would be modest and, therefore, that the accrual of a strong *Daphnia* population would require some time (De Stasio, 1990). In fact, *Daphnia* appeared quickly and dominated within one season. Further, the sequence of changes in the dominant cladoceran was as might be expected from the basic principles of size-selective predation (Brooks & Dodson, 1965): *Bosmina* yielded to *Diaphanosoma* which yielded to *Holopedium* and then large *Daphnia* species (Chapter 8; Carpenter *et al.*, 1987).

At the extreme end of the response scale, it appears that the zooplankton biomass of Peter Lake may have exceeded its resource base (Fig. 17.2). Given the magnitude of zooplankton standing stock in the premanipulation year (1984) and the variable but substantially smaller zooplankton biomass in Paul Lake, the amount of zooplankton in Peter Lake during 1989 and 1990 seems excessive. Low *Chaoborus* densities contributed some to that response but the zooplankton response to *Chaoborus* reduction in Tuesday Lake during 1988–90 was of a substantially lesser magnitude. A depensatory predation effect may be operating in this interaction and is clearly deserving of further evaluation. Similarly, the addition of trout may have created a 'nutrient pump' which facilitated accumulation of biomass at the trophic level immediately adjacent to that altered by trout predation effects.

Dramatic population responses are embedded in the lesser dynamics of an aggregate, ecosystem-scale variable such as total sestonic phosphorus. Selective predation and predator avoidance behaviors accelerate rates of change. Species disappear, invade and/or irrupt to dominance when the tenor of size-specific mortality changes. Numerical responses and population recoveries depend on founder effects derived from a seed bank of resting stages. Those are most evident in the algal blooms of Peter Lake and in the demise and slow recovery of dinoflagellates in Tuesday Lake (Chapter 11). At the ecosystem scale, the properties of species can play a key role. Not all species are important, but the aggregation of trophic levels offers only limited hope of precise forecasting (Carpenter *et al.*, 1992a; Power, 1992; Strong, 1992). We join

Schindler (1987, 1990) in the conclusion that predictability will improve in proportion to our mechanistic understanding of species interactions and life-history characteristics.

We found that ecosystem responses of Tuesday Lake were partially reversible. A further test of the lake's response capacity is now underway as the doctoral research of Russell Wright. A cohort of 4000 young-of-the-year largemouth bass (average mass was 3 g at stocking) was established there during July 1990. Overwinter survival of bass was excellent. The resident minnow population declined rapidly during 1991 as bass became piscivorous. Again, piscivory and predator avoidance behaviors resulted in virtual elimination of the planktivorous fishes. Cannibalism reduced the abundance of the smaller members of the bass cohort. The plastic growth potential of the dominant predator was evident as the largest of the survivors weighed in excess of 200 g by the end of 1991. These age 1 + bass were comparable in size to age 5 + bass in Paul Lake.

By August of 1991, large *Daphnia* were again prominent in Tuesday Lake and chlorophyll levels were similar to those observed in 1986. All fish were removed by applying rotenone in late autumn of 1991. Again, a modest population of adult minnows was reestablished in spring of 1992. The rate of response to reversal of the manipulation and the hysteresis of food-web effects will be tested again over the next few years. If the seed bank hypothesis is correct, then the recovery rate will be delayed as much as, if not more than, that observed during 1987–90. Key indicators of recovery will be the abundances of dinoflagellates, rotifers, copepods, *Bosmina* and *Chaoborus*.

Microbial interactions

This project benefitted from the recent flourish in effort on microbes and their role in food-web interactions (Chapter 14). Studies conducted by Pace and his co-workers took advantage of our manipulations as a testing ground for ideas about the role of bacteria and protozoa in these study lakes. Evidence accumulated to date suggests that the dynamics of microbial interactions are coupled to the trophic cascade by indirect and multiple pathways. Bacteria are strongly limited by phosphorus, which at times is supplied mainly by zooplankton excretion. Bacteria also depend upon organic carbon provided by phytoplankton, with which they compete for phosphorus. Crustacean zooplankton graze bacteria as well as algae and small bacterivores. Compensatory community changes in the phytoplankton may mitigate effects of grazers on organic carbon supply to bacteria. Thus trophic cascades, and their interactions with phosphorus input rates, do not have immediately obvious consequences for microbial processes in lakes. Because of the fundamental role of the microbiota in the biogeochemistry of lakes, these linkages are worth resolving through further study.

Littoral–pelagic interactions

Our original hypotheses focused exclusively on the pelagic system. That was a concession to parsimony and to the limitations of time and money. Owing to their steep-sided morphometry, the kettle lakes we studied had only modest littoral zones. As made evident in the refuge-seeking behavior of small fishes in response to piscivore additions, the pelagic system can be quickly and crucially linked to the littoral. Similarly, the diel migrations of zooplankton (Chapter 9) and the responses of meta-limnetic populations (Chapter 12) emphasize that vertical linkages within the pelagic zone can be highly responsive to changes in food-web structure. Other obvious links include the benthic invertebrates which are heavily represented in the diets of fishes, the sediment–water interactions involved in nutrient flux, the role that periphyton and their grazers play in competition with their ecological equivalents among the plankton, and the fact that fishes are highly mobile potential sources of local nutrient remineralization (Schindler, 1992).

In a conceptual sense the littoral zone is a component of variable significance. It accounts for a substantial share of total lake area and metabolism in some systems (Gasith, 1991; Wetzel, 1990). It can assume a brief but critical role in the life histories of species that have great impact on the pelagic system. Integration of the littoral zone in the whole lake context requires recognition of its relative role as habitat, refuge and nutrient sink or source (Lodge et al., 1988).

Shallow lakes may possess alternate stable states not found in the deep lakes we studied (Scheffer, 1990). Evidence is accumulating that shallow, productive lakes can be either macrophyte-dominated or phytoplankton-dominated, and that substantial manipulation can shift a lake between these alternative states (Jeppeson et al., 1990; Moss, 1990). Planktivore removal can lead to intense grazing that clears the water and creates conditions for macrophyte dominance (Jeppeson et al., 1990; Spencer & King, 1984). The refuge created by the macrophytes has important implications for sustainability of fish and plankton communities (Moss, 1990) and the pathways of nutrient cycling (Boers et al., 1991).

Do trophic cascades interact with nutrient loading?

Nutrient pools and the nutrient cycling rates that regulate production are profoundly influenced by food-web structure and the properties of dominant species. The extensive changes seen in Peter and Tuesday Lakes amply demonstrate this point. The relatively modest changes seen in Paul Lake suggest that variability in allochthonous nutrient supply was minor in our experimental systems. These lakes have small, fully vegetated and undisturbed drainage basins. They are unlike most lakes in that regard. They are also unusual in the absence of fishery exploitation that amplifies food-web effects. Most lakes are fished and have variable nutrient loading as a consequence of human activity in their drainage basins. In our efforts to understand variability in lakes, we must sort out the interactions of food-web effects, the lags associated with them, and the variability of nutrient loading (Fig. 17.3).

Several different (but not mutually exclusive) hypotheses have been proposed for the interaction of nutrient and food-web effects in lakes. A **Nutrient Attenuation Hypothesis** states that food-web effects diminish as nutrient enrichment increases (McQueen et al., 1986; McQueen, 1990). There may be a threshold level of nutrient supply, above which grazers cannot effectively regulate phytoplankton (Benndorf, 1990). This threshold may be related to the maximum size of algal colonies that can be controlled by herbivores (Gliwicz, 1990) and is indirectly supported by the evidence that large Daphnia can often constrain the initiation of blooms but cannot control them once they are under way (Carpenter, 1992). A **Mesotrophic Maximum Hypothesis** states that food-web effects are greatest in lakes with intermediate levels of nutrient supply (Elser & Goldman, 1991). In very oligotrophic lakes, predator–prey encounter rates may be so low that food-web effects are moot. In very rich systems, grazing-resistant algae may not be controlled by grazers, as in the Nutrient Attenuation Hypothesis. A **Chain Length Hypothesis** posits a direct relationship between food chain length and nutrient supply (Fretwell, 1977; Oksanen et al., 1981; Persson et al., 1988; Power, 1992). In unproductive and fishless lakes (chain length 2) and in productive lakes with abundant piscivores that suppress planktivorous fish (chain length 4), large zooplankton may regulate algae by grazing. In the intermediate condition (chain length 3), planktivorous fish suppress zooplankton and phytoplankton are regulated by resource competition. A **Destabilization Hypothesis** argues that nutrient enrichment destabilizes trophic interactions, but that the level of nutrients that induces variability depends on food-web structure (Carpenter, 1992). Food webs that foster stable populations of large herbivores are destabilized at higher nutrient inputs than are food webs that suppress large herbivores.

Present evidence supports all of these ideas to some extent, though none is supported unequivocally and each suffers from inadequacies. Because many (but not all) of the predictions of the hypotheses are the same, they can be discriminated only by evidence carefully targeted on areas where predictions differ (Sager & Richman, 1990; McQueen, 1990; Benndorf, 1990; Carpenter, 1992; Persson et al., 1992; Power, 1992). It may turn out that different hypotheses hold at different time scales, are highly system-specific, and/or that the assumptions embedded in erecting criteria for a critical test determine the outcome (Strong, 1992). Because of the overlap among the ideas, a new synthesis may emerge that represents a hybrid of them. Regardless of that, the ferment of ideas and the tenor of recent activity suggests a high degree of interest by ecologists and the consequent likelihood of important advances in the next few years.

One way to evaluate the relative importance of nutrient loading and food-web effects and their interactions is to conduct whole-lake experimental tests by adding nutrients to lakes of substantially different food-web structures. Based on our experimental results, nutrient additions to a piscivore-dominated system would be quickly passed from phytoplankton to their grazers. Herbivores would increase in direct proportion to the rate of nutrient loading until that rate exceeded their numerical response capacity. These kinds of systems should be most resistant to the effects of nutrient additions. In lakes dominated by planktivorous fishes, the small-bodied zooplankton should be less capable of controlling algal populations and, therefore, less likely to prevent the instabilities of algal blooms. This simplistic portrayal ignores the feedback of light penetration, its effects on metalimnetic algae, and the changes in efficiency of visual predators (Mazumder et al., 1990). It assumes that the current community composition will persist while changing only in relative proportions. Invaders are not anticipated a priori. It also ignores the accumulation of nutrients in an increasingly anaerobic hypolimnion and the prospect that storm-driven entrainment events can cause substantial, stochastic increases in epilimnetic nutrient levels. It also ignores the prospect that an enhanced nutrient load might elicit a much larger and more dynamic response from the microbial community. However, all of these prospects are experimentally testable at the ecosystem scale.

As this manuscript is coming to print, we are currently pursuing an experimental assessment of these interactions. Four lakes are employed in a 2×2 factorial design; two are piscivore-dominated and two are planktivore-dominated. After a two-year period of premanipulation study, one of each of the two kinds of food webs will be subjected to a regular and cumulative addition of nutrients for each of three years. The remaining pair of lakes will serve as the unenriched reference systems. Our primary hypothesis is that the piscivore-dominated system will be more resistant to nutrient additions. We expect that algal blooms will appear first and most frequently in the system where zooplankton are small. Microbial responses are to be evaluated through continued collaboration with Pace and his co-workers. A separate, collaborative effort on littoral zone coupling with the pelagic food web is also being pursued by David Lodge and his co-workers at the University of Notre Dame.

In view of our experiences with tests of the original cascade idea, we expect to be surprised by many of the results.

Where are trophic cascades expected?

Trophic cascades are intriguing because they can evoke big effects from small causes (Ricker, 1963; Walters, 1986). Equally intriguing are the converse cases where massive community change has little effect on ecosystem processes (Strong, 1992). Perturbations of keystone species that interact strongly with other important species are likely to yield trophic cascades (Paine, 1980). Ecosystem consequences are most likely where one of these species performs a unique and irreplaceable role in an ecosystem process (Schindler, 1990; Vitousek, 1990), where a predator with a broad diet is capable of regulating an entire ecosystem component (Carpenter et al., 1992a), when the constraints of competitive bottlenecks are released (Neill, 1988), or when exotics drop from a passing bird or boat and flourish (Lodge, 1992).

Because there are few ecosystem-scale studies of trophic cascades, it is easy to make generalizations such as 'trophic cascades do/do not/cannot occur in this lake/class of lakes/ecosystem(s)'. We expect that limnologists and ecologists will make many such pronouncements over the next few years, and that many of them will be questioned as additional case studies come to light. One particularly curious challenge arose from the analysis of DeMelo *et al.* (1992), whose review concluded that trophic cascades were demonstrated and of variable strength in most lakes but not in all. In an unusual extension of logic, they then concluded that the trophic cascade hypothesis must be rejected. We disagreed (Carpenter & Kitchell, 1992). Similarly, Strong (1992) erected stringent criteria (removal of vegetation) for a 'true cascade' then concluded that those were present but not ubiquitous and therefore, indicative of unique ecological conditions. Our current view is that the evidence of food-web effects is common, not necessarily universal, and of variable importance. Many if not most predators are selective; that causes changes in the population dynamics of their prey and the interactions expressed as community structure and ecosystem function. Arguing the opposite requires some disregard for the principles of natural selection. The important and unresolved questions have to do with the extent and relative significance of food web effects. As stated in Chapter 1, that is why we conducted the studies represented in this volume.

Comparisons at the ecosystem scale demonstrate cascading effects and considerable variability over broad gradients (Mills & Schiavone, 1982; Pace, 1984; Quiros, 1990; Carpenter *et al.*, 1991) (Fig. 1.2, Fig. 17.1). Ecosystem manipulations have caused trophic cascades in a surprising variety of lakes (Gulati *et al.*, 1990; Henrikson *et al.*, 1980; Kitchell, 1992; Reinertsen *et al.*, 1990; Shapiro & Wright, 1984). Exceptions are equally varied, and probably underreported because the results are less spectacular (McQueen *et al.*, 1989; Gulati *et al.*, 1990; Kitchell, 1992); this bias is unfortunate because the evidence of compensatory processes will be less well documented. At smaller scales (e.g. bottles, bags, buckets, mesocosms, limnocorrals, horse troughs, stock tanks, small ponds, etc.) various experimental approaches yield contradictory and sometimes bewildering conclusions, which, unfortunately, are constrained by a host of problems that limit interpretation at the ecosystem scale (Frost *et al.*, 1988; DeMelo *et al.*, 1992; Carpenter & Kitchell, 1992). Our experience, and what we offer to others as advice (Carpenter & Kitchell, 1988), is that strong treatments in the form of simply removing or adding a trophic level or a keystone species (Paine, 1966) are the most instructive.

We can conclude that trophic cascades occur in a wide variety of lake ecosystems, sometimes with spectacular consequences, and other times with little or no apparent effect. We think that explanation of these variable outcomes will come from breakthroughs on one (or more) of three fronts. First, the growing number of case studies may allow progress by inductive or comparative methods. Second, a combination of modeling, ecosystem experiments and comparative studies may reveal how food web and nutrient effects interact. Third, progress at the interface of community and ecosystem ecology may show how compensatory interactions among populations magnify or suppress ecosystem responses (Carpenter *et al.*, 1992a; Strong, 1992) and allow us to better predict the circumstances under which community perturbations have ecosystem effects.

Terrestrial ecosystems, like aquatic ones, exhibit generally positive but highly variable relations between primary and secondary production (McNaughton *et al.*, 1991). To what extent is the unexplained variability due to regulation of prey by predators, or to its failure? In complex systems, trophic ontogeny and prey switching (Polis, 1991) or compensatory species replacement may preclude trophic cascades (Strong, 1992). On the other hand, strong interactions are known between carnivores (including humans), mammalian herbivores, and the vegetation of forests and grasslands (e.g., Flader, 1974; McNaughton, 1979; Pastor *et al.*, 1988; Detling, 1988; Botkin, 1990). The trophic cascade argument was advanced to account for some of the variability in ecosystem process rates. The short generation times of plankton made it easier to test in lakes. With appropriate modifications of scale, we assert that this hypothesis bears further attention from terrestrial ecologists. They might profit from considering the frequency and extent of overgrazing or intense browsing by mammals and the effects of insect outbreaks. Those are the functional analogs of piscivore removal and the recruitment of a strong year class of fish.

Perhaps more relevant is the fact that most terrestrial and aquatic ecosystems are now dominated by humans, who impact and manage both populations and ecosystem processes. Widespread changes in the atmosphere and our landscapes are clearly evident in the concerns about global change issues. Aquatic ecosystems are similarly influenced by the nearly ubiquitous process of fishery exploitation and the fact that our effluent descends with rivers and streams to influence extensive areas of the coastal environment. In our search for general ecological principles, disregarding places where human effects are expressed (Strong, 1992) seems a waste of time and opportunity (Botkin, 1990).

Ecologists must accept the reality that species populations are increasingly exploited, enhanced, imported and extirpated. Management objectives emphasize species that are prized for economic or esthetic reasons and despised for lessening that. In the case of some that are endangered, species are protected as a caution against unknown consequences. Ecosystems are irrigated, eutrophied, acidified and polluted while management objectives include erosion control, improved water quality, restoration of native communities, and improved fishery yields. Finding places where none of the above applies seems a formidable challenge.

Our message in the preceding is that strong feedbacks, including trophic cascades, link populations and ecosystem processes. If ecology is to contribute to resource management, ecosystem restoration or rehabilitation, and environmental mitigation, it must address the links between populations and ecosystems. These feedbacks are the nexus of variability and the occasional remarkable surprise. Surprises should be fewer as our experience and understanding grow. Until then, the flexibility to learn from surprise (plus some modesty and a sense of humor) may be the most valuable adaptive tools for both ecological researchers and those charged with stewardship of our environmental resources.

Acknowledgements

We thank Tom Martin, Mike Pace and Lars Rudstam for helpful criticisms of this chapter.

References

Adams, M. S. & Breck, J. E. (1990). Bioenergetics. In *Methods of Fish Biology*, ed. C. B. Schreck & P. B. Moyle, pp. 389–415. Bethesda, Maryland: American Fisheries Society.

Andersen, T. & Hessen, D. O. (1991). Carbon, nitrogen, and phosphorus content of freshwater zooplankton. *Limnology and Oceanography*, **36**, 807–14.

Azam, F., Fenchel, T., Field, J. G., Ray, J. S., Meyer-Reil, L. A. & Thingstad, F. (1983). The ecological role of water-column microbes in the sea. *Marine Ecology Progress Series*, **10**, 257–63.

Bagenal, T. B. (1978). *Methods for Assessment of Fish Production in Fresh Waters.* Oxford, England: Blackwell.

Baker, A. L. & Brook, A. J. (1971). Optical density profiles as an aid to the study of microstratified phytoplankton populations in lakes. *Archiv für Hydrobiologie*, **69**, 214–33.

Balcer, M. D., Korda, N. L. & Dodson, S. I. (1984). *Zooplankton of the Great Lakes.* Madison, Wisconsin: University of Wisconsin Press.

Barlow, R. G., Burkill, P. H. & Mantoura, R. F. C. (1988). Grazing and degradation of algal pigments by marine protozoan *Oxyrrhis marina. Journal of Experimental Marine Biology and Ecology*, **119**, 119–29.

Bartell, S. M., Brenkert, A. L. & Carpenter, S. R. (1988). Parameter uncertainty and the behavior of a size-dependent plankton model. *Ecological Modelling*, **40**, 85–95.

Battarbee, R. W., Mason, J., Renberg, I. & Talling, J. F., eds. (1990). *Palaeolimnology and Lake Acidification.* London: The Royal Society.

Bayly, I. A. E. (1986). Aspects of diel vertical migration in zooplankton and its enigma variations. In *Limnology in Australia*, ed. P. De Deckker & W. D. Williams, pp. 349–68. Dordrecht: W. Junk.

Beck, M. B. (1983). Uncertainty, system identification, and the prediction of water quality. In *Uncertainty and Forecasting of Water Quality*, ed. M. B. Beck & G. van Straten, pp. 3–68. New York: Springer-Verlag.

Beckel, A. L. (1987). *Breaking New Waters: A Century of Limnology at the University of Wisconsin.* (*Transactions of the Wisconsin Academy of Sciences, Arts and Letters*, Special Issue.) Madison, Wisconsin: Wisconsin Academy of Sciences, Arts & Letters.

Becker, G. C. (1983). *Fishes of Wisconsin.* Madison, Wisconsin: University of Wisconsin Press.

Beddington, J. R. & May, R. M. (1977). Harvesting natural populations in a randomly fluctuating environment. *Science*, **197**, 463–5.

Bell, R. T. (1988). Thymidine incorporation and estimates of bacterioplankton production: are the conversion factors valid? *Archiv für Hydrobiologie, Ergebnisse der Limnologie*, **31**, 163–71.

Bender, E. A., Case, T. J. & Gilpin, M. E. (1984). Perturbation experiments in community ecology: theory and practice. *Ecology*, **65**, 1–13.

Benndorf, J. (1987). Food web manipulation without nutrient control: useful strategy in lake restoration? *Schweizerische Zeitschrift für Hydrologie*, **49**, 237–48.

Benndorf, J. (1990). Conditions for effective biomanipulation: conclusions derived from whole-lake experiments in Europe. *Hydrobiologia*, **200/201**, 187–203.

Bergquist, A. M. (1985). *Effects of Herbivory on Phytoplankton Community Composition, Size Structure, and Primary Production*. Ph.D. dissertation, University of Notre Dame, Notre Dame, Indiana.

Bergquist, A. M. & Carpenter, S. R. (1986). Limnetic herbivory: effects on phytoplankton populations and primary production. *Ecology*, **67**, 1351–60.

Bergquist, A. M., Carpenter, S. R. & Latino, J. C. (1985). Shifts in phytoplankton size structure and community composition during grazing by contrasting zooplankton assemblages. *Limnology and Oceanography*, **30**, 1037–45.

Bevelhimer, M. S., Stein, R. A. & Carline, R. F. (1985). Assessing significance of physiological differences among three esocids with a bioenergetics model. *Canadian Journal of Fisheries and Aquatic Sciences*, **42**, 57–69.

Bianchi, T. S. & Findlay, S. (1990). Plant pigments as tracers of emergent and submergent macrophytes from the Hudson River. *Canadian Journal of Fisheries and Aquatic Sciences*, **47**, 492–4.

Binford, M. W., Deevey, E. S. & Crisman, T. L. (1983). Paleolimnology: an historical perspective on aquatic ecosystems. *Annual Review of Ecology and Systematics*, **14**, 255–86.

Black, A. R. & Dodson, S. I. (1990). Demographic costs of *Chaoborus*-induced phenotypic plasticity in *Daphnia pulex*. *Oecologia*, **83**, 117–22.

Black, R. W. & Hairston, N. G., Jr. (1988). Predator driven changes in community structure. *Oecologia*, **77**, 468–79.

Bloesch, J. & Burns, N. (1980). A critical review of sedimentation trap technique. *Schweizerische Zeitschrift für Hydrologie*, **42**, 15–55.

Boers, P., van Ballegooijen, L. Uunk, J. (1991). Changes in phosphorus cycling in a shallow lake due to food web manipulations. *Freshwater Biology*, **25**, 9–20.

Bogdan, K. G. & Gilbert, J. J. (1984). Body size and food size in freshwater zooplankton. *Proceedings of the National Academy of Sciences, U.S.A.*, **81**, 6427–31.

Bollens, S. M. & Frost, B. W. (1989). Predator-induced diel vertical migration in a planktonic copepod. *Journal of Plankton Research*, **11**, 147–65.

Botkin, D. (1990). *Discordant Harmonies: a New Ecology for the Twenty-First Century*. Oxford University Press.

Bower, P. M., Kelly, C. A., Fee, E. J., Shearer, J. A., DeClerq, D. R. & Schindler, D. W. (1987). Simultaneous measurement of primary production by whole-lake and bottle radiocarbon additions. *Limnology and Oceanography*, **32**, 299–312.

Box, G. E. P., Hunter, W. G. & Hunter, J. S. (1978). *Statistics for Experimenters*. New York: Wiley.

Box, G. E. P. & Tiao, G. C. (1975). Intervention analysis with applications to economic and environmental problems. *Journal of the American Statistical Association*, **70**, 70–9.

Brahce, M. (1980). *The Vertical Distribution of Phytoplankton and Primary Production in Central Lake Michigan, July 1977*. M.S. thesis, Michigan Technological University, Houghton, Michigan.

Brenner, M., Leyden, B. & Binford, M. W. (1990). Recent sedimentary histories of shallow lakes in Guatemalan savannas. *Journal of Paleolimnology*, **4**, 239–52.

Brooks, J. L. & Dodson, S. I. (1965). Predation, body size, and composition of plankton. *Science*, **150**, 28–35.

Brown, S. R. (1969). Paleolimnological evidence from fossil pigments. *Mitteilungen – Internationale Vereinigung für theoretische und angewandte Limnologie*, **17**, 95–103.

Brown, S. R., Daley, R. J. & McNeely, R. N. (1977). Composition and stratigraphy of the fossil phorbin derivatives of Little Round Lake, Ontario. *Limnology and Oceanography*, **22**, 336–48.

Brown, S. R., McIntosh, H. J. & Smol, J. P. (1984). Recent paleolimnology of a meromictic lake: fossil pigments of photosynthetic bacteria. *Verhandlungen – Internationale Vereinigung für theoretische und angewandte Limnologie*, **22**, 1357–60.

Brylinsky, M. & Mann, K. H. (1973). An analysis of factors governing productivity in lakes and reservoirs. *Limnology and Oceanography*, **18**, 1–14.

Burns, C. W. (1968). The relationship between body size of filter-feeding Cladocera and the maximum size of particle ingested. *Limnology and Oceanography*, **13**, 675–8.

Carbonel, P., Colin, J., Danielopol, D. L., Loffler, H. & Neustrueva, I. (1988). Paleoecology of limnic ostracodes: a review of some major topics. *Palaeogeography, Palaeoclimatology, Palaeoecology*, **62**, 413–61.

Carpenter, S. R. (1983). Lake geometry: implications for production and sediment accretion rates. *Journal of Theoretical Biology*, **105**, 273–86.

Carpenter, S. R., ed. (1988a). *Complex Interactions in Lake Communities*. New York: Springer-Verlag.

Carpenter, S. R. (1988b). Transmission of variance through lake food webs. In *Complex Interactions in Lake Communities*, ed. S. R. Carpenter, pp. 119–38. New York: Springer-Verlag.

Carpenter, S. R. (1989). Replication and treatment strength in whole-lake experiments. *Ecology*, **70**, 453–63.

Carpenter, S. R. (1991). Large-scale perturbations: opportunities for innovation. *Ecology*, **71**, 2038–43.

Carpenter, S. R. (1992). Destabilization of planktonic ecosystems and blooms of blue-green algae. In *Food Web Management: A Case Study of Lake Mendota*, ed. J. F. Kitchell, pp. 461–81. New York: Springer-Verlag.

Carpenter, S. R. & Bergquist, A. M. (1985). Experimental tests of grazing indicators based on chlorophyll *a* degradation products. *Archiv für Hydrobiologie*, **102**, 303–17.

Carpenter, S. R., Elser, M. M. & Elser, J. J. (1986). Chlorophyll production, degradation, and sedimentation: implications for paleolimnology. *Limnology and Oceanography*, **31**, 112–24.

Carpenter, S. R., Frost, T. M., Heisey, D. & Kratz, T. K. (1989). Randomized intervention analysis and the interpretation of whole ecosystem experiments. *Ecology*, **70**, 1142–52.

Carpenter, S. R., Frost, T. M., Kitchell, J. F. & Kratz, T. K. (1992*a*). Species dynamics and global environmental change: a perspective from ecosystem experiments. In *Biotic Interactions and Global Change*, ed. J. Kingsolver, P. Kareiva & R. Huey, pp. 267–79. Sunderland, Massachusetts: Sinauer Associates.

Carpenter, S. R., Frost, T. M., Kitchell, J. F., Kratz, T. K., Schindler, D. W., Shearer, J., Sprules, W. G., Vanni, M. J. & Zimmerman, A. P. (1991). Patterns of primary production and herbivory in 25 North American lake ecosystems. In *Comparative Analyses of Ecosystems: Patterns, Mechanisms, and Theories*, ed. J. Cole, G. Lovett & S. Findlay, pp. 67–96. New York: Springer-Verlag.

Carpenter, S. R. & Kitchell, J. F. (1984). Plankton community structure and limnetic primary production. *American Naturalist*, **124**, 159–72.

Carpenter, S. R. & Kitchell, J. F. (1987). The temporal scale of variance in limnetic primary production. *American Naturalist*, **129**, 417–33.

Carpenter, S. R. & Kitchell, J. F. (1988). Consumer control of lake productivity. *BioScience*, **38**, 764–9.

Carpenter, S. R. & Kitchell, J. F. (1992). Trophic cascade and biomanipulation: interface of research and management. *Limnology and Oceanography*, **37**, 208–13.

Carpenter, S. R., Kitchell, J. F. & Hodgson, J. R. (1985). Cascading trophic interactions and lake productivity. *BioScience*, **35**, 634–9.

Carpenter, S. R., Kitchell, J. F., Hodgson, J. R., Cochran, P. A., Elser, J. J., Elser, M. M., Lodge, D. M., Kretchmer, D., He, X. & von Ende, C. N. (1987). Regulation of lake primary productivity by food web structure. *Ecology*, **68**, 1863–76.

Carpenter, S. R., Kraft, C. E., Wright, R., He, X., Soranno, P. A. & Hodgson, J. R. (1992*b*). Resilience and resistance of a lake phosphorus cycle before and after food web manipulation. *American Naturalist*, **140**, 781–98.

Carpenter, S. R. & Leavitt, P. R. (1991). Temporal variation in a paleolimnological record arising from a trophic cascade. *Ecology*, **72**, 277–85.

Carpenter, S. R., Leavitt, P. R., Elser, J. J. & Elser, M. M. (1988). Chlorophyll budgets: response to food web manipulation. *Biogeochemistry*, **6**, 79–90.

Carpenter, S. R. & Lodge, D. M. (1986). Effects of submersed macrophytes on ecosystem processes. *Aquatic Botany*, **26**, 341–70.

Carter, J. C. H. & Kwik, J. K. (1977). Instar succession vertical distribution, and interspecific competition amoung four species of *Chaoborus*. *Journal of Fisheries Research Board of Canada*, **34**, 113–18.

Chapra, S. C. & Reckhow, K. H. (1983). *Engineering Approaches for Lake Management*, volume 2: *Mechanistic Modeling*. Butterworth, Boston.

Charles, D. F. & Smol, J. P. (1988). New methods for using diatoms and chrysophytes to infer past pH of low-alkalinity lakes. *Limnology and Oceanography*, **33**, 1451–62.

Chatfield, C. (1980). *The Analysis of Time Series*, 2nd edn. London: Chapman and Hall.

Chatfield, C. (1984). *The Analysis of Time Series*, 3rd edn. London: Chapman and Hall.

Cho, B. C. & Azam, F. (1988). Heterotrophic bacterioplankton production measurement by the tritiated thymidine incorporation method. *Archiv für Hydrobiologie*,

Ergebnisse der Limnologie, **31**, 153–62.

Clady, M. D. (1974). Food habits of yellow perch, smallmouth bass and largemouth bass in two unproductive lakes in Northern Michigan. *American Midland Naturalist*, **91**, 453–9.

Clark, C. W. & Levy, D. A. (1988). Diel vertical migrations by juvenile sockeye salmon and the antipredation window. *American Naturalist*, **131**, 271–90.

Cochran, P. A., Lodge, D. M., Hodgson, J. R. & Knapik, P. G. (1988). Diets of syntropic finescale dace, *Phoxinus neogaeus*, and northern redbelly dace, *Phoxinus eos*: a reflection of trophic morphology. *Environmental Biology of Fishes*, **22**, 235–40.

Cohen, J. (1989). Food webs and community structure. In *Perspectives in Ecological Theory*, ed. J. Roughgarden, R. M. May & S. A. Levin, pp. 181–202. Princeton, New Jersey: Princeton University Press.

Cohen, J. E., Briand, F. & Newman, C. M. (1990). *Community Food Webs: Data and Theory*. New York: Springer-Verlag.

Cole, J., Lovett, G. & Findlay, S., eds. (1991). *Comparative Analyses of Ecosystems: Patterns, Mechanisms, and Theories*. New York: Springer-Verlag.

Cole, J. J., Findlay, S. & Pace, M. L. (1988). Bacterial production in fresh and saltwater ecosystems: a cross-system overview. *Marine Ecology Progress Series*, **43**, 1–10.

Connell, J. H. (1980). Diversity and the coevolution of competitors, or the ghost of competition past. *Oikos*, **35**, 131–8.

Coon, T., Lopez, M., Richerson, T., Powell, P. & Goldman, C. (1987). Summer dynamics of the deep chlorophyll maximum in Lake Tahoe. *Journal of Plankton Research*, **9**, 327–44.

Coveney, M. F. & Wetzel, R. G. (1988). Experimental evaluation of conversion factors for the [^{3}H]thymidine incorporation assay of bacterial secondary productivity. *Applied and Environmental Microbiology*, **54**, 2018–26.

Crowder, L. B., Drenner, R. W., Kerfoot, W. C., McQueen, D. J., Mills, E. O., Sommer, U., Spencer, C. N. & Vanni, M. J. (1988). Food web interactions in lakes. In *Complex Interactions in Lake Communities*, ed. S. R. Carpenter, pp. 141–60. New York: Springer-Verlag.

Crumpton, W. G. (1987). A simple and reliable method for making permanent mounts of phytoplankton for light and fluorescence microscopy. *Limnology and Oceanography*, **32**, 1154–9.

Culver, D. C. (1980). Seasonal variation in the sizes at birth and at first reproduction in Cladocera. In *Ecology and Evolution of Zooplankton Communities*, ed. W. C. Kerfoot, pp. 358–66. Hanover, New Hampshire: University Press of New England.

Cummins, K. W. & Wuycheck, J. C. (1971). Caloric equivalents for investigations in ecological energetics. *Mitteilungen – Internationalen Vereinigung für Theoretische und Angewandte Limnologie*, **18**, 1–158.

Currie, D. J. (1991). Large-scale variability and interactions among phytoplankton, bacterioplankton, and phosphorus. *Limnology and Oceanography*, **35**, 1437–55.

Currie, D. J. & Kalff, J. (1984). The relative importance of bacterioplankton and phytoplankton in phosphorus uptake in freshwater. *Limnology and Oceanography*, **29**, 311–21.

Curtis, J. T. (1959). *Vegetation of Wisconsin*. Madison, Wisconsin: University of Wisconsin Press.

Cushing, D. H. (1982). *Climate and Fisheries*. London: Academic Press.

Cuvier, G. (1817). *Le Regne Animal*, volume 17 (Texte): *Les Crustaces*. Paris: Masson.

Daan, N. & Ringleberg, J. (1969). Further studies on the positive and negative phototactic reaction of *Daphnia magna* Straus. *Netherlands Journal of Zoology*, **19**, 525–40.

Dagg, M. J. (1985). The effects of food limitation on diel migratory behavior in marine zooplankton. *Archiv für Hydrobiologie, Ergebnisse der Limnologie*, **21**, 247–55.

Daley, R. J. (1973). Experimental characterization of lacustrine chlorophyll diagenesis. II. Bacterial, viral and herbivore grazing effects. *Archiv für Hydrobiologie*, **72**, 409–39.

Daro, M. H. (1988). Migratory and grazing behavior of copepods and vertical distribution of phytoplankton. *Bulletin of Marine Science*, **43**, 710–29.

Davidson, G. A. (1988). A modified tape-peel technique for preparing permanent qualitative microfossil slides. *Journal of Paleolimnology*, **1**, 229–34.

Davies, B. H. (1976). Carotenoids, In *Chemistry and Biochemistry of Plant Pigments*, volume 2, ed. T. W. Goodwin, pp. 38–165. New York: Academic Press.

Davis, M. B. (1989). Retrospective studies. In *An Ecosystem Approach to Aquatic Ecology: Mirror Lake and its Environment*, ed. G. E. Likens, pp. 71–89. New York: Springer-Verlag.

Davis, M. B., Moeller, R. E. & Ford, J. (1984). Sediment focusing and pollen influx. In *Lake Sediments and Environmental History*, ed. E. Y. Haworth & J. W. G. Lund, pp. 261–94. Bath, England: Leicester University Press.

DeAngelis, D. L. (1988). Strategies and difficulties of applying models to aquatic populations and food webs. *Ecological Modelling*, **43**, 57–73.

DeAngelis, D. L., Bartell, S. M. & Brenkert, A. M. (1989). Effects of nutrient recycling and food-chain length on resilience. *American Naturalist*, **134**, 778–805.

DeAngelis, D. L. & Waterhouse, J. C. (1987). Equilibrium and nonequilibrium concepts in ecological models. *Ecological Monographs*, **57**, 1–21.

de Bernardi, R. (1981). Biotic interaction in freshwater and effect on community structure. *Bollettino di Zoologia*, **48**, 353–71.

DeMeester, L. & Dumont, H. J. (1988). The genetics of phototaxis in *Daphnia magna*: existence of three phenotypes for vertical migration among parthenogenetic females. *Hydrobiologia*, **162**, 47–55.

DeMelo, R., France, R. & McQueen, D. J. (1992). Biomanipulation: hit or myth? *Limnology and Oceanography*, **37**, 192–207.

DeMott, W. R. & Kerfoot, W. C. (1982). Competition among cladocerans: nature of the interaction between *Bosmina* and *Daphnia*. *Ecology*, **63**, 1949–66.

De Stasio, B. T., Jr. (1990). The role of dormancy and emergence patterns in the dynamics of a freshwater zooplankton community. *Limnology and Oceanography*, **35**, 1079–90.

Detling, J. K. (1988). Grasslands and savannas: regulation of energy flow and nutrient cycling by herbivores. In *Concepts of Ecosystem Ecology*, ed. L. R. Pomeroy & J. J. Alberts, pp. 131–48. New York: Springer-Verlag.

Dice, L. R. (1914). The factors determining the vertical movements of *Daphnia*.

Journal of Animal Behavior, **4**, 229–65.

Dini, M. L. (1989). *The Adaptive Significance of Diel Vertical Migration in* Daphnia. Ph.D. thesis, University of Notre Dame, Notre Dame, Indiana.

Dini, M. L. & Carpenter, S. R. (1988). Variability in *Daphnia* behavior following fish community manipulations. *Journal of Plankton Research,* **10**, 621–35.

Dini, M. L. & Carpenter, S. R. (1991). The effect of whole-lake fish community manipulations on *Daphnia* migratory behavior. *Limnology and Oceanography,* **36**, 370–7.

Dini, M. L. & Carpenter, S. R. (1992). Fish predators, food availability and diel vertical migration in *Daphnia. Journal of Plankton Research,* **14**, 359–78.

Dini, M. L., O'Donnell, J., Carpenter, S. R., Elser, M. M., Elser, J. J. & Bergquist, A. M. (1987). *Daphnia* size structure, vertical migration and phosphorus redistribution. *Hydrobiologia,* **150**, 185–91.

Dodson, S. I. (1972). Mortality in a population of *Daphnia rosea. Ecology,* **53**, 1011–23.

Dodson, S. I. (1988). The ecological role of chemical stimuli for the zooplankton: predator-avoidance behavior in *Daphnia. Limnology and Oceanography,* **33**, 1431–9.

Dodson, S. I. (1990). Predicting diel vertical migration of zooplankton. *Limnology and Oceanography,* **35**, 1195–200.

Dorazio, R. M., Bowers, J. A. & Lehman, J. T. (1987). Food-web manipulations influence grazer control of phytoplankton growth rates in Lake Michigan. *Journal of Plankton Research,* **9**, 891–9.

Douglas, M. S. V. & Smol, J. P. (1987). Siliceous protozoan plates in lake sediments. *Hydrobiologia,* **154**, 13–23.

Downing, J. A. & Rigler, F. H. (1984). *A Manual on Methods for the Assessment of Secondary Productivity in Fresh Waters.* Oxford, England: Blackwell Scientific Publications.

Draper, N. & Smith, H. (1981). *Applied Regression Analysis,* 2nd edn. New York: Wiley.

Driver, E. A. (1981). Caloric values of pond invertebrates eaten by ducks. *Freshwater Biology,* **11**, 579–81.

Driver, E. A., Sugden, L. G. & Kovach, R. J. (1974). Calorific, chemical and physical values of potential duck foods. *Freshwater Biology,* **4**, 281–92.

Dumont, H. J. (1972). A competition-based approach of the reverse vertical migration in zooplankton and its implications, chiefly based on a study of the interactions of the rotifer *Asplanchna priodonta* (Gorse) with several Crustacea Entomostraca. *Internationale Revue der Gesamten Hydrobiologie,* **57**, 1–38.

Dumont, H. J., Guisez, Y., Carels, I. & Verheye, H. M. (1985). Experimental isolation of positively and negatively phototactic phenotypes from a natural population of *Daphnia magna* Straus: a contribution to the genetics of vertical migration. *Hydrobiologia,* **126**, 121–7.

Dumont, H. J., Van de Velde, I. & Dumont, S. (1975). The dry weight estimate of biomass in a selection of Cladocera, Copepoda and Rotifera from plankton, periphyton and benthos of continental waters. *Oecologia,* **19**, 75–97.

Edmondson, W. T. (1960). Reproductive rates of rotifers in natural populations. *Memorie del Istituto Italiano di Idrobiologie,* **12**, 21–77.

Edmondson, W. T. (1979). Lake Washington and the predictability of limnological events. *Archiv für Hydrobiologie Beiheft*, **13**, 234–41.

Edmondson, W. T. (1991). *The Uses of Ecology: Lake Washington and Beyond.* Seattle: University of Washington Press.

Edmondson, W. T. & Litt, A. H. (1982). Daphnia in Lake Washington. *Limnology and Oceanography*, **27**, 272–93.

Edmondson, W. T. & Litt, A. H. (1987). Conochilus in Lake Washington. *Hydrobiologia*, **147**, 157–72.

Elser, J. J. (1987). Evaluation of size-related changes in chlorophyll-specific light extinction in some north temperate lakes. *Archiv für Hydrobiologie*, **111**, 171–82.

Elser, J. J. & Carpenter, S. R. (1988). Predation-driven dynamics of zooplankton and phytoplankton communities in a whole-lake experiment. *Oecologia*, **76**, 148–54.

Elser, J. J., Elser, M. M. & Carpenter, S. R. (1986a). Size fractionation of algal chlorophyll, carbon fixation, and phosphatase activity: relationships with species-specific size distributions and zooplankton community structure. *Journal of Plankton Research*, **8**, 365–83.

Elser, J. J., Elser, M. M., MacKay, N. & Carpenter, S. R. (1988). Zooplankton-mediated transitions between N- and P-limited algal growth. *Limnology and Oceanography*, **33**, 1–14.

Elser, J. J., Goff, N. C., MacKay, N. A., St. Amand, A. L., Elser, M. M. & Carpenter, S. R. (1987a). Species-specific algal responses to zooplankton: experimental and field observations in three nutrient-limited lakes. *Journal of Plankton Research*, **9**, 699–717.

Elser, J. J. & Goldman, C. R. (1991). Zooplankton effects on phytoplankton in lakes of contrasting trophic status. *Limnology and Oceanography*, **36**, 64–90.

Elser, J. J. & MacKay, N. A. (1989). Experimental evaluation of effects of zooplankton biomass and size distribution on algal biomass and productivity in three nutrient limited lakes. *Archiv für Hydrobiologie*, **114**, 481–96.

Elser, M. M., Elser, J. J. & Carpenter, S. R. (1986b). Paul and Peter lakes: a liming experiment revisited. *American Midland Naturalist*, **116**, 282–95.

Elser, M. M., von Ende, C. N., Soranno, P. A. & Carpenter, S. R. (1987b). Chaoborus populations: response to food web manipulation and potential effects on zooplankton communities. *Canadian Journal of Zoology*, **65**, 2846–52.

Enright, J. T. (1977). Diurnal vertical migration: adaptive significance and timing. I. Selective advantage: a metabolic model. *Limnology and Oceanography*, **22**, 856–72.

Estes, J. A. & Palmisano, J. F. (1974). Sea otters: their role in structuring nearshore communities. *Science*, **185**, 1058–60.

Fagerstrom, T. (1987). On theory, data and mathematics in ecology. *Oikos* **50**, 258–61.

Fahnenstiel, G. L. & Scavia, D. (1987). Dynamics of Lake Michigan phytoplankton: the deep chlorophyll layer. *Journal of Great Lakes Research*, **13**, 285–95.

Fedorenko, A. Y. (1975). Feeding characteristics and predation impact of Chaoborus (Diptera, Chaoboridae) larvae in a small lake. *Limnology and Oceanography*, **20**, 250–8.

Fee, E. J. (1976). The vertical and seasonal distribution of chlorophyll in lakes of the Experimental Lakes Area, northwestern Ontario: implications for primary production estimates. *Limnology and Oceanography*, **21**, 767–83.

Fee, E. J. (1979). A relation between lake morphometry and primary productivity and its use in interpreting whole-lake eutrophication experiments. *Limnology and Oceanography*, **24**, 401–16.

Fenchel, T. (1986). The ecology of heterotrophic microflagellates. *Advances in Microbial Ecology*, **9**, 57–97.

Findlay, S. E. G., Meyer, J. L. & Edwards, R. T. (1984). Measuring bacterial production via rate of incorporation of [3H]thymidine into DNA. *Journal of Microbiological Methods*, **2**, 57–72.

Findlay, S. E. G., Pace, M. L., Lints, D., Cole, J. J., Caraco, N. F. & Peierls, B. (1991). Weak coupling of bacterial and algal production in a heterotrophic ecosystem, the Hudson estuary. *Limnology and Oceanography*, **36**, 268–78.

Flader, S. (1974). *Thinking Like a Mountain*. Lincoln, Nebraska: University of Nebraska Press.

Forel, F. A. (1876). *Faune profonde du Lac Leman*. (*Bulletin de la Société Vaudoise des Sciences Naturelles*, no. 13.)

Fretwell, S. D. (1977). The regulation of plant communities by food chains exploiting them. *Perspectives in Biology and Medicine*, **20**, 169–85.

Frey, D. G. (1959). The late-glacial cladoceran fauna of a small lake. *Archiv für Hydrobiologie*, **54**, 209–75.

Frey, D. G. (1969). The rationale of paleolimnology. *Mitteilungen – Internationale Vereinigung für theoretische und angewandte Limnologie*, **20**, 95–123.

Frey, D. G. (1986). Cladocera analysis. In *Handbook of Holocene Palaeoecology and Palaeohydrology*, ed. B. E. Berglund & M. Ralska-Jasiewiczowa, pp. 667–92. New York: John Wiley and Sons.

Frey, D. G. (1988). Littoral and offshore communities of diatoms, cladocerans and dipterous larvae, and their interpretation in paleolimnology. *Journal of Paleolimnology*, **1**, 179–92.

Frost, T. M., DeAngelis, D. L., Bartell, S. M., Hall, D. J. & Hurlbert, S. H. (1988). Scale in the design and interpretation of aquatic community research. In *Complex Interactions in Lake Communities*, ed. S. R. Carpenter, pp. 229–60. New York: Springer-Verlag.

Fuhrmann, O. (1900). Beitrage zur Biologie des Neuenbergersees. *Biologisches Zentralblatt*, **20**, 85–96, 120–8.

Gabriel, W. & Thomas, B. (1988). Vertical migration of zooplankton as an evolutionarily stable strategy. *American Naturalist*, **132**, 199–216.

Gálvez, J. A., Niell, F. X. & Lucena, J. (1988). Description and mechanism of formation of a deep chlorophyll maximum due to *Ceratium hirundinella* (O. F. Muller) Bergn. *Archiv für Hydrobiologie*, **112**, 143–55.

Gardner, W. D. (1980). Sediment trap dynamics and calibration. A laboratory evaluation. *Journal of Marine Research*, **38**, 17–39.

Gasith, A. (1991). Can littoral resources influence ecosystem processes in large, deep lakes? *Verhandlungen – Internationale Vereinigung für theoretische und angewandte Limnologie*, **24**, 1073–76.

Gauch, H. G. (1982). *Multivariate Analysis in Community Analysis*. Cambridge University Press.

Geller, W. (1986). Diurnal vertical migration of zooplankton in a temperate great lake (L. Constance): a starvation avoidance mechanism? *Archiv für Hydrobiologie*

360 · **References**

(*Supplement*), **74**, 1–60.

George, D. G. (1983). Interrelations between the vertical distribution of *Daphnia* and chlorophyll *a* in two large limnetic enclosures. *Journal of Plankton Research*, **5**, 457–75.

George, E. L. & Hadley, W. F. (1979). Food and habitat partitioning between rock bass (*Ambloplites rupestris*) and smallmouth bass (*Micropterus dolomieui*) young of the year. *Transactions of the American Fisheries Society*, **108**, 253–61.

Giguere, L. A. & Dill, L. M. (1980). Seasonal patterns of vertical migration: a model for *Chaoborus trivittatus*. In *Evolution and Ecology of Zooplankton Communities*, ed. W. C. Kerfoot, pp. 122–8. Hanover, New Hampshire: University Press of New England.

Gilbert, J. J. (1988). Suppression of rotifer populations by *Daphnia*: a review of the evidence, the mechanisms, and the effects on zooplankton community structure. *Limnology and Oceanography*, **33**, 1286–303.

Gilbert, J. J. & Williamson, C. E. (1978). Predator-prey behavior and its effect on rotifer survival in associations of *Mesocyclops edax*, *Asplanchna girodi*, *Polyarthra vulgaris*, and *Keratella cochlearis*. *Oecologia*, **37**, 13–22.

Gillen, M. J. (1939). *Will of Martin J. Gillen, deceased*. Bessemer, Michigan: Probate Court of Gogebic County.

Giussani, G., de Bernardi, R. & Ruffoni, T. (1990). Three years of experience in biomanipulating a small eutrophic lake: Lago di Candia (Northern Italy). *Hydrobiologia*, **200/201**, 357–66.

Glew, J. R. (1988). A portable extruding device for close interval sectioning of unconsolidated core samples. *Journal of Paleolimnology*, **1**, 235–9.

Gliwicz, Z. M. (1969). Studies on the feeding of pelagic zooplankton in lakes with varying trophy. *Ekologia Polska, Seria A*, **17**, 663–707.

Gliwicz, Z. M. (1980). Filtering rates, food size selection, and feeding rates in cladocerans – another aspect of competition in filter-feeding zooplankton. In *Evolution and Ecology of Zooplankton Communities*, ed. W. C. Kerfoot, pp. 282–91. Hanover, New Hampshire: University Press of New England.

Gliwicz, Z. M. (1985). Predation or food limitation: an ultimate reason for extinction of planktonic cladoceran species. *Archive für Hydrobiologie, Beiheft*, **21**, 419–30.

Gliwicz, Z. M. (1986). A lunar cycle in zooplankton. *Ecology*, **67**, 883–97.

Gliwicz, Z. M. (1990). Why do cladocerans fail to control algal blooms? *Hydrobiologia*, **200/201**, 83–97.

Gliwicz, M. Z. & Pijanowska, J. (1988). Effect of predation and resource depth distribution on vertical migration of zooplankton. *Bulletin of Marine Science*, **43**, 695–709.

Gliwicz, Z. M. & Pijanowska, J. (1989). The role of predation in zooplankton succession. In *Plankton Ecology*, ed. U. Sommer, pp. 253–96. Berlin: Springer-Verlag.

Goldman, C. R., Jassby, A. & Powell, T. (1989). Interannual fluctuations in primary production: meteorological forcing at two subalpine lakes. *Limnology and Oceanography*, **34**, 310–23.

Gorham, E. & Sanger, J. (1975). Fossil pigments in Minnesota lake sediments and their bearing upon the balance between terrestrial and aquatic inputs to sedimentary organic matter. *Verhandlungen – Internationale Vereinigung für theoretische und angewandte Limnologie*, **19**, 2267–73.

Gran, G. (1952). Determination of the equivalance point in potentiometric titrations. *Analyst*, **77**, 661–71.

Green, J. (1957). Carotenoids in *Daphnia*. *Proceedings of the Royal Society of London*, **B147**, 392–401.

Gulati, R. D., Lammens, E. H. R. R., Meijer, M. L. & van Donk, E., eds. (1990). *Biomanipulation – Tool for Water Management*. (Proceedings of an international conference held in Amsterdam, The Netherlands.) Dordrecht, The Netherlands: Kluwer Academic Publishers.

Haas, L. (1982). Improved epifluorescence microscopy for observing planktonic micro-organisms. *Annales de l'Institut Océanographique, Paris*, **58** (Supplement), 261–6.

Haberyan, K. A. (1985). The role of copepod fecal pellets in the deposition of diatoms in Lake Tanganyika. *Limnology and Oceanography*, **30**, 1010–23.

Haberyan, K. A. (1990). The misrepresentation of the planktonic diatom assemblage in traps and sediments: Southern Lake Malawi, Africa. *Journal of Paleolimnology*, **3**, 35–44.

Hairston, N. G., Jr. (1980). The vertical distribution of diaptomid copepods in relation to body pigmentation. In *Evolution and Ecology of Zooplankton Communities*, ed. W. C. Kerfoot, pp. 98–110. Hanover, New Hampshire: University Press of New England.

Hairston, N. G., Sr. (1989). *Ecological Experiments: Purpose, Design, and Execution*. Cambridge University Press.

Hairston, N. G., Smith, F. E. & Slobodkin, L. B. (1960). Community structure, population control and competition. *American Naturalist*, **94**, 421–5.

Hall, D. J. (1964). An experimental approach to the dynamics of a natural population of *Daphnia galeata mendotae*. *Ecology*, **45**, 94–112.

Hall, D. J., Cooper, W. E. & Werner, E. E. (1970). An experimental approach to the production dynamics and structure of freshwater animal communities. *Limnology and Oceanography*, **15**, 839–928.

Hall, D. J. & Ehlinger, T. J. (1989). Perturbation, planktivory and pelagic community structure: the consequence of winterkill in a small lake. *Canadian Journal of Fisheries and Aquatic Sciences*, **46**, 2203–9.

Hall, D. J., Threlkeld, S. T., Burns, C. W. & Crowley, P. H. (1976). The size-efficiency hypothesis and the size structure of zooplankton communities. *Annual Review of Ecology and Systematics*, **7**, 177–208.

Hambright, K. D. (1991). Experimental analysis of prey selection by largemouth bass: role of predator mouth width and prey body depth. *Transactions of the American Fisheries Society*, **120**, 500–8.

Hanazato, T. & Yasuno, M. (1989). Zooplankton community structure driven by vertebrate and invertebrate predators. *Oecologia*, **81**, 450–58.

Haney, J. F. & Hall, D. J. (1975). Diel vertical migration and filter-feeding activities of *Daphnia*. *Archiv für Hydrobiologie*, **75**, 413–41.

Hardy, A. C. (1956). *The Open Sea – Its Natural History: The World of Plankton*. Boston: Houghton Mifflin.

Hardy, A. C. & Gunther, E. R. (1935). The plankton of the South Georgia whaling grounds and adjacent waters, 1926–27. *'Discovery' Reports,* **11,** 1–146.

Harvey, J. M. (1930). The action of light on *Calanus finmarchicus* (Gunner) as determined by its effect on heartrate. *Contributions to Canadian Biology,* **5,** 83–92.

Hasler, A. D., Brynildson, O. M. & Helm, W. T. (1951). Improving conditions for fish in brown-water lakes by alkalization. *Journal of Wildlife Management,* **15,** 347–52.

Havel, J. E. (1987). Predator-induced defenses: a review. In *Predation: Direct and Indirect Impacts on Aquatic Communities,* ed. W. C. Kerfoot & A. Sih, pp. 263–78. Hanover, New Hampshire: University Press of New England.

Havel, J. E. & Dodson, S. I. (1984). *Chaoborus* predation on typical and spined morphs of *Daphnia pulex*: behavioral observations. *Limnology and Oceanography,* **29,** 487–94.

Havel, J. E. & Dodson, S. I. (1985). Environmental cues for cyclomorphosis in *Daphnia retrocurva* Forbes. *Freshwater Biology,* **15,** 469–78.

He, X. (1986). *Population Dynamics of Northern Redbelly Dace (Phoxinus eos), Finescale Dace (Phoxinus neogaeus), and Central Mudminnow (Umbra limi), in Two Manipulated Lakes.* M.S. thesis (Oceanography and Limnology), University of Wisconsin–Madison.

He, X. (1990). *Effects of Predation on a Fish Community: a Whole Lake Experiment.* Ph.D. thesis (Oceanography and Limnology), University of Wisconsin–Madison.

He, X. & Kitchell, J. F. (1990). Direct and indirect effects of predation on a fish community: a whole lake experiment. *Transactions of the American Fisheries Society,* **119,** 825–35.

He, X. & Wright, R. A. (1992). Piscivore-planktivore interactions in an experimental lake: population and community responses to predation. *Canadian Journal of Fisheries and Aquatic Sciences,* **44,** 1176–83.

Healey, F. P. (1983). Effect of temperature and light intensity on the growth rate of *Synura sphagnicola. Journal of Plankton Research,* **5,** 767–74.

Healey, F. P. & Hendzel, L. L. (1980). Physiological indicators of nutrient deficiency in lake phytoplankton. *Canadian Journal of Fisheries and Aquatic Sciences,* **37,** 42–53.

Henrikson, L., Nyman, H. G., Oscarson, H. G. & Stenson, J. A. E. (1980). Trophic changes without changes in external loading. *Hydrobiologia,* **68,** 257–63.

Hessen, D. O. (1988). *Carbon, Nitrogen, and Phosphorus Content of Common Crustacean Zooplankton Species.* Dissertation, University of Oslo.

Hessen, D. O. (1990). Carbon nitrogen, and phosphorus status in Daphnia at varying food conditions. *Journal of Plankton Research,* **12,** 1239–50.

Hessen, D. O., Andersen, T. & Lyche, A. (1990). Carbon metabolism in a humic lake: pool sizes and cycling through zooplankton. *Limnology and Oceanography,* **35,** 84–99.

Hewett, S. W. & Johnson, B. L. (1987). *A Generalized Bioenergetics Model of Fish Growth for Microcomputers.* UW Sea Grant Technical Report No. WIS-SG-87-245. Madison: University of Wisconsin Sea Grant Institute.

Hewett, S. W. & Johnson, B. L. (1992). *Fish Bioenergetics Model 2, the Second Version of A Generalized Bioenergetics Model of Fish Growth for Microcomputers.* UW Sea Grant Technical Report. Madison: University of Wisconsin Sea Grant Institute, in press.

Hickman, M. & Schweger, C. E. (1991). Oscillaxanthin and myxoxanthophyll in two cores from Lake Wabamun, Alberta, Canada. *Journal of Paleolimnology*, **5**, 127–37.

Hickman, M., Schweger, C. E. & Klarer, D. M. (1990). Baptiste Lake, Alberta – a late Holocene history of changes in a lake and its catchment in the southern boreal forest. *Journal of Paleolimnology*, **4**, 253–68.

Hill, M. O. & Gauch, H. G. (1980). Detrended correspondence analysis, an improved ordination technique. *Vegetatio*, **42**, 47–58.

Hilsenhoff, W. L. (1975). *Aquatic Insects of Wisconsin*. Wisconsin Department of Natural Resources Technical Bulletin 89. Madison: Wisconsin Department of Natural Resources.

Hobbie, J. E., Daley, R. J. & Jasper, S. (1977). Use of Nuclepore filters for counting bacteria by fluorescence microscopy. *Applied and Environmental Microbiology*, **33**, 1225–8.

Hodgson, J. R. (1987). Occurrence of small mammals in the diets of largemouth bass (*Micropterus salmoides*). *Jack-Pine Warbler*, **64**, 39–40.

Hodgson, J. R., Carpenter, S. R. & Gripentrog, A. P. (1989). Effect of sampling frequency on intersample variance and food consumption estimates of nonpiscivorous largemouth bass. *Transactions of the American Fisheries Society*, **118**, 11–19.

Hodgson, J. R. & Cochran, P. A. (1988). The effect of sample methodology on diet analysis in largemouth bass (*Micropterus salmoides*). *Verhandlungen – Internationale Vereinigung für Theoretische und Angewandte Limnologie*, **23**, 1670–5.

Hodgson, J. R., Hodgson, C. J. & Brooks, S. M. (1991). Trophic interaction and competition between largemouth bass (*Micropterus salmoides*) and rainbow trout (*Onchorhynchus mykiss*) in a manipulated lake. *Canadian Journal of Fisheries and Aquatic Sciences*, **48**, 1704–12.

Hodgson, J. R. & Kitchell, J. F. (1987). Opportunistic foraging by largemouth bass (*Micropterus salmoides*). *American Midland Naturalist*, **118**, 323–36.

Hoenicke, R. & Goldman, C. R. (1987). Resource dynamics and seasonal changes in competitive interactions among three cladoceran species. *Journal of Plankton Research*, **9**, 397–417.

Hrbáček, J. (1969). On the possibility of estimating predation pressure and nutrition level of poplations of *Daphnia* from their remains in sediments. *Mitteilungen – Internationale Vereinigung für theoretische und angewandte Limnologie*, **17**, 262–74.

Hrbáček, J., Dvorakova, M., Korinek, V. & Prochazkova, L. (1961). Demonstration of the effect of the fish stock on the species composition of zooplankton and the intensity of metabolism of the whole plankton assemblage. *Verhandlungen – Internationale Vereinigung für theoretische und angewandte Limnologie*, **14**, 192–5.

Hunter, M. D. & Price, P. W. (1992). Playing chutes and ladders: heterogeneity and the relative roles of bottom-up and top-down forces in natural communities. *Ecology*, **73**, 724–32.

Huntley, M. & Brooks, E. R. (1982). Effects of age and food availability on diel vertical migration of *Calanus pacificus*. *Marine Biology*, **71**, 23–31.

Hurlbert, S. H. (1984). Pseudoreplication and the design of ecological field experiments. *Ecological Monographs*, **54**, 187–211.

Hurlbert, S. H. & Mulla, M. S. (1981). Impacts of mosquitofish (*Gambusia affinis*) predation on plankton communities. *Hydrobiologia*, **83**, 125–51.

Hurley, J. P. & Armstrong, D. E. (1991). Pigment preservation in lake sediments: a comparison of sedimentary environments in Trout Lake, Wisconsin. *Canadian Journal of Fisheries and Aquatic Sciences*, **48**, 472–86.

Hutchinson, G. E. (1967). *A Treatise on Limnology*, volume 2: *Introduction to Lake Biology and the Limnoplankton*. New York: Wiley.

Jamart, B. M., Winter, D., Banse, K., Anderson, G. C. & Lam, R. K. (1977). A theoretical study of phytoplankton growth and nutrient distribution in the Pacific Ocean off the northwestern U.S. coast. *Deep-Sea Research*, **24**, 753–73.

Jassby, A. D. & Powell, T. M. (1990). Detecting change in ecological time series. *Ecology*, **71**, 2044–52.

Jassby, A. D., Powell, T. M. & Goldman, C. R. (1990). Interannual fluctuations in primary production: direct physical effects and the trophic cascade at Castle Lake, CA. *Limnology and Oceanography*, **35**, 1021–38.

Jeffrey, W. H. & Paul, J. H. (1988). Underestimation of DNA synthesis by [^{3}H]thymidine incorporation in marine bacteria. *Applied and Environmental Microbiology*, **54**, 3165–8.

Jeppeson, E., Jensen, J. P., Kristensen, P., Sondergaard, M., Mortensen, E., Sortkjaer, O. & Olrik, K. (1990). Fish manipulation as a lake restoration tool in shallow, eutrophic, temperate lakes 2: threshold levels, long-term stability and conclusions. *Hydrobiologia*, **200/201**, 219–27.

Johnsen, G. H. & Jakobsen, P. J. (1987). The effect of food limitation on vertical migration in *Daphnia longispina*. *Limnology and Oceanography*, **32**, 873–80.

Johnson, T. C. (1974). The dissolution of siliceous microfossils in surface sediments of eastern tropical Pacific. *Deep-Sea Research*, **21**, 851–64.

Johnson, W. D. & Hasler, A. D. (1954). Rainbow trout production in dystrophic lakes. *Journal of Wildlife Management*, **18**, 113–34.

Juday, C. A. (1904). The diurnal movement of plankton crustacea. *Transactions of the Wisconsin Academy of Sciences, Arts and Letters*, **14**, 534–68.

Kareiva, P. & Sahakian, R. (1990). Tritrophic effects of a single architectural mutation in pea plants. *Nature*, **345**, 433–4.

Karl, D. M. (1986). Determination of *in situ* microbial biomass, viability, metabolism, and growth. In *Bacteria in Nature*, volume 2, ed. J. S. Poindexter & E. R. Leadbetter, pp. 85–176. New York: Plenum.

Keast, A. (1978). Trophic and spatial interrelationships in the fish species of an Ontario temperate lake. *Environmental Biology of Fishes*, **3**, 7–31.

Keast, A. (1985). The piscivore feeding guild of fishes in small freshwater ecosystems. *Environmental Biology of Fishes*, **12**, 119–29.

Kerfoot, W. C. (1974). Net accumulation rates and history of cladoceran communities. *Ecology*, **35**, 384–98.

Kerfoot, W. C. (1975). Seasonal changes of *Bosmina* (Crustacea, Cladocera) in Frains Lake, Michigan: laboratory observations of phenotypic changes induced by inorganic factors. *Freshwater Biology*, **5**, 227–43.

Kerfoot, W. C. (1980). Perspectives on cyclomorphosis: separation of phenotypes and genotypes. In *Evolution and Ecology of Zooplankton Communities*, ed. W. C. Kerfoot, pp. 470–96. Hanover, New Hampshire: University Press of New England.

Kerfoot, W. C. (1981). Long-term replacement cycles in cladoceran communities: a history of predation. *Ecology*, **62**, 216–33.

Kerfoot, W. C. (1985). Adaptive value of vertical migration. In *Migration: Mechanisms and Adaptive Significance*, ed. M. A. Rankin, pp. 92–113. (*Contributions in Marine Science*, supplement volume 27.) Port Aransas, Texas: Marine Science Institute, University of Texas at Austin.

Kerfoot, W. C. (1987a). Cascading effects and indirect pathways. In *Predation: Direct and Indirect Impacts on Aquatic Communities*, ed. W. C. Kerfoot & A. Sih, pp. 57–70. Hanover, New Hampshire: University Press of New England.

Kerfoot, W. C. (1987b). Translocation experiments: *Bosmina* responses to copepod predation. *Ecology*, **68**, 596–610.

Kerfoot, W. C. & Sih, A., eds. (1987). *Predation: Direct and Indirect Impacts on Aquatic Communities*. Hanover, New Hampshire: University Press of New England.

Kikuchi, K. (1930). Diurnal migration of plankton Crustacea. *Quarterly Review of Biology*, **5**, 189–206.

Kingston, J. C. & Birks, H. J. B. (1990). Dissolved organic carbon reconstruction from diatom assemblages in PIRLA project lakes, North America. *Philosophical Transactions of the Royal Society of London*, B**327**, 279–88.

Kitchell, J. A. & Kitchell, J. F. (1980). Size-selective predation, light transmission, and oxygen stratification: evidence from the recent sediments of manipulated lakes. *Limnology and Oceanography*, **25**, 389–402.

Kitchell, J. F. (1983). Energetics. In *Fish Biomechanics*, ed. P. Webb & D. Weihs, pp. 312–38. New York: Praeger Publishers.

Kitchell, J. F., ed. (1992). *Food Web Management: A Case Study of Lake Mendota*. New York: Springer-Verlag.

Kitchell, J. F., Bartell, S. M., Carpenter, S. R., Hall, D. J., McQueen, D. J., Neill, W. E., Scavia, D. & Werner, E. E. (1988). Epistemology, experiments, and pragmatism. In *Complex Interactions in Lake Communities*, ed. S. R. Carpenter, pp. 263–80. New York: Springer-Verlag.

Kitchell, J. F. & Carpenter, S. R. (1987). Piscivores, planktivores, fossils, and phorbins. In *Predation: Direct and Indirect Impacts on Aquatic Communities*, ed. W. C. Kerfoot & A. Sih, pp. 132–46. Hanover, New Hampshire: University Press of New England.

Kitchell, J. F., O'Neill, R. V., Webb, D., Gallepp, G., Bartell, S. M., Koonce, J. F. & Ausmus, B. S. (1979). Consumer regulation of nutrient cycling. *BioScience*, **29**, 28–34.

Kitchell, J. F., Stewart, D. J. & Weininger, D. (1977). Application of a bioenergetics model to yellow perch (*Perca flavescens*) and walleye (*Stizostedion vitreum vitreum*). *Journal of the Fisheries Research Board of Canada*, **34**, 1922–35.

Klein, B., Gieskes, W. W. C. & Kraay, G. G. (1986). Digestion of chlorophylls and carotenoids by marine protozoan *Oxyrrhis marina* studied by h.p.l.c. analysis of algal pigments. *Journal of Plankton Research*, **8**, 827–36.

Knapik, P. G. & Hodgson, J. R. (1986). Life history note: *Storeria occipitomaculata* (redbelly snake) predation. *Herpetological Review*, **17**, 22.

Knoechel, R. & Holtby, L. B. (1986). Cladoceran filtering rate: body length relationships for bacterial and large algal particles. *Limnology and Oceanography*, **31**, 195–200.

Krueger, D. A. & Dodson, S. I. (1981). Embryological induction and predation ecology of *Daphnia pulex*. *Limnology and Oceanography*, **26**, 219–23.

Lampert, W., Fleckner, W., Rai, H. & Taylor, B. E. (1986). Phytoplankton control by grazing zooplankton: a study on the spring clear water phase. *Limnology and Oceanography*, **31**, 478–90.

Lampert, W. & Taylor, B. E. (1985). Zooplankton grazing in a eutrophic lake: implications of diel vertical migration. *Ecology*, **66**, 68–82.

Lampert, W. H. (1989). The adaptive significance of diel vertical migration in zooplankton. *Functional Ecology*, **3**, 21–7.

Lane, P. A. (1978). Role of invertebrate predation in structuring zooplankton communities. *Verhandlungen – Internationale Vereinigung für theoretische und angewandte Limnologie*, **20**, 480–5.

Lathrop, R. C. & Carpenter, S. R. (1992). Zooplankton and their relationship to phytoplankton. In *Food Web Management: A Case Study of Lake Mendota*, ed. J. F. Kitchell, pp. 129–52. New York: Springer-Verlag.

Leavitt, P. R. (1988). Experimental determination of carotenoid degradation. *Journal of Paleolimnology*, **1**, 215–28.

Leavitt, P. R. & Brown, S. R. (1988). Effects of grazing by *Daphnia* on algal carotenoids: implications for paleolimnology. *Journal of Paleolimnology*, **1**, 201–14.

Leavitt, P. R. & Carpenter, S. R. (1989). Effects of sediment mixing and benthic algal production on fossil pigment stratigraphies. *Journal of Paleolimnology*, **2**, 147–58.

Leavitt, P. R. & Carpenter, S. R. (1990a). Aphotic pigment degradation in the hypolimnion: implications for sedimentation studies and paleolimnology. *Limnology and Oceanography*, **35**, 520–35.

Leavitt, P. R. & Carpenter, S. R. (1990b). Regulation of pigment sedimentation by photo-oxidation and herbivore grazing. *Canadian Journal of Fisheries and Aquatic Sciences*, **47**, 1166–76.

Leavitt, P. R., Carpenter, S. R. & Kitchell, J. F. (1989). Whole-lake experiments: the annual record of fossil pigments and zooplankton. *Limnology and Oceanography*, **34**, 700–17.

LeCren, E. D. & Lowe-McConnell, R., eds. (1981). *Functioning of Freshwater Ecosystems*. London: Cambridge University Press.

Lee, S. & Fuhrman, J. A. (1987). Relationships between biovolume and biomass of naturally derived marine bacterioplankton. *Applied and Environmental Microbiology*, **53**, 1298–303.

Legendre, L., Demers, S., Yentsch, C. M. & Yentsch, C. S. (1983). The ^{14}C method: patterns of dark CO_2 fixation and DCMU correction to replace the dark bottle. *Limnology and Oceanography*, **28**, 996–1003.

Lehman, J. T. (1980). Release and cycling of nutrients between planktonic algae and herbivores. *Limnology and Oceanography*, **25**, 620–32.

Lehman, J. T. & Sandgren, C. (1985). Species-specific rates of growth and grazing loss among freshwater algae. *Limnology and Oceanography*, **30**, 34–46.

Lei, C. H. & Armitage, K. B. (1980). Growth, development and body size of field and laboratory populations of *Daphnia ambigua*. *Oikos*, **35**, 31–48.

Leibold, M. A. (1990). Resources and predators can affect the vertical distribution of zooplankton. *Limnology and Oceanography*, **35**, 938–44.

Lewis, W. M. (1977). Feeding selectivity of a tropical *Chaoborus* population.

Freshwater Biology, **7**, 311–25.

Lewis, W. M. (1979). *Zooplankton Community Analysis: Studies on a Tropical System.* New York: Springer-Verlag.

Likens, G. E. (1985). An experimental approach for the study of ecosystems. *Journal of Ecology*, **73**, 381–96.

Likens, G. E., ed. (1989). *Long-Term Studies in Ecology.* New York: Springer-Verlag.

Likens, G. E., Borman, F. H., Johnson, N. M., Fisher, D. W. & Pierce, R. S. (1970). Effects of forest cutting and herbicide treatment on nutrient budgets in Hubbard Brook watershed-ecosystem. *Ecological Monographs*, **40**, 23–47.

Lindeman, R. L. (1942). The trophic-dynamic aspect of ecology. *Ecology*, **23**, 399–418.

Liu, K. (1990). Holocene paleoecology of the boreal forest and Great Lakes-St. Lawrence forest in Northern Ontario. *Ecological Monographs*, **60**, 179–212.

Lodge, D. M. (1992). Species invasions and deletions: community effects and responses to climate and habitat change. In *Biotic Interactions and Global Change*, ed. J. Kingsolver, P. Kareiva & R. Huey, 367–87. Sunderland, Massachusetts: Sinauer.

Lodge, D. M., Barko, J. W., Strayer, D., Melack, J. M., Mittelbach, G. G., Howarth, R. W., Menge, B. & Titus, J. E. (1988). Spatial heterogeneity and habitat interactions in lake communities. In *Complex Interactions in Lake Communities*, ed. S. R. Carpenter, pp. 181–208. New York: Springer-Verlag.

Longhurst, A. R. (1976). Vertical migration. In *The Ecology of the Seas*, ed. D. H. Cushing & J. J. Walsh, pp. 116–37. Philadelphia: W. B. Saunders.

Luecke, C. (1986). A change in the pattern of vertical migration of *Chaoborus flavicans* after the introduction of trout. *Journal of Plankton Research*, **8**, 649–57.

Luecke, C. & Litt, A. H. (1987). Effects of predation by *Chaoborus flavicans* on crustacean zooplankton of Lake Lenore, Washington. *Freshwater Biology*, **18**, 185–92.

Lynch, M. (1979). Predation, competition, and zooplankton community structure: an experimental study. *Limnology and Oceanography*, **24**, 253–72.

Lynch, M. (1980). The evolution of cladoceran life histories. *Quarterly Review of Biology*, **55**, 23–42.

Lynch, M. & Shapiro, J. (1981). Predation, enrichment, and phytoplankton community structure. *Limnology and Oceanography*, **26**, 86–102.

MacIsaac, H. J. & Gilbert, J. J. (1989). Competition between rotifers and cladocerans of different body sizes. *Oecologia*, **81**, 295–301.

MacKay, N. A., Carpenter, S. R., Soranno, P. A. & Vanni, M. J. (1990). The impact of two *Chaoborus* species on a zooplankton community. *Canadian Journal of Zoology*, **68**, 981–5.

Magnuson, J. J., Beckel, A. L., Mills, K. & Brandt, S. B. (1985). Surviving winter hypoxia: behavioral adaptations of fishes in a northern Wisconsin winterkill lake. *Environmental Biology of Fishes*, **14**, 241–50.

Magnuson, J. J. & Bowser, C. J. (1990). A network for long-term ecological research in the United States. *Freshwater Biology*, **23**, 137–43.

Makarewicz, J. C. & Likens, G. E. (1975). Niche analysis of a zooplankton community. *Science*, **190**, 1000–3.

Malone, T. C. (1980). Algal size. In *Physiological Ecology of Phytoplankton*, ed. I

Morris, pp. 433–64. Berkeley: University of California Press.

Malueg, K. W. (1963). *A Study on the Vertical Distribution of Phytoplankton with Respect to the Effects of Lime-Treatment.* M.S. thesis (Zoology-Botany), University of Wisconsin–Madison.

Mann, K. H. & Breen, P. A. (1972). The relation between lobster abundance, sea urchins, and kelp beds. *Journal of the Fisheries Research Board of Canada*, **29**, 603–5.

Mantoura, R. F. C. & Llewellyn, C. A. (1983). The rapid determination of algal chlorophyll and carotenoid pigments and their breakdown products in natural waters by reverse phase high-performance liquid chromatography. *Analytica Chimica Acta*, **151**, 297–314.

Marker, A. F. H., Crowther, C. A. & Gunn, R. J. M. (1980). Methanol and acetone as solvents for estimating chlorophyll *a* and phaeopigments by spectrophotometry. *Archiv für Hydrobiologie, Ergebnisse der Limnologie*, **14**, 52–69.

Matson, P. A. & Carpenter, S. R., eds. (1990). Statistical analysis of ecological response to large-scale perturbations. *Ecology*, **71**, 2037–68.

Mazumder, A. D., McQueen, D. J., Taylor, W. D. & Lean, D. R. S. (1988). Effects of fertilization and planktivorous fish (yellow perch) predation on size distribution of particulate phosphorus and assimilated phosphate: large exclosure experiments. *Limnology and Oceanography*, **33**, 421–30.

Mazumder, A., Taylor, W. D., McQueen, D. J. & Lean, D. R. S. (1990). Effects of fish and plankton on lake temperature and mixing depth. *Science*, **247**, 312–15.

McAllister, C. D. (1969). Aspects of estimating zooplankton production from phytoplankton production. *Journal of the Fisheries Research Board of Canada*, **26**, 199–220.

McIntosh, R. P. (1985). *The Background of Ecology.* Cambridge University Press.

McLaren, I. A. (1963). Effects of temperature on growth of zooplankton, and the adaptive value of vertical migration. *Journal of the Fisheries Research Board of Canada*, **20**, 685–722.

McLaren, I. A. (1974). Demographic strategy of vertical migration by a marine copepod. *American Naturalist*, **108**, 91–102.

McNaught, D. C. (1978). Spatial heterogeneity and niche differentiation in zooplankton of Lake Huron. *Verhandlungen – Internationale Vereinigung für theoretische und angewandte Limnologie*, **20**, 341–6.

McNaughton, S. J. (1979). Grazing as an optimization process: grass-ungulate relationships in the Serengeti. *American Naturalist*, **113**, 691–703.

McNaughton, S. J., Oesterheld, M., Frank, D. A. & Williams, K. J. (1991). Primary and secondary production in terrestrial ecosystems. In *Comparative Analyses of Ecosystems: Patterns, Mechanisms, and Theories*, ed. J. Cole, G. Lovett & S. Findlay, pp. 120–39. New York: Springer-Verlag.

McQueen, D. J. (1990). Manipulating lake community structure: where do we go from here? *Freshwater Biology*, **23**, 613–20.

McQueen, D. J., Johannes, M.R.S., Post, J. R., Stewart, T. J. & Lean, D.R.S. (1989). Bottom-up and top-down impacts on freshwater pelagic community structure. *Ecological Monographs*, **59**, 289–309.

McQueen, D. J., Post, J. R. & Mills, E. L. (1986). Trophic relationships in freshwater pelagic ecosystems. *Canadian Journal of Fisheries and Aquatic Sciences*, **43**, 1571–81.

Menge, B. A. & Sutherland, J. P. (1976). Species diversity gradients: synthesis of the roles of predation, competition, and temporal heterogeneity. *American Naturalist,* **110**, 351–69.

Menge, B. A. & Sutherland, J. P. (1987). Community regulation: variation in disturbance, competition, and predation in relation to environmental stress and recruitment. *American Naturalist,* **130**, 730–57.

Mills, E. L. & Schiavone, A. (1982). Evaluation of fish communities through assessment of zooplankton populations and measures of lake productivity. *North American Journal of Fisheries Management,* **2**, 14–27.

Moll, R. & Stoermer, E. (1982). A hypothesis relating trophic status and subsurface chlorophyll maxima of lakes. *Archiv für Hydrobiologie,* **94**, 425–40.

Moll, R. A., Brahce, M. R. & Peterson, T. P. (1984). Phytoplankton dynamics within the subsurface chlorophyll maximum of Lake Michigan. *Journal of Plankton Research,* **6**, 751–66.

Moore, M. & Gilbert, J. J. (1987). Age-specific *Chaoborus* predation on rotifer prey. *Freshwater Biology,* **17**, 223–36.

Morgan, T. H. (1903). *Evolution and Adaptation.* London: Macmillan.

Moriarty, D. J. W. (1986). Measurement of bacterial growth rates in aquatic systems from rates of nucleic acid synthesis. *Advances in Microbial Ecology,* **9**, 245–92.

Moss, B. (1990). Engineering and biological approaches to the restoration from eutrophication of shallow lakes in which aquatic plant communities are important components. *Hydrobiologia,* **200/201**, 367–77.

Murdoch, W. W. (1966). 'Community structure, population control, and competition' – a critique. *American Naturalist,* **100**, 219–26.

Murdoch, W. W. & McCauley, E. (1985). Three distinct types of dynamic behavior shown by a single planktonic system. *Nature,* **316**, 628–30.

Murray, J. & Hjort, J. (1912). *The Depths of the Ocean.* London: Macmillan.

Naud, M. & Magnan, P. (1988). Diel onshore–offshore migrations in northern redbelly dace, *Phoxinus eos* (Cope), in relation to prey distribution in a small oligotrophic lake. *Canadian Journal of Zoology,* **66**, 1249–53.

Neill, W. E. (1981). Impact of *Chaoborus* predation upon the structure and dynamics of a crustacean zooplankton community. *Oecologia,* **48**, 164–77.

Neill, W. E. (1984). Regulation of rotifer densities by crustacean zooplankton in an oligotrophic montane lake in British Columbia. *Oecologia,* **61**, 174–81.

Neill, W. E. (1988). Complex interactions in oligotrophic lake food webs: responses to nutrient enrichment. In *Complex Interactions in Lake Communities,* ed. S. R. Carpenter, pp. 31–44. Berlin: Springer-Verlag.

Neill, W. E. (1990). Induced vertical migration in copepods as a defence against invertebrate predation. *Nature,* **345**, 524–6.

Neill, W. E. & Peacock, A. (1980). Breaking the bottleneck: interactions of invertebrate predators and nutrients in oligotrophic lakes. In *Evolution and Ecology of Zooplankton Communities,* ed. W. C. Kerfoot, pp. 715–24. Hanover, New Hampshire: University Press of New England.

Newell, S. Y., Fallon, R. D. & Tabor, P. S. (1986). Direct microscopy of natural assemblages. In *Bacteria in Nature,* volume 2, ed. J. S. Poindexter & E. R. Leadbetter, pp. 1–48. New York: Plenum.

Nilsson, M. & Renberg, I. (1990). Viable endospores of *Thermoactinomyces vulgaris* in lake sediments as indicators of agricultural history. *Applied Environmental Micro-*

biology, **56**, 2025–8.

Northcote, T. G. (1964). Use of a high-frequency echosounder to record distribution and migration of *Chaoborus* larvae. *Limnology and Oceanography*, **9**, 87–91.

Northcote, T. G. (1988). Fish in the structure and function of freshwater ecosystems: a 'top-down' view. *Canadian Journal of Fisheries and Aquatic Sciences*, **45**, 361–79.

Northcote, T. G., Walters, C. J. & Hume, J. M. B. (1978). Initial impacts of experimental fish introductions on the macrozooplankton of small oligotrophic lakes. *Verhandlungen – Internationale Vereinigung für theoretische und angewandte Limnologie*, **20**, 2003–12.

Odum, E. P. (1969). The strategy of ecosystem development. *Science*, **164**, 262–70.

Ohman, M. D., Frost, B. W. & Cohen, E. B. (1983). Reverse diel migration: an escape from invertebrate predators. *Science*, **220**, 1404–7.

Oksanen, L. (1983). Trophic exploitation and arctic phytomass patterns. *American Naturalist*, **122**, 45–52.

Oksanen, L. (1990). Predation, herbivory, and plant strategies along gradients of primary productivity. In *Perspectives on Plant Competition*, ed. D. Tilman & J. Grace, pp. 445–74. New York: Academic Press.

Oksanen, L., Fretwell, S. D., Arruda, J. & Niemela, P. (1981). Exploitation ecosystems in gradients of primary productivity. *American Naturalist*, **118**, 240–61.

Ortner, P. B., Wiebe, P. H. & Cox, J. L. (1980). Relationships between oceanic epizooplankton distributions and the seasonal deep chlorophyll maximum in the Northwestern Atlantic Ocean. *Journal of Marine Research*, **38**, 507–31.

O'Sullivan, P. E. (1983). Annually-laminated lake sediments and the study of Quaternary environmental change. *Quaternary Science Review*, **1**, 245–313.

Pace, M. L. (1984). Zooplankton community structure, but not biomass, influences the phosphorus-chlorophyll *a* relationship. *Canadian Journal of Fisheries and Aquatic Sciences*, **41**, 1089–96.

Pace, M. L. & Funke, E. B. (1991). Regulation of planktonic microbial communities by nutrients and herbivores. *Ecology*, **72**, 904–14.

Pace, M. L., McManus, G. B. & Findlay, S. E. G. (1990). Plankton community structure determines the fate of bacterial production in a temperate lake. *Limnology and Oceanography*, **35**, 795–808.

Paine, R. T. (1966). Food web complexity and species diversity. *American Naturalist*, **100**, 65–75.

Paine, R. T. (1980). Food webs: linkage interaction strength, and community infrastructure. *Journal of Animal Ecology*, **49**, 667–85.

Paine, R. T. (1988). Food webs: road maps of interactions or grist for the theoretical mill? *Ecology*, **69**, 1648–54.

Paloheimo, J. E. (1974). Calculation of instantaneous birth rate. *Limnology and Oceanography*, **19**, 692–4.

Parkin, T. B. & Brock, T. D. (1980). Photosynthetic bacterial production in lakes: the effects of light intensity. *Limnology and Oceanography*, **25**, 711–18.

Pastor, J., Naiman, R. J., Dewey, B. & McInnes, P. (1988). Moose, microbes, and the boreal forest. *BioScience*, **38**, 770–7.

Pastorok, R. A. (1980). Selection of prey by *Chaoborus* larvae: a review and new evidence for behavioral flexibility. In *Evolution and Ecology of Zooplankton*

Communities, ed. W. C. Kerfoot, pp. 538–54. Hanover, New Hampshire: University Press of New England.

Paull, R. K. & Paull, R. A. (1977). Geology of Wisconsin and Upper Michigan. Dubuque, Iowa: Kendall/Hunt Publishing Company.

Pedrós-Alió, C., Gasol, J. M. & Guerrero, R. (1987). On the ecology of a Cryptomonas phaseolus population forming a metalimnetic bloom in Lake Cisó, Spain: annual distribution and loss factors. Limnology and Oceanography, 32, 285–98.

Pennak, R. W. (1978). Fresh-Water Invertebrates of the United States. New York: John Wiley and Sons.

Persson, L., Andersson, G., Hamrin, S. F. & Johansson, L. (1988). Predator regulation and primary production along the productivity gradient of temperate lake ecosystems. In Complex Interactions in Lake Communities, ed. S. R. Carpenter, pp. 45–68. New York: Springer-Verlag.

Persson, L., Diehl, S., Johansson, L., Andersson, G. & Hamrin, S. F. (1992). Trophic interactions in temperate lake ecosystems – a test of food chain theory. American Naturalist, in press.

Persson, L. & Greenberg, L. A. (1990). Juvenile competitive bottlenecks: the perch (Perca fluviatilis)-roach (Rutilus rutilus) interaction. Ecology, 71, 44–56.

Peters, R. H. (1975). Phosphorus regeneration by natural populations of limnetic zooplankton. Verhandlungen – Internationale Vereinigung für theoretische und angewandte Limnologie, 19, 273–9.

Peters, R. H. (1983). The Ecological Implications of Body Size. Cambridge University Press.

Peters, R. H. & Downing, J. A. (1984). Empirical analysis of zooplankton filtering and feeding rates. Limnology and Oceanography, 29, 763–84.

Petterson, K. (1980). Alkaline phosphatase activity and algal surplus phosphorus as phosphorus-deficiency indicators in Lake Erken. Archiv für Hydrobiologie, 89, 54–87.

Pijanowska, J. & Dawidowicz, P. (1987). The lack of vertical migration in Daphnia: the effect of homogeneously distributed food. Hydrobiologia, 148, 175–81.

Pimm, S. L. (1982). Food Webs. New York: Chapman and Hall.

Pip, E. & Robinson, G. G. C. (1982). A study of the seasonal dynamics of three phycoperiphytic communities using nuclear track autoradiography. I. Inorganic carbon uptake. Archiv für Hydrobiologie, 94, 341–71.

Polis, G. A. (1991). Complex trophic interactions in deserts: an empirical critique of food-web theory. American Naturalist, 138, 123–55.

Porter, K. G. (1973). Selective grazing and differential digestion of algae by zooplankton. Nature, 244, 179–80.

Porter, K. G., Paerl, H., Hodson, R., Pace, M., Priscu, J., Riemann, B., Scavia, D. & Stockner, J. (1988). Microbial interactions in lake food webs. In Complex Interactions in Lake Communities, ed. S. R. Carpenter, pp. 209–28. New York: Springer-Verlag.

Power, M. E. (1990). Effects of fish in river food webs. Science, 250, 811–14.

Power, M. E. (1992). Top down and bottom up forces in food webs: do plants have primacy? Ecology, 73, 733–46.

Press, W. H., Flannery, B. P., Teukolsky, S. A. & Vetterling, W. T. (1989).

Numerical Recipes in C. Cambridge University Press.

Price, P. W., Bouton, C. E., Gross, P., McPheron, B. A., Thompson, J. N. & Weis, A. E. (1980). Interactions among three trophic levels: influence of plants on interactions between insect herbivores and natural enemies. *Annual Review of Ecology and Systematics*, **11**, 41–65.

Priscu, J. C. & Goldman, C. R. (1984). The effect of temperature on photosynthetic and respiratory electron transport system activity in the shallow and deep-living phytoplankton of a subalpine lake. *Freshwater Biology*, **14**, 143–55.

Quiros, R. (1990). Factors related to variance of residuals in chlorophyll-total phosphorus regressions in lakes and reservoirs of Argentina. *Hydrobiologia*, **200/201**, 343–55.

Rahel, F. J. (1986). Biogeographic influences on fish species composition of northern Wisconsin lakes with applications for lake acidification studies. *Canadian Journal of Fisheries and Aquatic Sciences*, **43**, 124–34.

Raven, J. A. & Geider, R. J. (1988). Temperature and algal growth. *New Phytologist*, **110**, 441–61.

Reasoner, M. A. & Healy, R. E. (1986). Identification and significance of tephras encountered in a core from Mary Lake, Yoho National Park, British Columbia. *Canadian Journal of Earth Sciences*, **24**, 1991–9.

Reckhow, K. H. & Chapra, S. C. (1983). *Engineering Approaches for Lake Management*, volume I: *Data Analysis and Empirical Modelling*. Boston: Butterworth.

Reinertsen, H., Jensen, A., Koksvik, J. I., Langelaand, A. & Olsen, Y. (1990). Effects of fish removal on the limnetic ecosystem of a eutrophic lake. *Canadian Journal of Fisheries and Aquatic Sciences*, **47**, 166–73.

Reissen, H. P. & Sprules, W. G. (1990). Demographic costs of antipredator defenses in *Daphnia pulex*. *Ecology*, **71**, 1536–46.

Renberg, I. & Wik, M. (1985). Soot particle counting in recent lake sediments: an indirect dating method. *Ecological Bulletins, Stockholm*, **37**, 53–7.

Repeta, D. J. & Gagosian, R. B. (1982). Carotenoid transformations in coastal marine waters. *Nature*, **295**, 51–4.

Repeta, D. J. & Gagosian, R. B. (1987). Carotenoid diagenesis in recent marine sediments – I. The Peru continental shelf (15°S, 75°W). *Geochimica et Cosmochimica Acta*, **51**, 1001–9.

Reynolds, C. S. (1984a). *The Ecology of Freshwater Phytoplankton*. Cambridge University Press.

Reynolds, C. S. (1984b). Phytoplankton periodicity: the interactions of form, function and environmental variability. *Freshwater Biology*, **14**, 111–42.

Rice, J. A. & Cochran, P. A. (1984). Independent evaluation of a bioenergetics model for largemouth bass. *Ecology*, **63**, 732–9.

Richerson, P. J., Lopez, M. & Coon, T. (1978). The deep chlorophyll maximum layer of Lake Tahoe. *Verhandlungen – Internationale Vereinigung für theoretische und angewandte Limnologie*, **20**, 426–33.

Richman, S., Bohon, S. A. & Robbins, S. E. (1980). Grazing interactions among freshwater calanoid copepods. In *Ecology and Evolution of Zooplankton Communities*, ed. W. C. Kerfoot, pp. 219–33. Hanover, New Hampshire: University Press of New England.

Ricker, W. E. (1963). Big effects from small causes: two examples from fish population dynamics. *Journal of the Fisheries Research Board of Canada*, **20**, 257–64.

Ricker, W. E. (1975). *Computation and Interpretation of Biological Statistics of Fish Populations*. (*Bulletin of the Fisheries Research Board of Canada*, No. 191.) Ottawa: Department of the Environment, Fisheries and Marine Service.

Riemann, B. & Søndergaard, M. (1986). *Carbon Dynamics in Eutrophic Temperate Lakes*. New York: Elsevier Science.

Ringleberg, J. (1964). The positively phototactic reaction of *Daphnia magna* Straus. A contribution to the understanding of diurnal vertical migration. *Netherlands Journal of Sea Research*, **2**, 319–406.

Ringleberg, J. (1991). A mechanism of predator-mediated induction of diel vertical migration in *Daphnia hyalina*. *Journal of Plankton Research*, **13**, 83–9.

Robbins, J. A., Keilty, T., White, D. S. & Edgington, D. N. (1989). Relationships among tubificid abundances, sediment composition, and accumulation rates in Lake Erie. *Canadian Journal of Fisheries and Aquatic Sciences*, **46**, 223–31.

Roth, J. C. (1968). *Chaoborus* species in a southern Michigan lake (Diptera, Chaoboridae). *Limnology and Oceanography*, **13**, 242–9.

Russell, F. S. (1927). The vertical distribution of plankton in the sea. *Biological Review*, **2**, 213–62.

Sager, P. E. & Richman, S. (1990). Patterns of phytoplankton-zooplankton interactions along a trophic gradient: I. Production and utilization. *Verhandlungen – Internationale Vereinigung für theoretische und angewandte Limnologie*, **24**, 393–6.

Sager, P. E. & Richman, S. (1991). Functional interaction of phytoplankton and zooplankton along the trophic gradient in Green Bay, Lake Michigan. *Canadian Journal of Fisheries and Aquatic Sciences*, **48**, 116–22.

Sanders, R. W. (1991). Mixotrophic protists in marine and freshwater ecosystems. *Journal of Protozoology*, **38**, 76–81.

Sandgren, C. D., ed. (1988). *Growth and Reproductive Strategies of Freshwater Phytoplankton*. Cambridge University Press.

Sandman, O., Lichu, A. & Simola, H. (1990). Drainage ditch history as recorded in the varved sediment of a small lake in east Finland. *Journal of Paleolimnology* **3**, 161–9.

Sanger, J. E. (1988). Fossil pigments in paleoecology and paleolimnology. *Palaeogeography, Palaeoclimatology, Palaeolimnology*, **62**, 343–59.

Sanni, S. & Waervagen, S. B. (1990). Oligotrophication as a result of planktivorous fish removal with rotenone in the small, eutrophic Lake Mosvatn, Norway. *Hydrobiologia*, **200/201**, 263–74.

SAS Institute (1985). *SAS User's Guide: Statistics*. Cary, North Carolina: SAS Institute, Inc.

SAS Institute (1988). *SAS/ETS Users Guide*. Cary, North Carolina: SAS Institute, Inc.

Scheffer, M. (1990). Multiplicity of stable states in freshwater systems. *Hydrobiologia*, **200/201**, 475–86.

Schindler, D. E. (1992). *The Role of Food Web Configuration and Migratory Behavior in Lake Phosphorus Cycling*. M.S. thesis (Zoology), University of Wisconsin–Madison.

Schindler, D. W. (1978). Factors regulating phytoplankton production and standing crop in the world's lakes. *Limnology and Oceanography*, **23**, 478–86.

Schindler, D. W. (1987). Detecting ecosystem response to anthropogenic stress. *Canadian Journal of Fisheries and Aquatic Sciences*, **44**, 6–25.

Schindler, D. W. (1988). Experimental studies of chemical stressors on whole lake ecosystems. *Verhandlungen Internationale Vereinigung Limnologie*, **23**, 11–41.

Schindler, D. W. (1990). Experimental perturbations of whole lakes as tests of hypotheses concerning ecosystem structure and function. *Oikos*, **57**, 25–41.

Schindler, D. W., Beaty, K. G., Fee, E. J., Cruikshank, D. R., DeBruyn, E. R., Findlay, D. L., Linsey, G. A., Shearer, J. A., Stainton, M. P. & Turner, M. A. (1990). Effects of climatic warming on lakes of the central boreal forest. *Science*, **250**, 967–70.

Schmidt, K. (1978). Biosynthesis of carotenoids. In *The Photosynthetic Bacteria*, ed. R. K. Clayton & W. R. Sistrom, pp. 729–50. New York: Plenum Press.

Schmitz, W. R. (1958). *Artificially Induced Circulation in Thermally Stratified Lakes*. Ph.D. thesis (Zoology), University of Wisconsin–Madison.

Schoenly, K. & Cohen, J. E. (1991). Temporal variation in food web structure: 16 empirical cases. *Ecological Monographs*, **61**, 267–98.

Scott, W. B. & Crossman, E. J. (1973). *Freshwater Fishes of Canada*. (*Bulletin of the Fisheries Research Board of Canada*, No. 184.) Ottawa: Fisheries Research Board of Canada.

Scrope-Howe, S. & Jones, D. A. (1986). The vertical distribution of zooplankton in the Western Irish Sea. *Estuarine and Coastal Shelf Science*, **22**, 785–802.

Seaburg, K. G. (1957). A stomach sampler for live fish. *Progressive Fish-Culturist*, **19**, 137–9.

Shapiro, J. (1990). Biomanipulation: the next phase – making it stable. *Hydrobiologia*, **200/201**, 13–27.

Shapiro, J., Lamarra, V. & Lynch, M. (1975). Biomanipulation: an ecosystem approach to lake restoration. In *Proceedings of a Symposium on Water Quality Management Through Biological Control*, ed. P. L. Brezonik & J. L. Fox, pp. 85–96. Gainesville: University of Florida.

Shapiro, J. & Wright, D. I. (1984). Lake restoration by biomanipulation: Round Lake, Minnesota the first two years. *Freshwater Biology*, **14**, 371–83.

Sherr, E. B. (1988). Direct use of high molecular weight polysaccharide by heterotrophic flagellates. *Nature*, **335**, 348–51.

Sherr, E. B. & Sherr, B. F. (1991). Planktonic microbes: tiny cells at the base of the ocean's food web. *Trends in Ecology and Evolution*, **6**, 50–4.

Sherr, E. B., Sherr, B. F. & McDaniel, J. (1991). Clearance rates of <6 μm fluorescently labeled algae (FLA) by estuarine protozoa: potential grazing impact of flagellates and ciliates. *Marine Ecology Progress Series*, **69**, 81–92.

Siebeck, O. (1978). UV Toleranz und Photoreaktiviergen bei Daphnien aus Biotopen verschiedener Hohenregionen. *Naturwissenschaften*, **65**, 390–1.

Simon, M. & Azam, F. (1989). Protein content and protein synthesis rates of planktonic marine bacteria. *Marine Ecology Progress Series*, **51**, 201–13.

Smith, C. & Reay, P. (1991). Cannibalism in teleost fish. *Reviews in Fish Biology and Fisheries*, **1**, 41–64.

Smith, F. E. (1969). Effects of enrichment in mathematical models. In *Eutrophication: Causes, Consequences, Correctives*, pp. 631–45. Washington, D.C.: National Academy of Sciences.

Smith, V. H. (1979). Nutrient dependence of primary production in lakes. *Limnology and Oceanography*, **24**, 1051–64.

Smits, J. D. & Riemann, B. (1988). Calculation of cell production from [³H]thymidine incorporation with freshwater bacteria. *Applied and Environmental Microbiology*, **54**, 2213–19.

Smol, J. P. (1987). Freshwater algae. *Geosciences Canada*, **14**, 208–17.

Smol, J. P. (1990). Paleolimnology: recent advances and future challenges. In *Scientific Perspectives in Theoretical and Applied Limnology*, ed. R. de Bernardi, G. Giussani & L. Barbanti, pp. 253–84. (*Memorie dell'Istituto Italiano di Idrobiologia Dott. Marco de Marchi*, volume 47.) Verbania Pallanza, Italy: Istituto Italiano di Idrobiologia.

Snow, H. E. (1969). *Comparative growth of eight species of fish in thirteen northern Wisconsin lakes.* (*Research Report of the Wisconsin Department of Natural Resources*, No. 46.) Madison: Wisconsin Department of Natural Resources.

Sokal, R. R. & Rohlf, F. J. (1981). *Biometry.* New York: W. H. Freeman and Company.

Sommer, U. (1988). Phytoplankton succession in microcosm experiments under simultaneous grazing pressure and resource limitation. *Limnology and Oceanography*, **33**, 1037–54.

Sommer, U. (1989). *Plankton Ecology: Succession in Plankton Communities.* Berlin: Springer-Verlag.

Sommer, U., Gliwicz, Z. M., Lampert, W. & Duncan, A. (1986). The PEG-Model of seasonal succession of planktonic events in fresh waters. *Archiv für Hydrobiologie*, **106**, 433–71.

Soranno, P. A., ed. (1990). *Methods of the Cascading Trophic Interactions Project*, 2nd edn. Madison: Center for Limnology, University of Wisconsin.

Soto, D. & Hurlbert, S. H. (1991). Long-term experiments on calanoid-cyclopoid interactions. *Ecological Monographs*, **61**, 245–65.

Sournia, A. (1978). *Phytoplankton Manual.* Paris: Unesco.

Spencer, C. N. & King, D. L. (1984). Role of fish in regulation of plant and animal communities in eutrophic ponds. *Canadian Journal of Fisheries and Aquatic Sciences*, **41**, 1851–5.

Sprules, W. G., Carter, J. C. H. & Ramcharan, C. W. (1984). Phenotypic associations in the Bosminidae (Cladocera): phenotypic patterns. *Limnology and Oceanography*, **29**, 161–9.

St. Amand, A. L. (1990). *Mechanisms Controlling Metalimnetic Communities and the Importance of Metalimnetic Phytoplankton to Whole Lake Primary Productivity.* Ph.D. dissertation, University of Notre Dame, Notre Dame, Indiana.

St. Amand, A. L., Soranno, P. A., Carpenter, S. R. & Elser, J. J. (1989). Algal nutrient deficiency: growth bioassays versus physiological indicators. *Lake and Reservoir Management*, **5**, 27–35.

Stainton, M. P., Capal, M. J. & Armstrong, F. A. J. (1977). *The Chemical Analysis of Fresh Water*, 2nd edn. (Canada Fisheries and Marine Service Miscellaneous Special Publication 25.) Ottawa, Canada.

Stasiak, R. H. (1978). Reproduction, age and growth of finescale dace, *Chrosomos neogaeus*, in Minnesota. *Transactions of the American Fisheries Society*, **107**, 720–3.

Stemberger, R. S. & Gilbert, J. J. (1987). Defenses of planktonic rotifers against predators. In *Predation: Direct and Indirect Impacts on Aquatic Communities*, ed. W. C. Kerfoot & A. Sih, pp. 227–39. Hanover, New Hampshire: University Press of New England.

Sterner, R. W. (1986). Herbivores' direct and indirect effects on algal populations. *Science*, **231**, 605–7.

Sterner, R. W. (1989). The role of grazers in phytoplankton succession. In *Plankton Ecology: Succession in Plankton Communities*, ed. U. Sommer, pp. 107–70. Berlin: Springer-Verlag.

Sterner, R. W. (1990). The ratio of nitrogen to phosphorus resupplied by zooplankton: grazers and the algal competitive arena. *American Naturalist*, **136**, 209–29.

Sterner, R. W., Elser, J. J. & Hessen, D. O. (1992). Stoichiometric relationships among producers, consumers and nutrient cycling in pelagic ecosystems. *Biogeochemistry*, **17**, 49–67.

Stewart-Oaten, A., Murdoch, W. & Parker, K. (1986). Environmental impact assessment: 'pseudoreplication' in time? *Ecology*, **67**, 929–40.

Stich, H.-B. & Lampert, W. (1981). Predator evasion as an explanation of diurnal vertical migration by zooplankton. *Nature*, **293**, 396–8.

Stich, H.-B. & Lampert, W. (1984). Growth and reproduction of migrating and non-migrating *Daphnia* species under simulated food and temperature conditions of diurnal vertical migration. *Oecologia*, **61**, 192–6.

Stockner, J. G. & Porter, K. G. (1988). Microbial food webs in freshwater planktonic ecosystems. In *Complex Interactions in Lake Ecosystems*, ed. S. R. Carpenter, pp. 69–83. New York: Springer-Verlag.

Stoecker, D. K., Michaels, A. E. & Davis, L. H. (1987). Large proportion of marine planktonic ciliates found to contain functional chloroplasts. *Nature*, **326**, 260–2.

Strayer, D. J., Glitzenstein, S., Jones, C. G., Kolasa, J., Likens, G. E., McDonnell, M. J., Parker, G. G. & Pickett, S. T. A. (1986). *Long-Term Ecological Studies: an Illustrated Account of Their Design, Operation, and Importance to Ecology*. (Occasional Publication No. 1.) Millbrook, New York: Institute of Ecosystem Studies.

Strong, D. R. (1983). Natural variability and the manifold mechanisms of ecological communities. *American Naturalist*, **122**, 636–60.

Strong, D. R. (1992). Are trophic cascades all wet? Differentiation and donor-control in speciose ecosystems. *Ecology*, **73**, 747–54.

Stross, R. G. (1958). *Experimentally Induced Changes in Lakes: a) Environmental Changes Following Lime-Application to Stained Lakes; b) Changes in the Planktonic Crustacea Following the Introduction of Trout to a Fish-Free Lake*. Ph.D. thesis, University of Wisconsin–Madison.

Stross, R. G. & Hasler, A. D. (1960). Some lime-induced changes in lake metabolism. *Limnology and Oceanography*, **5**, 265–72.

Stroud, R. H. & Clepper, H., eds. (1979). *Predator-Prey Systems in Fisheries Management*. Washington, D.C.: Sport Fishing Institute.

Sugihara, G., Schoenly, K. & Trombia, A. (1989). Scale-invariance in food web properties. *Science*, **245**, 48–52.

Swain, E. B. (1985). Measurement and interpretation of sedimentary pigments. *Freshwater Biology*, **15**, 53–75.

Tessier, A. J. (1986). Comparative population regulation of two planktonic cladocera (*Holopedium gibberum* and *Daphnia catawba*). *Ecology*, **67**, 285–302.

Thienemann, A. (1919). Ueber die vertikale Schichtung des Planktons in Ulmener

Maar und die Planktonproduktion der anderen Eifelmann. *Verhandlungen Naturwissenschaften Verein preussi. Rheinlands und Westfalens*, **74**.

Thomas, G. (1977). Effects of near ultraviolet light on microorganisms. *Photochemistry and Photobiology*, **26**, 669–73.

Thorp, J. H. (1986). Two distinct roles for predators in freshwater assemblages. *Oikos*, **47**, 75–82.

Threlkeld, S. T. (1979). The midsummer dynamics of two *Daphnia* species in Wintergreen Lake, Michigan. *Ecology*, **60**, 165–79.

Tilman, D. (1982). *Resource Competition and Community Structure*. Princeton, New Jersey: Princeton University Press.

Tilman, D. (1989). Ecological experimentation: strengths and conceptual problems. In *Long-Term Studies in Ecology: Approaches and Alternatives*, ed. G. E. Likens, pp. 136–57. Berlin: Springer-Verlag.

Tolonen, K., Alasaarela, E. & Liehu, A. (1988). Retention of mercury in sediments of the Baltic Sea near Oulu, Finland. *Journal of Paleolimnology*, **1**, 133–40.

Tonn, W. M. (1985). Density compensation in Umbra-Perca fish assemblages of northern Wisconsin lakes. *Ecology*, **66**, 415–29.

Tonn, W. M. & Magnuson, J. J. (1982). Patterns in the species composition and richness of fish assemblages in northern Wisconsin lakes. *Ecology*, **63**, 1149–66.

Tonn, W. M., Magnuson, J. J., Rask, M. & Toivonen, J. (1990). Intercontinental comparison of small lake fish assemblages: the balance between local and regional processes. *American Naturalist*, **136**, 345–75.

Toolan, T., Wehr, J. D. & Findlay, S. (1991). Inorganic phosphorus stimulation of bacterioplankton production in a meso-eutrophic lake. *Applied and Environmental Microbiology*, **57**, 2074–8.

Turner, A. M. & Mittlebach, G. G. (1990). Predator avoidance and community structure: interactions among piscivores, planktivores, and plankton. *Ecology*, **71**, 2241–54.

UNDA (1988). *Gillen Land o'Lakes File UBVG*. Notre Dame, Indiana: University of Notre Dame Archives.

Uutala, A. J. (1990). *Chaoborus* (Diptera: Chaoboridae) mandibles – paleolimnological indicators of the historical status of fish populations in acid sensitive lakes. *Journal of Paleolimnology*, **4**, 139–52.

van Donk, E., Grimm, M. P., Gulati, R. D., Heuts, P. G. M., de Kloet, W. A. & van Liere, E. (1990). First attempt to apply whole-lake food web manipulation on a large scale in the Netherlands. *Hydrobiologia*, **200/201**, 291–302.

Vanni, M. J. (1986). Competition in zooplankton communities: suppression of small species by *Daphnia pulex*. *Limnology and Oceanography*, **31**, 1039–56.

Vanni, M. J. (1987). Effects of nutrients and zooplankton size on the structure of a phytoplankton community. *Ecology*, **68**, 624–35.

Vanni, M. J., Carpenter, S. R. & Luecke, C. (1992a). A simulation model of the interactions among nutrients, phytoplankton and zooplankton in Lake Mendota. In *Food Web Management: A Case Study of Lake Mendota*, ed. J. F. Kitchell, pp. 231–9. New York: Springer-Verlag.

Vanni, M. J. & Temte, J. (1990). Seasonal patterns of grazing and nutrient limitation of phytoplankton in a eutrophic lake. *Limnology and Oceanography*, **35**, 697–709.

Vanni, M. J., Temte, J., Allen, Y., Dodds, R., Howard, P. J., Leavitt, P. R. & Luecke, C. (1992b). Herbivory, nutrients, and phytoplankton dynamics in Lake Mendota,

1987–89. In *Food Web Management: A Case Study of Lake Mendota*, ed. J. F. Kitchell, pp. 241–71. New York: Springer-Verlag.

Vaqué, D. & Pace, M. L. (1992). Grazing on bacteria by flagellates and cladocerans in lakes of contrasting food web structure. *Journal of Plankton Research*, 14, 307–21.

Venrick, E. L., McGowan, J. A. & Mantyla, A. W. (1973). Deep maxima of photosynthetic chlorophyll in the Pacific Ocean. *Fisheries Bulletin*, 71, 41–52.

Vincent, W. F. (1981). Rapid physiological assays for nutrient demand by the plankton. I. Nitrogen. *Journal of Plankton Research*, 3, 685–97.

Vitousek, P. M. (1990). Biological invasions and ecosystem processes: towards an integration of population biology and ecosystem studies. *Oikos*, 57, 7–13.

von Ende, C. N. (1979). Fish predation, interspecific predation, and the distribution of two *Chaoborus* species. *Ecology*, 60, 119–28.

von Ende, C. N. (1982). Phenology of four *Chaoborus* species. *Environmental Entomology*, 11, 9–16.

Walker, I. R. (1987). Chironomidae (Diptera) in paleoecology. *Quaternary Science Reviews*, 6, 29–40.

Walker, I. R. & Mathewes, R. W. (1989). Chironomidae (Diptera) remains in surficial lakes sediments from the Canadian Cordillera: analysis of the fauna across an altitudinal gradient. *Journal of Paleolimnology*, 2, 61–80.

Walker, I. R., Smol, J. P., Engstrom, D. R. & Birks, H. J. B. (1991). An assessment of Chironomidae as quantitative indicators of past climate change. *Canadian Journal of Fisheries and Aquatic Sciences*, 48, 975–8.

Wallace, R. L. (1987). Coloniality in the phylum Rotifera. *Hydrobiologia*, 147, 141–55.

Walls, M., Caswell, H. & Ketola, M. (1991). Demographic costs of *Chaoborus*-induced defenses in *Daphnia pulex*: a sensitivity analysis. *Oecologia*, 87, 43–50.

Walsby, A. F. & Reynolds, C. S. (1980). Sinking and floating. In *Physiological Ecology of Phytoplankton*, ed. I. Morris, pp. 371–412. Berkeley: University of California Press.

Walters, C. (1986). *Adaptive Management of Renewable Resources*. New York: Macmillan.

Walters, C. J., Krause, E., Neill, W. E. & Northcote, T. G. (1987). Equilibrium models for seasonal dynamics of plankton biomass in four oligotrophic lakes. *Canadian Journal of Fisheries and Aquatic Sciences*, 44, 1002–17.

Walters, C. J., Robinson, D. C. E. & Northcote, T. G. (1990). Comparative population dynamics of *Daphnia rosea* and *Holopedium gibberrum* in four oligotrophic lakes. *Canadian Journal of Fisheries and Aquatic Sciences*, 47, 401–9.

Waters, T. F. (1957). The effects of lime application to acid bog lakes in northern Michigan. *Transactions of the American Fisheries Society*, 86, 329–44.

Wei, W. W. S. (1990). *Time Series Analysis*. New York: Addison-Wesley.

Weider, L. J. (1984). Spatial heterogeneity of *Daphnia* genotypes: vertical migration and habitat partitioning. *Limnology and Oceanography*, 29, 225–35.

Weismann, A. (1874). Ueber Bau und Lebenserscheinungen von *Leptodora hyalina* Lilljeborg. *Zeitschrift für wissenschaftliche Zoologie*, 24, 349–418.

Weismann, A. (1877). Das Tierleben im Bodensee. *Schriften für Geschichte des Bodensees und seiner Umgebung*, 7, 1–31.

Welschmeyer, N. A. & Lorenzen, C. J. (1985a). Chlorophyll budgets: zooplankton grazing and phytoplankton growth in a temperate fjord and the Central Pacific Gyres. *Limnology and Oceanography*, **30**, 1–21.

Welschmeyer, N. A. & Lorenzen, C. J. (1985b). Role of herbivory in controlling phytoplankton abundance: annual pigment budget for a temperate fjord. *Marine Biology*, **90**, 75–86.

Werner, E. E. (1979). Niche partitioning by food size in fish communities. In *Predator-Prey Systems in Fisheries Management*, ed. R. H. Stroud & H. Clepper, pp. 311–22. Washington, D.C.: Sport Fishing Institute.

Werner, E. E. (1986). Species interactions in freshwater fish communities. In *Community Ecology*, ed. J. Diamond & T. Case, pp. 344–58. New York: Harper and Row.

Werner, E. E. & Gilliam, J. F. (1984). The ontogenetic niche and species interactions in size-structured populations. *Annual Review of Ecology and Systematics*, **15**, 393–425.

Werner, E. E., Gilliam, J. F., Hall, D. J. & Mittelbach, G. G. (1983). An experimental test of the effects of predation risk on habitat use in fish. *Ecology*, **64**, 1540–8.

Wetzel, R. G. (1990). Land-water interfaces: metabolic and limnological regulators. *Verhandlungen – Internationale Vereinigung für theoretische und angewandte Limnologie*, **24**, 6–24.

Wetzel, R. G. & Likens, G. E. (1979). *Limnological Analyses*. Philadelphia: W. B. Saunders Co.

Wetzel, R. G. & Likens, G. E. (1991). *Limnological Analyses*, 2nd edn. New York: Springer-Verlag.

Williamson, C. E. (1980). The predatory behavior of *Mesocyclops edax*: Predator preferences, prey defenses, and starvation-induced changes. *Limnology and Oceanography*, **25**, 903–9.

Wissing, T. E. & Hasler, A. D. (1971). Intraseasonal change in caloric content of some freshwater invertebrates. *Ecology*, **52**, 371–3.

Wootton, R. J. (1990). *Ecology of Teleost Fishes*. London: Chapman and Hall.

Worthington, E. B. (1931). Vertical movements of freshwater macroplankton. *Internationale Revue der gesamten Hydrobiologie/Hydrogeographie*, **25**, 394–436.

Wright, D. I. & Shapiro, J. (1984). Nutrient reduction by biomanipulation: an unexpected phenomenon and its possible cause. *Verhandlungen – Internationale Vereinigung für theoretische und angewandte Limnologie*, **22**, 518–24.

Wright, J. C. (1965). The population dynamics and production of *Daphnia* in Ferry Canyon Reservoir, Montana. *Limnology and Oceanography*, **10**, 583–90.

Wright, R. A. & Kitchell, J. F. (1993). Prey size and growth dynamics of juvenile largemouth bass: an experimental and individual-based modeling approach. Unpublished manuscript.

Yan, N. D., Keller, W., MacIsaac, H. J. & McEachern, L. J. (1991). Regulation of zooplankton community structure of an acidified lake by *Chaoborus*. *Ecological Applications*, **1**, 52–65.

Yentsch, C. M., Yentsch, C. S. & Strube, L. R. (1977). Variations in ammonium enhancement, an indication of nitrogen deficiency in New England coastal phytoplankton populations. *Journal of Marine Research*, **35**, 537–55.

Yodzis, P. (1988). The indeterminacy of ecological interactions as perceived through perturbation experiments. *Ecology*, **69**, 508–15.

Zaret, T. M. (1980). *Predation and Freshwater Communities*. New Haven, Connecticut: Yale University Press.

Zaret, T. M. & Suffern, J. S. (1976). Vertical migration in zooplankton as a predator avoidance mechanism. *Limnology and Oceanography*, **21**, 804–13.

Zimmerman, G., Goetz, M. & Meilke, P. W., Jr. (1985). Use of an improved statistical method for group comparison to study effects of prairie fire. *Ecology*, **66**, 606–11.

Züllig, H. (1989). Role of carotenoids in lake sediments for reconstructing trophic history during the late Quaternary. *Journal of Paleolimnology*, **2**, 23–40.

Index

Printed in the United States
By Bookmasters